HYDRAULI

in
Civil Engineering

TITLES OF RELATED INTEREST

HYDRAULICS
in
Civil Engineering

A. J. Chadwick
B.Sc. M.Sc. C.Eng. MICE MIWES

J. C. Morfett
Dip.Eng. M.Sc. C.Eng. MIMechE

Department of Civil Engineering, Brighton Polytechnic

London
ALLEN & UNWIN
Boston　　　　　Sydney

Allen & Unwin (Publishers) Ltd,
40 Museum Street, London WC1A 1LU, UK

Allen & Unwin (Publishers) Ltd,
Park Lane, Hemel Hempstead, Herts HP2 4TE, UK

Allen & Unwin Inc.,
8 Winchester Place, Winchester, Mass. 01890, USA

George Allen & Unwin (Australia) Ltd,
8 Napier Street, North Sydney, NSW 2060, Australia

First published in 1986

British Library Cataloguing in Publication Data

Chadwick, A.J.
  Hydraulics in civil engineering.
1. Hydraulic engineering
I. Title   II. Morfett, J.C.
627   TC145
ISBN 0-04-627003-5
ISBN 0-04-627004-3 Pbk

Library of Congress Cataloging in Publication Data
Chadwick, A. J. (Andrew John)
  Hydraulics in civil engineering.
Bibliography: p.
Includes index.
1. Hydraulics. 2. Hydraulic engineering. 3. Civil
engineering. I. Morfett, J. C. (John C.) II. Title.
TC160.C47 1986   627   85-22918
ISBN 0-04-627003-5 (alk. paper)
ISBN 0-04-627004-3 (pbk. : alk. paper)

Set in 10 on 12 point Times with mathematics in 9 point by
Mathematical Composition Setters Ltd., Salisbury, UK
and printed in Great Britain by Butler & Tanner Ltd,
Frome and London

Preface

Our aim in writing this book has been to provide a comprehensive coverage of hydraulics that is appropriate to a degree course in Civil Engineering. It should also be suitable for practising civil engineers as a guide and source of reference to some of the more specialist aspects of hydraulic engineering.

The material falls into two distinct sections. Part I presents the fundamental theoretical concepts. Part II exemplifies some of the ways in which basic concepts may be applied to the design of hydraulic systems. A substantial number of worked examples are included in the text, since it is our experience that these are essential if the student is to gain confidence in applying theory to typical engineering problems. In certain respects, the content of this book differs from many of its predecessors. The coverage has been broadened to include Flood Hydrology, Sediment Transport and the cognate areas of River and Coastal Engineering. The presentation also takes account of the widespread availability of the microcomputer. There are examples of various numerical techniques, together with some flow charts and an appendix (Appendix B) containing some programs.

All engineering is a synthesis, and a book such as this owes much to the many engineers and researchers in the field of hydraulics. We have tried to give due acknowledgement to all our principal sources, and to trace ownership of copyright where required. If any attribution is inaccurate, we would be glad to be informed so that future editions may be corrected. If any errors are discovered by readers we would also be grateful to be informed.

ANDREW CHADWICK and JOHN MORFETT
Brighton Polytechnic
Department of Civil Engineering
March 1985

Acknowledgements

We are extremely grateful to a number of colleagues who have contributed in various ways during the drafting of the manuscript:

F. W. Matthews wrote Appendix A, and also assisted with some of the computational work.
Colin Prior produced all of the line illustrations.
Hanna Chadwick undertook the considerable task of word-processing the manuscript.
Dr B. O. Hilson, Professor D. M. McDowell (formerly of Manchester University) and Mr M. R. Rees, all of Brighton Polytechnic, read various sections of the draft manuscript and gave us the benefit of their constructive criticism.

We are also indebted to Mr J. H. Loveless of Kings College London, Mr R. D. Faulkner of Loughborough University of Technology and Dr G. Fleming of Strathclyde University for reviewing the complete typescript. Their many constructive criticisms have undoubtedly improved the final text.

Any remaining errors are, of course, the sole responsibility of the authors.

Finally, we gratefully acknowledge the following individuals and organisations who have given permission for the reproduction of illustrative and tabular material:

Figure 4.5 reproduced from *Charts for the hydraulic design of channels and pipes*, 5th edn, courtesy of Hydraulics Research Limited, Wallingford, UK. Figures 8.12 and 8.14 reproduced from H. Darbyshire and L. Draper, 'Forecasting wind-generated sea waves', *Engineering* (15 April 1983), by permission of L. Draper and the Design Council, London. Figures 10.3 and 10.4 reproduced from *Flood studies report* vol. I, and Tables 10.2 and 10.3 from *Flood studies report* vol. IV, by permission of the Institute of Hydrology, Wallingford, UK. Figure 10.18 and Table 10.8 reproduced from *The Wallingford procedure* vol. I, courtesy of Hydraulics Research Limited, Wallingford, UK. Figures 13.1, 13.4, 13.5 and 13.8 are extracted from *BS3680* and are reproduced by permission of the British Standards Institution: complete copies of the document can be obtained from BSI, Linford Wood, Milton Keynes MK14 6LE, UK. Figure 13.13 reproduced from D. A. Ervine, *Proc. Instn Civ. Engrs* Part 2 **61**, 383–400 (June) by permission of the author and the Institution of Civil Engineers.

Photograph 1 courtesy of Armfield Technical Education Co. Ltd, Ringwood, Hampshire, UK. Photographs 2, 5 and 7 courtesy of Hydraulics Research Limited, Wallingford, UK. Photograph 3 courtesy of M. Dawes, Esq. Photograph 4 courtesy of the Cement and Concrete Association (photography), the Welsh Water Authority (client) and Halcrow Water (consulting engineers on the project). Photograph 6 courtesy of the Central Electricity Generating Board (photography by John Mills, Liverpool). Photograph 8 courtesy of Shephard, Hill and Co. Ltd., Hillingdon, Middlesex.

Contents

Preface   *page*   v

Acknowledgements   vii

List of principal symbols   xiii

A short history of hydraulics   xix

Introductory notes   xxii

Part I   Principles and basic applications

1   Hydrostatics   3
   1.1   Pressure   3
   1.2   Pressure measurement   5
   1.3   Pressure forces on submerged bodies   8
   1.4   Flotation   14

2   Principles of fluid flow   20
   2.1   Introduction   20
   2.2   Classification of flows   20
   2.3   Visualisation of flow patterns   21
   2.4   The fundamental equations of fluid dynamics   23
   2.5   Application of the conservation laws to fluid flows   24
   2.6   Application of the energy equation   31
   2.7   Application of the momentum equation   35
   2.8   Velocity and discharge measurement   41
   2.9   Potential flows   46
   2.10 Some typical flow patterns   50
       References and further reading   61

3   Behaviour of real fluids   62
   3.1   Real and ideal fluids   62
   3.2   Viscous flow   63
   3.3   The stability of laminar flows and the onset of turbulence   66
   3.4   Shearing action in turbulent flows   69
   3.5   The boundary layer   74
   3.6   Some implications of the boundary layer concept   82
   3.7   Cavitation   85

3.8 Surface tension effects 86
3.9 Summary 86
References and further reading 87

4 Flow in pipes and closed conduits 88
4.1 Introduction 88
4.2 The historical context 88
4.3 Fundamental concepts of pipe flow 91
4.4 Laminar flow 93
4.5 Turbulent flow 96
4.6 Local head losses 108
4.7 Partially full pipes 112
References and further reading 117

5 Open channel flow 118
5.1 Flow with a free surface 118
5.2 Flow classification 119
5.3 Natural and artificial channels and their properties 120
5.4 Velocity distributions, energy and momentum coefficients 122
5.5 Laminar and turbulent flow 123
5.6 Uniform flow 125
5.7 Rapidly varied flow: the use of energy principles 131
5.8 Rapidly varied flow: the use of momentum principles 143
5.9 Critical depth meters 148
5.10 Gradually varied flow 151
5.11 Unsteady flow 170
References and further reading 179

6 Pressure surge in pipelines 180
6.1 Introduction 180
6.2 Effect of 'rapid' valve closure 183
6.3 Unsteady compressible flow 183
6.4 Analysis of more-complex problems 192
6.5 The method of characteristics 197
6.6 Characteristic equations for system boundaries 200
6.7 Concluding remarks 209
References and further reading 211

7 Hydraulic machines 212
7.1 Classification of machines 212
7.2 Continuous flow pumps 212
7.3 Performance data for continuous flow pumps 218
7.4 Pump selection 219
7.5 Hydro-power turbines 220
7.6 Turbine selection 223
7.7 Cavitation in hydraulic machines 223
References and further reading 224

8 Wave theory 225
  8.1 Wave motion 225
  8.2 Airy waves 226
  8.3 Wave energy and wave power 229
  8.4 Refraction and shoaling 230
  8.5 Standing waves 236
  8.6 Wave diffraction 238
  8.7 Wave prediction 239
  8.8 Analysis of wave records 243
  8.9 Extension of wave records 246
      References and further reading 246

9 Sediment transport 247
  9.1 Introduction 247
  9.2 The threshold of movement 247
  9.3 A general description of the mechanics of sediment transport 251
  9.4 Sediment transport equations 255
  9.5 Concluding notes on transport formulae 267
      References and further reading 268

10 Flood hydrology 269
  10.1 Classifications 269
  10.2 Methods of flood prediction for rural catchments 270
  10.3 Catchment characteristics 271
  10.4 Frequency analysis 273
  10.5 Unit hydrograph theory 280
  10.6 Summary of design flood procedures for rural catchments 288
  10.7 Flood routing 294
  10.8 Design floods for reservoir safety 302
  10.9 Methods of flood prediction for urban catchments 303
      References and further reading 307

11 Dimensional analysis and the theory of physical models 309
  11.1 Introduction 309
  11.2 The idea of 'similarity' 310
  11.3 Dimensional homogeneity and its implications 311
  11.4 Dimensional analysis 312
  11.5 Dimensional analysis involving more variables 316
  11.6 Applications of dynamic similarity 317
  11.7 Hydraulic models 326
      References and further reading 335

Part II Aspects of hydraulic engineering

12 Pipeline systems 339
  12.1 Introduction 339
  12.2 Design of a simple pipe system 339

12.3	Series, parallel and branched pipe systems	342
12.4	Distribution systems	346
12.5	Design of pumping mains	353
12.6	Surge protection	358
	References and further reading	368

13 Hydraulic structures — 369

13.1	Introduction	369
13.2	Thin plate (sharp-crested) weirs	369
13.3	Long-based weirs	379
13.4	Flumes	384
13.5	Spillways	391
13.6	Energy dissipators	403
13.7	Control gates	406
13.8	Lateral discharge structures	412
13.9	Outlet structures	417
13.10	Concluding remarks	417
	References and further reading	418

14 Canal and river engineering — 419

14.1	Introduction	419
14.2	Optimisation of a channel cross section	419
14.3	Unlined channels	421
14.4	Morphology of natural channels	428
14.5	River engineering	431
14.6	Concluding note	433
	References and further reading	433

15 Coastal engineering — 435

15.1	The action of waves on beaches	435
15.2	Littoral drift	436
15.3	Natural bays	438
15.4	Sea defence or coast protection?	439
15.5	Sea walls	439
15.6	Groynes and beach replenishment	440
15.7	Permeable breakwaters and rip-rap	442
15.8	Artificial headlands	443
	References and further reading	445

Postscript	446
Problems	447
Appendix A The matrix approach to dimensional analysis	463
Appendix B Computer programs	471
Appendix C Moments of area	479
Index	487

List of tables

1   The six primary units in the SI system
4.1   The chronological development of pipe flow theories
4.2   Typical k_S values
4.3   Local head loss coefficients
5.1   Geometric properties of some common prismatic channels
5.2   Typical values of Manning's n
5.3   Computer solution of Example 5.7
5.4   Computer solution of Example 5.8
5.5   Computer solution of Example 5.9
8.1   Wave generation: to show group wave speed
8.2   Tabular solution for breaking waves
10.1   Tabular and matrix methods of convolution
10.2   Annual maximum discharges for the River Nidd at Hunsingore weir
10.3   Catchment characteristics for the River Nidd at Hunsingore weir
10.4   Frequency analysis for the River Nidd at Hunsingore weir
10.5   Synthetic unit hydrograph analysis for the River Nidd at Hunsingore weir
10.6   Relationships between stage, volume and outflow for Example 10.2
10.7   Tabular solution for reservoir routing
10.8   Methods and models included in the WASSP package
12.1   Pipe materials and joints for water supply pipes
13.1   Solution to Example 13.7

List of principal symbols

A area (m^2)
A catchment area (km^2)
A_h plan area of reservoir at stage h (m^2)
A_p areal packing of grains
A_R ratio of partially open valve area to fully open area
A_S cross-sectional area of sediment particle (m^2)
A_{ST} cross-sectional area of surge tower (m^2)

A_V cross-sectional area of flow at valve (m^2)

B*   sediment transport parameter
B   centre of buoyancy
B   surface width of channel (m)
b   width (m)

c   celerity (m/s)
C   Chézy coefficient $(m^{1/2}/s)$
C   concentration (mg/l)
C_c contraction coefficient
C_d discharge coefficient
C_D drag coefficient
C_f friction coefficient
C_G group celerity of waves (m/s)
C_L lift coefficient
C_0 celerity of waves in deep water (m/s)
C_v velocity coefficient

d   water depth under a wave (m)
d   diameter (m)
D   diameter (m)
D   duration (h)
d_B water depth for a breaking wave (m)
D_m hydraulic mean depth (m)

e   pipe wall thickness (m)
e   sediment transport 'efficiency'
E   Young's modulus (N/m^2)
E   energy (J)
E_K kinetic energy of wave (J)
E_P potential energy of wave (J)
E_S specific energy (m)

F   fetch length (km)
F   force (N)
F_D drag force (N)
Fr   Froude Number

g   gravitational acceleration (m/s^2)
G   centre of gravity

H   head (m)
H   wave height (m)
H_0 wave height in deep water (m)
H_B wave height at breaker line (m)
H_C cavitation head (m)

h_f	frictional head loss (m)
h_L	local head loss (m)
H_p	pump head (m)
H_t	turbine head (m)
i	rainfall intensity (mm/h)
I	inflow (to reservoir) (m^3/s)
I	second moment of area (m^4)
K	bulk modulus (N/m^2)
K	von Kármán constant
k_L	local loss coefficient
K_N	dimensionless specific speed
K_R	refraction coefficient for waves
k_S	surface roughness (m)
K_S	shoaling coefficient for waves
l	length (m)
L	length (dimension)
L	wave length (m)
$\bar{l}$	distance to centroid of area (m)
l'	distance to centre of pressure (m)
m	mass (kg)
M	mass (dimension)
M	metacentre
M	momentum (kg/m s)
n	Manning's roughness coefficient
N	rotational speed (1/s)
N_S	specific speed
O	outflow (from reservoir) (m^3/s)
P	power (kW)
P	rainfall depth (mm)
p^*	piezometric pressure (N/m^2)
p	pressure (N/m^2)
P	probability
P	wetted perimeter
P_S	sill or crest height (m)
Q	discharge (m^3/s)
q	discharge per unit channel width $(m^3/m\ s)$
Q_p	peak runoff (m^3/s)
Q_S	sediment discharge (m^3/s)
q_S	sediment discharge per unit channel width $(m^3/m\ s)$
Q_{Tr}	flood discharge of return period T_r (m^3/s)

r radius (m)
R radius (m)
R hydraulic radius (m)
R_p reading on pressure gauge
Re   Reynolds' Number

S_c slope of channel bed to give critical flow
S_{CR} slope of channel bed to give critical shear stress for sediment transport
S_f slope of hydraulic gradient
S_0 slope of channel bed
St   Strouhal Number

T tension force (N)
t time (s)
T wave period (s)
T_B time base of the unit hydrograph (h)
t_c time of concentration (min)
t_e time of entry into drainage system (min)
t_f time of flow through a drainage pipe (min)
T_p periodic time (s)
T_p time to peak runoff of the unit hydrograph (h)
T_r return period (years)
T_T periodic time of tide (h)
T_S surface tension (N/m)

u velocity (m/s)
U wind speed (m/s)
$\bar{u}$ average velocity (m/s)
u_0 initial velocity (m/s)
U_∞ 'free stream' velocity (m/s)
u_* friction velocity (m/s)

V velocity (usually mean velocity of flow)(m/s)
V volume (m^3)
V_A absolute velocity (m/s)
V_F radial velocity in hydraulic machine (m/s)
V_{FS} fall velocity of a sediment particle (m/s)
V_I tangential velocity of pump impeller (m/s)
V_R relative velocity (m/s)
V_S volume of sediment particle (m^3)
V_{ST} velocity of flow in a surge tower (m/s)
V_W tangential (whirl) velocity in hydraulic machine

W weight (N)
W' immersed weight of a particle (N)
We   Weber Number

y water depth (m)
y_c critical depth (m)
y_n normal depth (m)

z height above datum (m)

α	angle (degree or rad)
α	velocity coefficient
β	momentum coefficient
Γ	circulation
δ	boundary layer thickness (m)
δ_*	displacement thickness (m)
ϵ	eddy viscosity
ϵ	spectral width
ζ	vorticity
η	efficiency
η	height of wave profile (m)
θ	angle (degree or rad)
θ	momentum thickness (m)
λ	friction factor
λ	scale factor
μ	absolute viscosity (kg/m s)
ν	kinematic viscosity (m^2/s)
Π	dimensionless group
ρ	density of liquid (kg/m^3)
ρ_S	density of sediment (kg/m^3)
σ	hoop stress (N/m^2)
τ	shear stress (N/m^2)
τ_{CR}	critical shear stress for sediment transport (N/m^2)
τ_0	shear stress at a solid boundary (N/m^2)
ϕ	velocity potential
Φ	sediment transport parameter
ψ	stream function
Ψ	sediment transport parameter
ω	rotational speed (1/s)

A short history of hydraulics

Hydraulics is a very ancient science. The Egyptians and Babylonians constructed canals, both for irrigation and for defensive purposes. No attempts were made at that time to understand the laws of fluid motion. The first notable attempts to rationalise the nature of pressure and flow patterns were undertaken by the Greeks. The laws of hydrostatics and buoyancy were enunciated, Ctesibius and Hero designed hydraulic equipment such as the piston pump and water clock, and (of course) there was the Archimedes screw pump. The Romans appear, like the Egyptians, to have been more interested in the practical and constructional aspects of hydraulics than in theorising. Thus, development continued slowly until the time of the Renaissance, when men such as Leonardo Da Vinci began to publish the results of their observations. Ideas which emerged then, respecting conservation of mass (continuity of flow), frictional resistance and the velocity of surface waves, are still in use, though sometimes in a more refined form. The Italian School became famous for their work. Toricelli *et al.* observed the behaviour of water jets. They compared the path traced by a free jet with projectile theory, and related the jet velocity to the square root of the pressure generating the flow. Guglielmini *et al.* published the results of observations on river flows. The Italians were hydraulicians in the original sense of the word, i.e. they were primarily empiricists. Up to this point, mathematics had played no significant part in this sort of scientific work. Indeed, at that time mathematics was largely confined to the principles of geometry, but this was about to change.

In the 17th century, several brilliant men emerged. Descartes, Pascal, Newton, Boyle, Hooke and Leibnitz laid the foundations of modern mathematics and physics. This enabled researchers to perceive a logical pattern in the various aspects of mechanics. On this basis, four great pioneers – Bernoulli, Euler, Clairaut and D'Alembert – developed the academic discipline of hydrodynamics. They combined a sound mathematical framework with an acute perception of the physical phenomena which they were attempting to represent. In the 18th century, further progress was made, both in experimentation and in analysis. In Italy, for example, Poleni investigated the concept of discharge coefficients.

However, it was French and German thinkers who now led the way. Henri de Pitot constructed a device which could measure flow velocity. Antoine Chézy (1718–98), followed by Eytelwein and Woltmann, developed a rational equation to describe flow in streams. Men such as Borda, Bossut and du Buat not only extended knowledge, but took pains to see that the available knowledge was disseminated. Woltmann and Venturi used Bernoulli's work as a basis for developing the principles of flow measurement.

The 19th century was a period of further advance. Hagen (1797–1884) constructed experiments to investigate the effects of temperature on pipe flow. His understanding of the nature of fluid viscosity was limited to Newton's ideas, yet so careful was his work that the results were within 1% of modern measurements. He injected sawdust into the fluid for some of his experiments, in order to visualise the motion. He was probably the first person knowingly to observe turbulence, though he was unable to grasp its significance fully. At almost the same time, a French doctor (Poiseuille) was also making observations on flow in pipes (in an attempt to understand the flow of blood in blood vessels), which led to the development of equations for laminar flow in pipes. Further contributions were made by Weisbach, Bresse and Henri Darcy, who developed equations for frictional resistance in pipe and channel flows (the first attempts to grapple with this problem coincided with signs of an incipient awareness of the existence of the 'boundary layer'). During the later part of the century, important advances were made in experimentation. The first practical wind tunnel, the first towing tank for model testing of ships and the first realistic attempt to model a tidal estuary (by O. Reynolds) were all part of this flowering of knowledge. These techniques are still used today. Reynolds also succeeded in defining the different types of flow, observing cavitation and explaining Darcy's friction law in greater detail.

Even at this stage, studies of fluid flows were subdivided into 'Classical Hydrodynamics' (which was a purely mathematical approach with little interest in experimental work) and experimental 'Hydraulics'. Navier, Stokes, Schwarz, Christoffel and other hydrodynamicists all contributed to the development of a formidable array of mathematical equations and methods (including the conformal transformation). Their work agreed only sporadically with the practitioners (the hydraulicians) and, indeed, there were frequently yawning disparities between the results suggested by the two schools. The rapid growth of industry in the 19th and 20th centuries was by now producing a demand for a better understanding of fluid flow phenomena. The real breakthrough came with the work of Prandtl. He proposed (in 1901) that flow was 'divided into two interdependent parts. There is on the one hand the free fluid which can be treated as inviscid (i.e. which obeyed the laws of hydrodynamics) ... and on the other hand the transition layer at the fixed boundaries' (the transition layer is the thin layer of fluid within which frictional forces dominate). With this brilliant insight, Prandtl effectively fused together the two disparate schools of thought and laid the foundation for the development of the unified science of Fluid Mechanics.

The 20th century has, in consequence, seen tremendous advances in the understanding and application of fluid mechanics in almost every branch of engineering. It is only possible to give the barest outline here. Prandtl and Th. von Kármán published a series of papers in the 1920s and 1930s, covering various aspects of boundary layer theory and turbulence. Their work was supplemented by increasingly sophisticated laboratory research (e.g. the work of Dryden and his colleagues at NACA in the USA). These efforts had an impact on every aspect of engineering fluid mechanics. In the 1930s, the efforts of Nikuradse (in Germany), Moody (in America), Colebrook (Great Britain) and others resulted in a clearer understanding of pipe flows and, in particular, of the factors affecting pipe friction. This led directly to the modern methods for estimating flows in pipes and channels. Since 1945 the advent of the computer and electronic instrumentation systems has revolutionised many aspects of hydraulics. Our understanding of steady and unsteady flows, and of sediment transport, has improved continuously. In consequence, the range of measurement techniques and analytical tools now available to engineers make it possible to design more effectively and economically. Research continues apace worldwide, and there is every reason to suppose that the last part of the 20th century will be an exciting period of advance, change and challenge for the engineer.

Introductory notes

Before formally embarking on a study of the engineering aspects of the behaviour of fluids, it is worth pausing to consider the area to be studied, the nature of the substance, and one or two other basic points. Students who already have some grounding in fluid mechanics may wish to proceed directly to the main body of the text.

Civil engineering hydraulics

'Fluid mechanics' is the general title given to the study of all aspects of the behaviour of fluids which are relevant to engineers. Within this very broad discipline, a number of subsections have developed. Of these subsections, hydraulics is the branch which concentrates on the study of liquids. Civil engineers are largely, though not exclusively, concerned with one liquid, namely water. The development of the industrial society rests largely on the ability of civil engineers to provide adequate water services, such as the supply of potable water, drainage, flood control, etc.

The nature of fluids

A fluid is a substance which can readily flow, i.e. in which there can be a continuous relative motion between one particle and another. A fluid is inelastic in shear, and therefore continuously deforms under application of a shear force without the possibility of a return to its original disposition.
Fluids are subdivided into the following:

Liquids which have a definite volume for a given mass, i.e. they cannot readily be altered (say, compressed) due to changes of temperature or pressure – if liquid is poured into a container, a clearly defined interface is established between the liquid and the atmosphere;
Gases generally exhibit no clear interface, will expand to fill any container, and are readily compressible.

Balance of forces
on immersed molecule

(a)

Imbalance of forces on molecule
at surface of liquid

Meniscus effect

(b)

Figure 1 Surface tension.

Properties of fluids

Density is the ratio of mass of a given quantity of a substance to the volume occupied by that quantity.

Specific weight is the ratio of weight of a given quantity of a substance to the volume occupied by that quantity. An alternative definition is that specific weight equals the product of density and gravitational acceleration.

Relative density (otherwise called 'specific gravity') is the ratio of the density of a substance to some standard density (usually the standard density is that of water at $4°C$).

Viscosity represents the susceptibility of a given fluid to shear deformation, and is defined by the ratio of the applied shear stress to rate of shear strain. This property is discussed in detail in Chapter 3.

Surface tension is the tensile force per unit length at the free surface of a liquid. The reason for the existence of this force arises from inter-molecular attraction (Fig. 1a). In the body of the liquid, a molecule is surrounded by other molecules and intermolecular forces are symmetrical and in equilibrium. At the surface of the liquid (Fig. 1b), a molecule has this force acting only through $180°$. This imbalance of forces means that the molecules at the surface tend to be drawn together, and they act rather like a very thin membrane under tension. This causes a slight deformation at the surface of a liquid (the meniscus effect).

Bulk modulus is a measure of the compressibility of a liquid, and is the ratio of the change in pressure to the volumetric strain caused by the pressure change. This property is discussed in Chapter 6.

Units

In order to define the magnitudes of physical quantities (such as mass or velocity) or properties (density, etc.), it is necessary to set up a standardised framework of units. In this book, the Système Internationale d'Unités (SI system) will generally be used. There are six primary units from which all

Table 1 The six primary units in the SI system.

Quantity	Symbol of quantity	SI unit	Symbol of unit
length	L	metre	m
mass	M	kilogramme	kg
time	T	second	s
electric current	i	ampere	A
luminous intensity	I	candela	Cd
temperature	t	kelvin	K

others are derived, and these are listed in Table I. For a complete specification of a magnitude, a number and a unit are required (e.g. 5 m, 3 kg).
The principal derived units are as follows:

velocity is the distance travelled in unit time and therefore has the dimensions of LT^{-1} and units m/s;

acceleration is the change of velocity in unit time and has dimensions LT^{-2} (= velocity per second) and units m/s^2;

force is derived by the application of Newton's second law, which may be summarised as 'force = mass × acceleration' − it is measured in newtons, where 1 newton (N) is that force which will accelerate a mass of 1 kg at 1 m/s^2, and has dimensions MLT^{-2} and units kg m/s^2;

energy (or 'work', or heat) is measured in joules, 1 joule being the work equivalent of a force of 1 N acting through a distance of 1 m, and has dimensions ML^2T^{-2} and units N m;

power, the rate of energy expenditure, is measured in watts (= joules per second), and has dimensions ML^2T^{-3} and units N m/s.

The dimensions and units of the properties of fluids are the following:

density, which by definition is mass/volume = ML^{-3}, has units of kg/m^3;

specific weight is the force applied by a body of given mass in a gravitational field − its dimensions are $ML^{-2}T^{-2}$ and its units are N/m^3;

relative density is a ratio, and is therefore dimensionless;

$$viscosity \quad = \frac{\text{shear stress}}{\text{rate of shear strain}} = \frac{(\text{force/area})}{(\text{velocity/distance})}$$

has dimensions $ML^{-1}T^{-1}$ and its units may be expressed as either N s/m^2 or kg/m s;

surface tension is defined as force per unit length, so its dimensions are MT^{-2} and its units are N/m.

Dimensional homogeneity

If any equation is to represent some physical situation accurately, then it must achieve numerical equality and dimensional equality (or equality of units).

It is strongly recommended that when constructing equations the numerical value and the units of each term are written down. It is then possible to check that both of the required equalities are, in fact, achieved. Thus, if at the end of a calculation one has

$$55 \text{ kg m} = 55 \text{ s}$$

then a mistake has been made, even though there is numerical equality. This point is especially important when equations incorporate coefficients. Unless the coefficients are known to be dimensionless, a check should be made to ascertain precisely what the units relevant to a given numerical value are. If an engineer applies a particular numerical value for a coefficient without checking units, it is perfectly possible to design a whole system incorrectly and waste large sums of money. Subsequent excuses ('I didn't know it was in centimetres') are not usually popular!

PART I

Principles and basic applications

1

Hydrostatics

1.1 Pressure

Hydrostatics is the study of fluids at rest, and is therefore the simplest aspect of hydraulics. The main characteristic of a stationary fluid is the force which it brings to bear on its surroundings. A fluid force is frequently specified as a pressure, p, which is the force exerted on a unit area. Pressure is measured in N/m^2 or in 'bar' (1 bar = 10^5 N/m^2).

Pressure is not constant everywhere in a body of fluid. In fact, if pressure is measured at a series of different depths below the upper surface of the fluid, it will be found that the pressure reading increases with increasing depth. An exact relationship can be developed between pressure, p, and depth, y, as follows. Suppose there is a large body of liquid (a lake or swimming pool, for example), then take any imaginary vertical column of liquid within that main body (Fig. 1.1). The column of fluid is at rest, therefore all of the forces acting on the column are in equilibrium. If this statement is to be true for any point on the boundary surfaces of the column, the action and reaction forces must be perpendicular to the boundary surface. If any forces were not perpendicular to the boundary, then a shear force

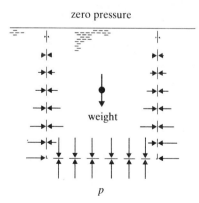

Figure 1.1 Pressure distribution around a column of liquid.

component would exist; this condition arises only for fluids in motion. It follows that the only force which is supporting the column of fluid is the force acting upwards due to the pressure on the base of the column. For the column to be in equilibrium, the upward force must exactly equal the weight force acting downward.

The weight of the column is $\rho g V$ where V is the volume of the column. The volume is the product of the horizontal cross-sectional area and height, $V = Ay$, and thus the weight is given by $\rho g Ay$.

The force acting upwards is the product of pressure and horizontal cross-sectional area, i.e. pA. Therefore

$$pA = \rho g Ay$$

and so

$$p = \rho g y \qquad (1.1a)$$

This is the basic hydrostatic equation or 'law'. By way of example, in fresh water (which has a density of 1000 kg/m^3) the pressure at a depth of 10 m is

$$p(\text{N/m}^2) = 1000 \ (\text{kg/m}^3) \times 9.81 \ (\text{m/s}^2) \times 10 \ (\text{m})$$

$$= 98\ 100 \ \text{N/m}^2$$

$$= 98.1 \ \text{kN/m}^2$$

The equation is correct both numerically and in terms of its units. For all practical purposes, the value of g ($= 9.81$ m/s^2) is constant on the Earth's surface. The product ρg will therefore also be constant for any homogeneous incompressible fluid, and (1.1a) then indicates that pressure varies linearly with the depth y (Fig. 1.2).

Gauge pressure and absolute pressure

An important case of pressure variation is that of a liquid with a gaseous atmosphere above its free surface. The pressure of the gaseous atmosphere

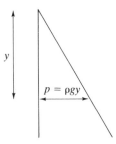

Figure 1.2   Pressure variation with depth.

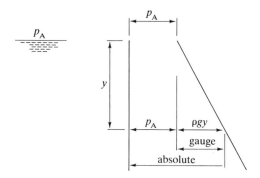

Figure 1.3 Gauge and absolute pressure.

immediately above the free surface is p_A (Fig. 1.3). For equilibrium, the pressure in the liquid at the free surface is p_A, and therefore at any depth y below the free surface the absolute pressure p_{ABS} (i.e. the pressure with respect to absolute zero) must be

$$p_{ABS} = p_A + \rho g y$$

The gauge pressure is the pressure with respect to p_A (i.e. p_A is treated as the pressure 'datum'):

$$p = \rho g y = \text{gauge pressure}$$

It is possible for gauge pressure to be positive (above p_A) or negative (below p_A). Negative gauge pressures are usually termed vacuum pressures. Virtually every civil engineering project is constructed on the Earth's surface, so it is customary to take atmospheric pressure as the datum. Most pressure gauges read zero at atmospheric pressure.

1.2 Pressure measurement

The argument so far has centred upon variation of pressure with depth. However, suppose that a pipeline is filled with liquid under pressure (Fig. 1.4a). At one point the pipe has been pierced and a vertical transparent tube has been attached. The liquid level would rise to a height y, and since (1a) may be rearranged to read

$$p/\rho g = y \tag{1.1b}$$

this height will indicate the pressure. The term $p/\rho g$ is often called 'pressure head' or just 'head'. A vertical tube pressure indicator is known as a piezometer. The piezometer is of only limited use. Even to record quite

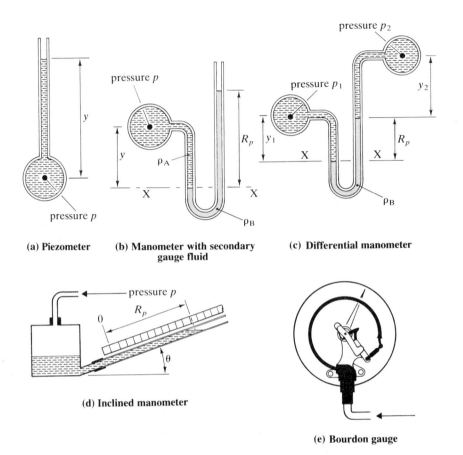

(a) Piezometer   (b) Manometer with secondary   (c) Differential manometer
                        gauge fluid

(d) Inclined manometer

(e) Bourdon gauge

Figure 1.4   Pressure measuring devices.

moderate water pressures, the height of a piezometer becomes imprac-
ticably large. For example, a pressure of water of $100 \, kN/m^2$ corresponds
to a height of roughly 10 m, and civil engineers commonly design systems
for higher pressure than this. The piezometer is obviously useless for gas
pipelines.

To overcome this problem, engineers have developed a range of devices.
From a purely hydrostatic viewpoint, perhaps the most significant of these
is the manometer (Fig. 1.4b), which is a U-shaped transparent tube. This
permits the use of a second gauge fluid having a density (ρ_B) which differs
from the density (ρ_A) of the primary fluid in the pipeline. The two fluids
must be immiscible, and must not react chemically with each other. When
the pressure in the pipeline is p, the gauge fluid will be displaced so that
there is a difference of R_p metres between the gauge fluid levels in the left-
and right-hand vertical sections. To evaluate the pressure p, we proceed as
follows.

Set a horizontal datum X–X at the lower gauge fluid level, i.e. at the interface between fluid A and fluid B in the left-hand vertical section. The pressure in the left-hand section at X–X is the sum of the pressure p at the centre of the pipe and the pressure due to the height y of fluid A, i.e.

$$p_{\text{X--X}} = p + \rho_A g y$$

In the right-hand section the pressure at X–X is due to the height R_p of fluid B (atmospheric pressure is ignored, so the final answer will be a gauge pressure), i.e.

$$p_{\text{X--X}} = \rho_B g R_p$$

In any continuous homogeneous fluid, pressure is constant along any horizontal datum. Since X–X passes through the interface on the left-hand side and through fluid B on the right-hand side, $p_{\text{X--X}}$ must be the same on both sides. Therefore

$$p_{\text{X--X}} = p + \rho_A g y = \rho_B g R_p$$

so

$$p = \rho_B g R_p - \rho_A g y \qquad (1.2a)$$

Thus, if ρ_A, ρ_B and y are known, the pressure p may be calculated for any reading R_p on the manometer. Equation (1.2a) is often written in terms of pressure head:

$$\frac{p}{\rho_A g} = \frac{\rho_B}{\rho_A} R_p - y \qquad (1.2b)$$

The manometer may also be connected up to record the pressure difference between two points (Fig. 1.4c).

Other devices include the following:

(a) The sloping tube manometer (Fig. 1.4d), which works on the same essential principle as the U-tube. However, the right-hand limb comprises a transparent tube sloping at an angle θ while the left-hand limb comprises a tank whose horizontal cross section is much larger than that of the tube. The primary fluid extends from the pipeline through a flexible connecting tube into the top of the tank. The pressure p forces fluid from the tank into the transparent sloping tube, giving the reading R_p. The level in the tank falls only by a relatively small amount, so the reading is usually taken from a fixed datum. The equation of pressure may be adapted from (1.2a) if it is realised that $y = R_p \sin \theta$.

(b) The Bourdon Gauge is a semi-mechanical device (Fig. 1.4e). It consists of a tube formed into the shape of a question mark, with its upper end sealed. The upper end is connected by a linkage to a pointer. An increase in pressure causes a slight deformation of the tube. This is transmitted to the pointer, which rotates through a corresponding angle. Bourdon gauges can be designed for a much wider range of pressures than can manometers. On the other hand, they are somewhat less precise and require a certain amount of maintenance.

(c) Pressure transducers are devices which convert a pressure into a corresponding electrical output. They are particularly applicable for use in conjunction with automatic (computer-controlled) systems, or where variable or pulsating pressures need to be measured. For details of such devices, the manufacturer's literature is usually the best source of information.

1.3   Pressure forces on submerged bodies

The omnidirectional nature of pressure

In developing the equation of hydrostatic pressure distribution, it was emphasised that pressure forces were assumed to be perpendicular to the imaginary surfaces of the column of liquid. However, this, taken to its logical conclusion, must mean that at any point in the fluid the pressure acts equally in all directions. That this is true may be formally demonstrated as follows. Imagine a small sphere of fluid which is stationary within a larger body of the same fluid (Fig. 1.5). Action and reaction forces must be in equilibrium at all points round the sphere. If the sphere diameter is reduced, then even though $d \to 0$ (and therefore the weight of the sphere tends to zero) the equilibrium statement must still be true. Therefore, in an uninterrupted fluid continuum, pressure at any point acts equally in all directions.

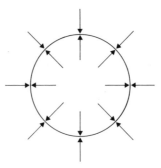

Figure 1.5   Pressure at a point in a liquid.

Pressure on plane surfaces

If a body is immersed in a fluid, it must be subject to the external pressure of the fluid and must provide the reaction necessary for equilibrium. Once again, it is emphasised that in the absence of fluid motion the action and reaction forces must be perpendicular to the surface of the body at any point. Generally, in design calculations it is the total force of the fluid on the body that is required, rather than the pressure at any given location. To show how this is evaluated, take an arbitrary plane surface which is immersed in a liquid (Fig. 1.6). Somewhere on the surface, take a small element of area δA. The force δF on that area, due to the pressure of the liquid, is $p\delta A = \delta F$. Obviously, the total force acting on the whole plane surface must be the sum of all the products ($p\delta A$). However, p is not a constant. Thus if the element is a distance y below the free surface of the liquid, then $p = \rho g y$ and $p\delta A = \delta F = \rho g y \, \delta A$ and therefore the total force is $F = \rho g \int y \, dA$ (ρ and g, being constants, are outside the integral). But

$$y = l \sin \theta$$

therefore

$$F = \rho g \int l \sin \theta \, dA = \rho g (\sin \theta) \int l \, dA$$

The quantity $\int l \, dA$ is a geometrical characteristic of the shape, and is known as the first moment of area. This integral can be evaluated for any shape and equals the product $A\bar{l}$, where A is the area of the plane surface and $\bar{l}$ is the distance from the origin O to the centroid of the plane. Therefore

$$F = \rho g (\sin \theta) A \bar{l} \tag{1.3}$$

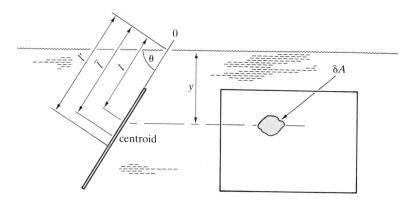

Figure 1.6 Pressure on a plane surface.

In addition to the magnitude of the force F, it is usually necessary to know the precise location and angle of its line of action. The pressure is, of course, distributed over the whole of the immersed body, but it is almost always more convenient to regard F as a concentrated point load.

Returning to the element of area, the force δF produces a moment $\delta F\,l$ about the origin. This may be written as

$$\text{moment} = \delta F\,l = \rho g y\,\delta A\,l$$

$$= \rho g l(\sin\theta)\,\delta A\,l$$

$$= \rho g(\sin\theta)l^2\,\delta A$$

In the limit, $dF l = \rho g(\sin\theta)l^2 dA$, so for the whole surface the moment is $\rho g(\sin\theta)\int l^2 dA$.

The quantity $\int l^2 dA$ is another geometrical characteristic, known as the second moment of area, which has the symbol I (see Appendix C). Therefore

$$\text{moment about origin} = \rho g(\sin\theta)I$$

Therefore, the distance from the origin to the point of action of F is

$$l' = \frac{\text{moment}}{\text{force}} = \frac{\rho g(\sin\theta)I}{\rho g(\sin\theta)A\bar{l}} = \frac{I}{A\bar{l}} \tag{1.4}$$

I is usually evaluated by use of the Parallel Axes theorem.

Example 1.1   Pressure forces on a dam

A rockfill dam is to have the cross section shown in Figure 1.7a. The reservoir design depth is to be 10 m. Estimate:

(a) the force on the dam per unit width;
(b) the location of the centre of pressure.

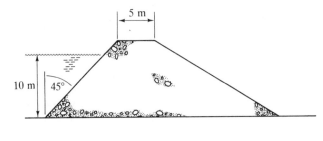
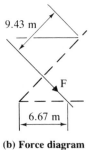

    (a) Detail of dam   **(b) Force diagram**

Figure 1.7   Dam section (Example 1.1).

Solution

Using trigonometry, the wetted slant height is 14.14 m.

$$\text{Force on dam} = \rho g(\sin\theta)\int l\,dA = \rho g(\sin\theta)A\bar{l}$$

$$= \rho g \sin\theta \text{ area} \times \left(\frac{\text{wetted height}}{2}\right)$$

$$= 1000 \times 9.81 \times \sin 45°(14.14 \times 1) \times 14.14/2$$

$$= 693\,460 \text{ N/m width}$$

The moment of the force is $\rho g \sin\theta \times$ (2nd moment of area). Therefore

$$M = 1000 \times 9.81 \times \sin\theta \ [I_0 + A(\text{wetted height}/2)^2]$$

where $I_0 =$ second moment of area about centroid. Hence,

$$M = 1000 \times 9.81 \times \sin 45° \left(\frac{1 \times 14.14^3}{12} + (1 \times 14.14)\left(\frac{14.14}{2}\right)^2\right)$$

$$= 6\,537\,040 \text{ N m}$$

So

$$l' = 6\,537\,040/693\,460 = 9.43 \text{ m}$$

(see Fig 1.7b). Therefore the slant height from base to the centre of pressure is 4.71 m.

Pressure on curved surfaces

Not all immersed structures are flat. Some dams have an upstream face which may be curved in both the vertical and the horizontal planes. Certain types of adjustable hydraulic structures (used to control water levels), such as gates, may also be curved.

It has already been shown that pressure is perpendicular to an immersed surface. The pressure on a curved surface would be distributed as shown in Figure 1.8a. To calculate the total force directly is inconvenient. At any small area δA, the force $\delta F(=p\delta A)$ will be perpendicular to the surface. This force can be resolved into two components (Fig. 1.8b), one vertical, δF_y, and one horizontal, δF_x. It can easily be shown that $\sum \delta F_y$ is the total weight, W, of the volume of liquid above the curved surface. The horizontal force, $F_x = \sum \delta F_x$, is equal to the pressure force on a vertical plane surface equal in height to the projected height of the curved surface. The resultant

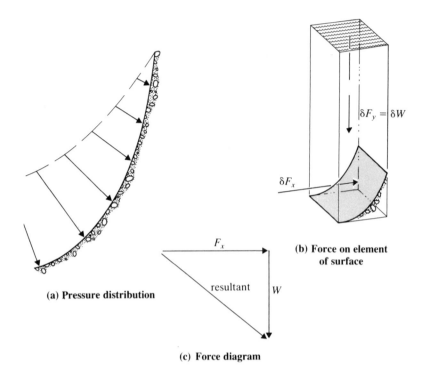

(a) Pressure distribution

F_x

resultant W

δF_x

$\delta F_y = \delta W$

(b) Force on element of surface

(c) Force diagram

Figure 1.8   Pressure force on curved surface.

force may be obtained from the triangle of forces (Fig. 1.8c). The weight component of the volume will act through the centre of gravity, while the horizontal component will act at the distance l' below the free surface of the liquid. The resultant force will act through the point of intersection of the lines of action of W and F_x.

Example 1.2   Hydrostatic force on a quadrant gate

Find the magnitude and direction of the resultant force of water on the quadrant gate shown in Figure 1.9a. The principal dimensions of the gate are:

$$\text{radius of gate} = 1 \text{ m}$$

$$\text{width of gate} = 3 \text{ m}$$

$$\text{water density} = 1000 \text{ kg/m}^3$$

The position of the centre of gravity is, as shown, $4R/3\pi$ horizontally from the origin.

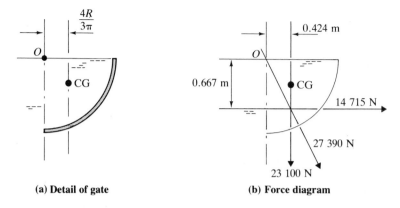

(a) Detail of gate   (b) Force diagram

Figure 1.9   Pressure on quadrant gate (Example 1.2).

Solution

(a) For vertical component,

$$\text{area of sector} = \pi R^2/4 = \pi \times 1^2/4 = 0.785 \text{ m}^2$$

$$\text{volume of water} = 0.785 \times 3 = 2.355 \text{ m}^3$$

$$\text{weight of water} = 9.81 \times 1000 \times 2.355 = 23\ 100 \text{ N}$$

$$\text{centre of gravity is } 4 \times 1/3\pi = 0.424 \text{ m from origin}$$

(b) For horizontal component, centroid of vertical projected area is 0.5 m below free surface of water. Therefore, from (1.3),

$$\text{force} = \rho g A \bar{l} = 1000 \times 9.81 \times 3 \times 1 \times 0.5 = 14\ 715 \text{ N}$$

The point of action of this component lies l' below the free surface:

$$l' = I/A\bar{l}, \text{ where } I = A\bar{l}^2 + I_0 = (3 \times 1)0.5^2 + (3 \times 1^3)/12.$$

$$l' = \frac{(3 \times 1) \times 0.5^2 + (3 \times 1^3)/12}{(3 \times 1) \times 0.5} = 0.667 \text{ m}$$

Since this is for the vertical projected surface, l' is a vertical distance. The resultant force, F, is given by

$$F = \sqrt{23\ 100^2 + 14\ 715^2} = 27\ 390 \text{ N}$$

This will act at an angle $\tan^{-1}$ (23 100/14 715), i.e. 57.5° below the horizontal. The force is therefore disposed as shown in Figure 1.9b, with its line of action passing through 'O'.

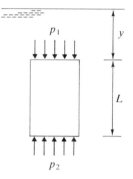

Figure 1.10   Pressure on immersed body.

1.4   Flotation

Buoyancy forces

Although civil engineers are not boat designers, they do have to deal with cases of buoyancy from time to time. Some typical examples are:

buried gas pipelines in waterlogged ground;
exploration rigs used by oil or gas corporations;
towing large steel dock/lock gates by sea or river (assuming that the structure can float, of course).

Figure 1.10 shows the vertical forces acting on an immersed cylinder of horizontal cross-sectional area A with its axis vertical.

The force acting downwards is due to pressure on the top surface i.e. $p_1A = \rho gyA$. The force acting upwards is due to pressure on the bottom surface, and is given by $p_2A = \rho g(y + L)A$. So

$$\text{total upthrust} = F_B = \rho g(y + L)A - \rho gyA = \rho gLA$$

Where LA is the volume of the cylinder. This leads to Archimedes' principle that 'the upthrust on a body is equal to the weight of the fluid displaced'. The upthrust acts through the centre of buoyancy B, which is the centre of gravity of the displaced fluid. For the case of a floating body, there must be equilibrium between F_B and the weight of the body.

Stable flotation

If a body is designed to float, it must do so in a stable fashion. This means that if the body suffers an angular displacement, it will automatically return to the original (correct) position. To appreciate how a floating body may exhibit this self-righting property, it must be realised that (with the

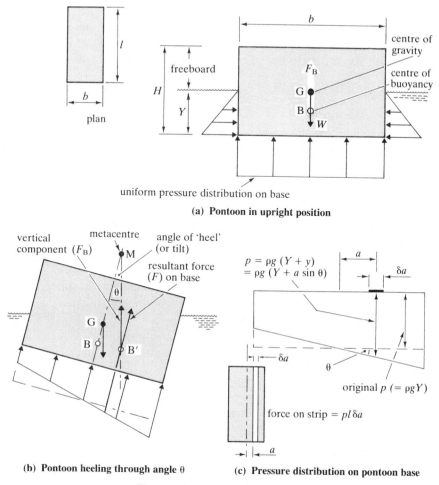

(a) **Pontoon in upright position**

(b) **Pontoon heeling through angle θ**   (c) **Pressure distribution on pontoon base**

Figure 1.11   Flotation of pontoon.

exception of circular sections) the angular displacement of a body causes a lateral displacement in the position of B. Consider, for example, the pontoon of rectangular section shown in Figure 1.11a. For upright flotation, the pressure distribution across the base is uniform. The centre of buoyancy lies on the vertical centreline. Since the pontoon is upright, the weight, W, must also be acting along the vertical centreline. If the vessel now rotates about its longitudinal axis ('heels over') through angle θ (Fig. 1.11b), the pressure distribution on the base becomes non-uniform, although still linear. This is, therefore, similar to the other pressure distribution patterns which have been examined. The shift from a uniform distribution (with the vessel upright) to a non-uniform distribution necessarily implies a corresponding shift in the position of the line of action of the resultant force. The buoyancy force now acts through B' rather than through the original centre of buoyancy. If a vertical line (representing the buoyancy force) is drawn through B', it intercepts the centreline of the vessel at point M, which is called the 'metacentre'. Using trigonometry, the distance $\overline{BB}'$ may be found:

$$\overline{BB}' = \overline{MB}' \sin \theta$$

Unfortunately, the distances $\overline{MB}'$ and $\overline{BB}'$ are both unknown. In order to evaluate them, use is made of the pressure distribution as follows.

The pressure, p, on the base at any distance a from the centreline is

$$p = \rho g (Y + a \sin \theta)$$

The corresponding force on an element of width δa and length l is

$$\delta F = p l \, \delta a = \rho g (Y + a \sin \theta) l \, \delta a$$

The vertical component of δF is the buoyancy component, so that

$$\delta F_B = \rho g (Y + a \sin \theta) l \, \delta a \cos \theta$$

The moment of this component about B is

$$\rho g (Y + a \sin \theta) l \, \delta a (\cos \theta) a$$

So, for the whole vessel the moment due to the buoyancy force is given by

$$\int_{-b/2}^{+b/2} \rho g (Y + a \sin \theta) l (\cos \theta) a \, da$$

to which the solution is

$$\text{moment} = \rho g (\sin \theta)(\cos \theta)(b^3 l / 12) = \rho g (\sin \theta)(\cos \theta) I$$

Remember, this is the moment due to the change in position of the centre of buoyancy. It has already been shown that buoyancy force $F_B = \rho g V$ and that $\overline{BB}' = \overline{MB}' \sin \theta$. Thus

$$\text{moment} = F_B \times \overline{BB}' = F_B \times \overline{MB}' \sin \theta$$

$$= \rho g (\sin \theta)(\cos \theta) I$$

Substituting for F_B and rearranging,

$$\overline{MB} = (I / V) \cos \theta$$

If θ is small, $\cos \theta \to 1$ and

$$\overline{MB} = I/V \tag{1.5}$$

The appearance of the second moment of area could have been anticipated in view of its association with linear non-uniform stress distributions.

The self-righting ability of a body is a function of the geometrical relationship between M and the centre of gravity G:

if M is above G, the body is stable (i.e. self-righting);
if M coincides with G, the body is neutrally stable and will neither capsize nor be self-righting;
if M is below G, the body is unstable and will capsize.

A glance at the disposition of the forces shows that the righting action is due to the moment of buoyancy force about the centre of gravity. This may be expressed as

$$\text{righting moment} = F_B \times \overline{MG}\, \theta$$

where θ is measured in radians. A number of simplifications are inherent in the derivation of these relationships. For example, the effect of the pressure distribution on each side has been ignored. However, as long as θ is small, the resulting errors are negligible.

Example 1.3   Buoyancy (uplift) force on quadrant gate

For a gate having the same dimensions as for Example 1.2, find the magnitude and direction of the force if the water is on the opposite side of the gate (Fig. 1.12).

Solution

(a) For vertical component,

area of sector (as before) = 0.785 m^2

volume of water displaced = 2.355 m^3

buoyancy force, $F_B = 9.81 \times 1000 \times 2.355 = 23\ 100$ N

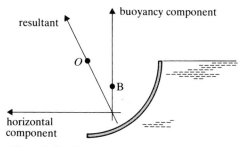

Figure 1.12   Pressure on quadrant gate (Example 1.3).

The centre of gravity of the sector is 0.424 m from the origin, so this will be the position of the centre of buoyancy.

(b) For horizontal component, centroid of vertical projected area is 0.5 m below the free surface of the water, so

$$\text{force} = \rho g A \bar{l} = 1000 \times 9.81 \times 3 \times 1 \times 0.5 = 14\,715 \text{ N}$$

The calculation of the position of the points of action is as before. Therefore, $l' = 0.667$ m in the vertical plane. The resultant force, F, is given by

$$F = \sqrt{23\,100^2 + 14\,715^2} = 27\,390 \text{ N}$$

The solution is numerically identical to that of Example 1.2. The only difference is that here each force is acting in the opposite direction to the corresponding force in Example 1.2.

Example 1.4 Pontoon stability

During serious river flooding, one span of a road bridge collapses. Government engineers decide to install a temporary pontoon bridge while repairs are in progress (Fig. 1.13). The river is 80 m wide and 7 m deep, and is not tidal. An outline specification of the temporary scheme is:

Clearance between pontoon base and river bed = 5.5 m
pontoon freeboard (i.e. distance from water line to deck) = 1.5 m
maximum pontoon self weight = 220 t
width of highway = 10 m
maximum side tilt due to 40 t vehicle load = 4°

Assume that the centre of gravity of the vehicle is 3 m above the bridge deck and 2 m off-centre. Assume that the pontoon centre of gravity lies at the geometrical centre of the pontoon.

Estimate the principal pontoon dimensions.

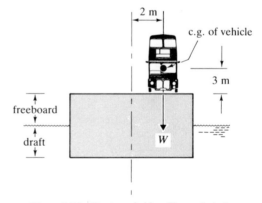

Figure 1.13 Pontoon bridge (Example 1.4).

Solution

$$\text{Pontoon draft (i.e. depth of immersion)} = 7 - 5.5 = 1.5 \text{ m}$$
$$\text{total height of pontoon} = \text{draft} + \text{freeboard} = 1.5 + 1.5 = 3 \text{ m}$$
$$\text{displacement due to pontoon} + \text{vehicle} = 220 + 40 = 260 \text{ t}$$

Therefore, the volume of water displaced is 260 m^3, since 1 m^3 of water has a mass of 1 t. So,

$$\text{pontoon length} = \frac{\text{volume displaced}}{\text{width} \times \text{draft}}$$

$$= \frac{260}{10 \times 1.5} = 17.333 \text{ m}$$

This is not a whole submultiple of the river width, so say that the pontoon length is 20 m, with a corresponding draft of 1.3 m.

Check that the pontoon stability is within specified limits:

$$\overline{\text{MB}} = \frac{I}{V} = \frac{(20 \times 10^3)/12}{260} = 6.41 \text{ m}$$

(from (1.5)). The pontoon centre of gravity is at its geometrical centre, 1.5 m above the base. The centre of gravity of the vehicle is 3 m above the deck, and therefore 6 m above the base of the pontoon. Taking moments about the pontoon base, the combined centre of gravity position is found:

$$\text{height of combined C.G. above the base} = \frac{(220 \times 1.5) + (40 \times 6)}{260}$$

$$= 2.192 \text{ m}$$

The centre of buoyancy, B, is at the centre of gravity of the displaced fluid. This lies at the geometrical centre of the immersed section of the pontoon, and is therefore 0.65 m above base.

Therefore, the distance $\overline{\text{MG}}$ between the metacentre and the centre of gravity is

$$\overline{\text{MG}} = 6.41 + 0.65 - 2.192 = 4.868 \text{ m}$$
$$\text{Overturning moment due to the vehicle} = 40 \times 2 = 80 \text{ tm}$$
$$\text{Righting moment} = 260 \times 4.868 \times \theta^c$$

For equilibrium, righting moment = overturning moment, so

$$\theta^c = \frac{80}{260 \times 4.868} = 0.062 \text{ rad} = 3.55°$$

which is within specification.

2

Principles of fluid flow

2.1 Introduction

The problems encountered in Chapter 1 involved only a small number of quantities, basically ρ, g and y. In consequence, the equations developed were simple and precise.

Thus, if the results of hydrostatic experimentation and calculation are compared, they are found to agree within the limits of experimental accuracy. Unfortunately, this combination of simplicity and accuracy does not apply when we turn our attention to problems involving fluid flows. Engineering flows are mostly very complex, and it is not usually possible to evolve precise theoretical models.

However, a great deal can be learnt about flows by adopting the techniques and equations which were developed by the hydrodynamicists. These equations are reasonably straightforward because they are developed for the case of an 'ideal' fluid. An ideal fluid has no viscosity (i.e. it is inviscid), has no surface tension and is incompressible. Viscosity and compressibility are the major reasons for the complexity of real fluid flows. Of course, no such substance as an 'ideal' fluid actually exists. Nevertheless, for certain types of problem the equations for 'ideal' flows are remarkably accurate.

2.2 Classification of flows

Flows can be classified in a number of ways. The system generally adopted is to consider the flow as being characterised by two parameters – time and distance. The class into which any particular flow falls is usually a reliable guide to the appropriate method of solution.

The first major subdivision is based on consideration of the timescale. This categorises all flows as either steady or unsteady.

A flow is steady if the parameters describing that flow do not vary with time. Typical parameters of a flow are velocity, discharge (volume per second passing a given point), pressure, or depth of flow (e.g. in a river or channel). Conversely, a flow is unsteady if these parameters do vary with time.

Because of the complex equations associated with unsteady flows, engineers often use steady flow equations even where a small degree of unsteadiness is present. A case in point is that of flows in rivers, where the discharge is rarely, if ever, absolutely steady. Nevertheless, if changes occur slowly, steady flow equations give quite accurate results.

The second subdivision relates to the scale of distance. This classifies flows as being uniform or non-uniform.

A flow is uniform if the parameters describing the flow do not vary with distance along the flow path. Conversely, for a non-uniform flow, the magnitude of the parameters varies from point to point along the flow path.

The existence of a uniform flow necessarily implies that the area of the cross section perpendicular to the direction of flow is constant. A typical example is that of the flow in a pipeline of constant diameter. By contrast, one would have a non-uniform flow in a tapered pipe.

The two sets of subdivisions or classifications (steady–unsteady, uniform–non-uniform) are not mutually exclusive. Some flows exhibit changes with respect to both time and distance, while others change with respect to time or distance only. However, the majority of flows will fall into one of the classifications listed below.

Steady uniform flow. For such a flow the discharge is constant with time, and the cross section through which the flow passes is of constant area. A typical example is that of constant flow through a long straight pipe of uniform diameter.

Steady non-uniform flow. The discharge is constant with time, but the cross-sectional area varies with distance. Examples are flow in a tapering pipe and flow with constant discharge in a river (the cross section of a river usually varies from point to point).

Unsteady uniform flow. The cross section through the flow is constant, but the discharge varies with time. This is a complex flow pattern. An example is that of pressure surge in a long straight pipe of uniform diameter.

Unsteady non-uniform flow, in which the cross section and discharge vary with both time and distance. This is typified by the passage of a flood wave in a natural channel, and is the most complex flow to analyse.

2.3   Visualisation of flow patterns

Streamlines

The fundamental method of visualising a flow pattern is by means of 'streamlines'. A streamline is constructed by drawing a line which is tangential to the velocity vectors of a connected series of fluid particles (Fig.

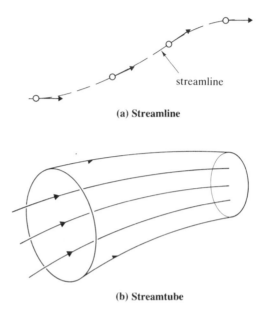

(a) Streamline

(b) Streamtube

Figure 2.1   Definitions of a streamline and a streamtube.

2.1a). The streamline is thus a line representing the direction of flow of the series of particles at a given instant. Because the streamline is always tangential to the flow, it follows that there is no flow across a streamline.

A set of streamlines may be arranged to form an imaginary pipe or tube. This is known as a 'streamtube' (Fig. 2.1b). Under certain specific circumstances, streamtubes can actually be identified. For example, the internal surface of a pipeline must also be a streamtube, since the vectors representing the flow adjacent to the surface must be parallel to that surface. The surface is therefore 'covered' with streamlines.

The hydrodynamicists developed a theoretical framework which makes it possible to construct streamlines for a variety of ideal flows, by using a graphical technique. The resulting diagrams are known as 'flow nets', and these are discussed in Section 2.9.

Streaklines

Clearly, it will be difficult to construct a streamline for a real flow, since individual particles of fluid are not visible to the eye. A simple method of obtaining approximate information regarding streamline patterns is to inject a dye into the flow. The dye will trace out a path known as a 'streakline', which may be photographed. Usually the characteristics of the dye (its density, etc.) will not be identical to those of the fluid. The streakline, therefore, will not necessarily be absolutely identical to the streamline.

One-, two- and three-dimensional flow

Most real flows are three-dimensional, in that velocity, pressure and other parameters may vary in three directions (x, y, z). There may also be variation of the parameters with time.

In practice it is nearly always possible to consider the flow to be one- or two-dimensional. This greatly simplifies the equations of flow. Minor adjustments to these simplified equations can often be incorporated to allow for a two- or three-dimensional flow. For example, steady uniform flow in a pipe is considered to be one-dimensional, the flow being characterised by a streamline along the centreline of the pipe. The velocity and pressure variations across the pipe are ignored, or are considered separately as secondary effects.

2.4   The fundamental equations of fluid dynamics

Description and physical basis

In order to develop the equations which describe a flow, hydrodynamicists assumed that fluids are subject to certain fundamental laws of physics. The pertinent laws are:

(a) conservation of matter (conservation of mass);
(b) conservation of energy;
(c) conservation of momentum.

These principles were initially developed for the case of a solid body, and it is worth expanding them a little before proceeding.

(a) The law of conservation of matter stipulates that matter can be neither created nor destroyed, though it may be transformed (e.g. by a chemical process). Since this study of the mechanics of fluids excludes chemical activity from consideration, the law reduces to the principle of conservation of mass.
(b) The law of conservation of energy states that energy may be neither created nor destroyed. Energy can be transformed from one guise to another (e.g. potential energy can be transformed into kinetic energy), but none is actually lost. Engineers sometimes loosely refer to 'energy losses' due to friction, but in fact the friction transforms some energy into heat, so none is really 'lost'. The basic equation may be derived from the First Law of Thermodynamics, though here it will be obtained in a slightly simplified manner.
(c) The law of conservation of momentum states that a body in motion cannot gain or lose momentum unless some external force is applied. The

classical statement of this law is Newton's Second Law of Motion, i.e.

force = rate of change of momentum.

Applying these laws to a solid body is relatively straightforward, since the body will be of measurable size and mass. For example, if the mass of a body is known, the force required to produce a given acceleration is easily calculated. However, for the case of a fluid the attempt to apply these laws presents a problem. A flowing fluid is a continuum – that is to say, it is not possible to subdivide the flow into separate small masses. How, then, can the three basic laws be applied? How can a suitable mass of fluid be identified so that we can investigate its momentum or its energy? The answer lies in the use of 'control volumes'.

Control volumes

A control volume is a purely imaginary region within a body of flowing fluid. The region is usually (though not always) at a fixed location and of fixed size. Inside the region, all of the dynamic forces cancel each other. Attention may therefore be focused on the forces acting externally on the control volume. The control volume may be of any shape. Therefore, a shape may be selected which is most convenient for any particular application. In the work which follows, a short streamtube (Fig 2.1b) will be used as a control volume. This may be visualised as a short transparent pipe or tube. Fluid enters through one end of the tube and leaves through the other. Any forces act externally along the boundaries of the control volume.

2.5 Application of the conservation laws to fluid flows

The continuity equation (principle of conservation of mass)

During any time interval δt, the principle of conservation of mass implies that for any control volume the mass flow entering minus the mass flow leaving equals the change of mass within the control volume.

If the flow is steady, then the mass must be entering (or leaving) the volume at a constant rate. If we further restrict our attention to incompressible flow, then the mass of fluid within the control volume must remain fixed. In other words, the change of mass within the control volume is zero. Therefore, during time δt,

mass flow entering = mass flow leaving

Since the flow is incompressible, the density of the fluid is constant throughout the fluid continuum. Mass flow entering may be calculated by taking the product

(density of fluid) × (volume of fluid entering per second)

Mass flow is therefore represented by the product ρQ, hence

$$\rho Q(\text{entering}) = \rho Q(\text{leaving})$$

But since flow is incompressible, the density is constant, so

$$Q(\text{entering}) = Q(\text{leaving}) \qquad (2.1a)$$

This is the 'continuity equation' for steady incompressible flow.

The dimensions of Q are L^3T^{-1} (SI units m^3/s). This can be expressed alternatively as L^2LT^{-1}, which is the product of an area and a velocity. Supposing that measurements are taken of the velocity of flow across the entry to the control volume, and that the velocity is constant at u_1 m/s. Then, if the cross-sectional area of the streamtube at entry is A_1,

$$Q(\text{entering}) = u_1 A_1$$

Note that the area must be that of a plane perpendicular to the direction of flow. Similar remarks apply to the flow leaving the volume. Thus, if the velocity of flow leaving the volume is u_2 and the area of the streamtube at exit is A_2, then

$$Q(\text{leaving}) = u_2 A_2$$

Therefore, the continuity equation may also be written as

$$u_1 A_1 = u_2 A_2 \qquad (2.1b)$$

The energy equation (principle of conservation of energy)

This may be simply derived by considering the forms of energy available to the fluid. Figure 2.2 shows the control volume used to develop the equation. The combination of a flow and a pressure implies that work is done. Thus, if pressure p acts on area A, the corresponding force is pA. If the fluid is flowing, then in travelling through a length L the work done equals the product of force and distance, i.e.

$$\text{'flow work' done} = pAL$$

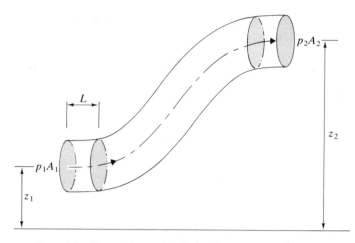

Figure 2.2   Streamtube used to derive the energy equation.

For the control volume under consideration, the fluid entering the system travels through distance L during time interval δt. The flow work done during this time is $p_1 A_1 L$. The mass, m, entering the system is $\rho_1 A_1 L$ during δt. Therefore the kinetic energy (K.E.) entering the system is

$$\text{K.E.} = \tfrac{1}{2} m u^2 = \tfrac{1}{2} \rho_1 A_1 L u_1^2$$

The potential energy of a body is mgz, where z is the height of the body above some arbitrary datum. The potential energy of the mass $\rho_1 A_1 L$ entering during δt is $\rho_1 A_1 Lgz$. The total energy entering is the sum of the flow work done and the potential and kinetic energies, i.e.

$$p_1 A_1 L + \tfrac{1}{2} \rho_1 A_1 L u_1^2 + \rho_1 A_1 lgz_1$$

It is more convenient to consider the energy per unit weight of fluid. Note that the weight of fluid entering the system is $\rho_1 g A_1 L$, so the energy per unit weight of fluid is

$$\frac{p_1 A_1 L}{\rho_1 g A_1 L} + \frac{1}{2}\frac{\rho_1 A_1 L u_1^2}{\rho_1 g A_1 L} + \frac{\rho_1 A_1 Lgz_1}{\rho_1 g A_1 L} = \frac{p_1}{\rho_1 g} + \frac{u_1^2}{2g} + z_1$$

Similarly, at the exit the total energy per unit weight of fluid leaving the system is

$$\frac{p_2}{\rho_2 g} + \frac{u_2^2}{2g} + z_2$$

If, during the passage from entry to exit, no energy is supplied to the fluid or extracted from the fluid, then clearly

$$\text{energy entering} = \text{energy leaving}$$

If the flow is incompressible, then $\rho_1 = \rho_2 = \rho$, hence

$$\frac{p_1}{\rho g} + \frac{u_1^2}{2g} + z_1 = \frac{p_2}{\rho g} + \frac{u_2^2}{2g} + z_2 = H = \text{constant} \qquad (2.2)$$

This is the 'Bernoulli equation' named after Daniel Bernoulli (1700–82), who published one of the first books on fluid flow in 1738.

Several points should be made about this equation.

(a) The statement that 'no energy is extracted from the fluid' implies that the fluid is frictionless. If this were not so, frictional forces would transform some of the energy into heat, then

$$\text{energy entering at } 1 > \text{energy leaving at } 2$$

i.e. there would be a 'loss' of energy between 1 and 2.

(b) The reader should check the dimensions of each separate component of the Bernoulli equation:

$$\frac{p}{\rho g} \equiv \frac{\text{N}}{\text{m}^2} \cdot \frac{\text{m}^3}{\text{kg}} \cdot \frac{\text{s}^2}{\text{m}} \equiv \frac{ML}{L^2 T^2} \cdot \frac{L^3}{M} \cdot \frac{T^2}{L} = L \text{ (m)}$$

$$\frac{u^2}{2g} \equiv \frac{\text{m}^2}{\text{s}^2} \cdot \frac{\text{s}^2}{\text{m}} \equiv \frac{L^2}{T^2} \cdot \frac{T^2}{L} = L \text{ (m)}$$

$$z \equiv L \text{ (m)}$$

Thus, all the constituent parts of the equation have units of metres. For this reason, each term may be regarded as a 'head': $p/\rho g = $ pressure head, $u^2/2g = $ kinetic, or velocity head, $z = $ potential or elevation head.

(c) The equation was developed for a streamtube of finite area A. However, A must be 'small', otherwise the height (z) of a streamline at the bottom of the tube is significantly less than that of a streamline at the top of the tube. Strictly, therefore, Bernoulli's equation applies along a single streamline. However, if the diameter of the appropriate control volume is small compared with its length, engineers apply Bernoulli's equation between two points without specific reference to a particular streamline.

It is possible to derive Bernoulli's equation in a mathematically rigorous fashion, as is now demonstrated.

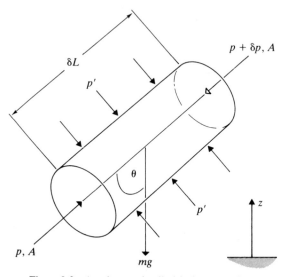

Figure 2.3   An elemental cylindrical streamtube.

Figure 2.3 shows a small (elemental) streamtube of cylindrical section. Taking this to be a control volume at an instant in time, the forces acting in the direction of flow along the streamtube are:

Pressure forces

upstream end	$+ pA$
downstream end	$-(p + \delta p)A$
circumference	zero (pressure forces p' cancel)

Weight force

$$- mg \cos \theta = - \rho A\, \delta L\, g \cos \theta$$

Newton's Second Law (force = mass × acceleration) may be used to relate these forces:

$$pA - (p + \delta p)A - \rho A\, \delta L\, g \cos \theta = \rho A\, \delta L\, (du/dt)$$

Setting $\delta L \cos \theta = \delta z$ and cancelling terms,

$$- A\, \delta p - \rho Ag\, \delta z = \rho A\, \delta L\, (du/dt)$$

Dividing by $\rho A\, \delta L$, and in the limit

$$\frac{1}{\rho}\frac{dp}{dL} + \frac{du}{dt} + g\frac{dz}{dL} = 0$$

For steady flow, velocity (u) only varies with distance (L). Therefore,

$$\frac{du}{dt} = \frac{dL}{dt}\frac{du}{dL} = u\frac{du}{dL}$$

Hence, substituting for du/dt,

$$\frac{1}{\rho}\frac{dp}{dL} + u\frac{du}{dL} + g\frac{dz}{dL} = 0 \tag{2.3}$$

Equation (2.3) is known as Euler's equation (in honour of the Swiss mathematician Leonhard Euler (1707–83)).

For incompressible fluids, Euler's equation may be integrated to yield

$$\frac{p}{\rho} + \frac{u^2}{2} + gz = \text{constant}$$

or, dividing by g,

$$\frac{p}{\rho g} + \frac{u^2}{2g} + z = \text{constant}$$

This is, of course, Bernoulli's equation (cf. (2.2)).

The momentum equation (principle of conservation of momentum)

Again, consider the streamtube shown in Fig. 2.2 and apply Newton's Second Law in the form force = rate of change of momentum (i.e. $F = d(mu)/dt$). In a time interval δt,

$$\text{momentum entering} = \rho\,\delta Q_1\,\delta t\,u_1$$

$$\text{momentum leaving} = \rho\,\delta Q_2\,\delta t\,u_2$$

and $\delta Q_1 = \delta Q_2 = \delta Q$ by the continuity principle. Hence, the force required to produce this change of momentum is

$$\delta F = \frac{\rho\,\delta Q\,\delta t(u_2 - u_1)}{\delta t} = \rho\,\delta Q(u_2 - u_1)$$

As the velocities may have components in the x-, y- and z-directions, it is more convenient to consider each component separately. Hence,

$$\delta F_x = \rho\,\delta Q(u_{2x} - u_{1x})$$

and similarly for δF_y and δF_z. Integrating over a region,

$$F_x = \rho Q(V_{2x} - V_{1x}) \qquad (2.4)$$

(similarly for F_y and F_z) provided that the velocity (V) is uniform across the collection of streamtubes.

Equation (2.4) is the momentum equation for steady flow for a region of uniform velocity. The momentum force is composed of the sum of all external forces acting on the streamtube (control volume) and may include pressure forces (F_P) and reaction forces (F_R), i.e.

$$F_x = \sum(F_P + F_R) \qquad (2.5)$$

This is explained more fully in Section 2.7, where the momentum equation is used to find the forces acting on pipe bends, etc.

Energy and momentum coefficients

In deriving the momentum equation, it was stressed that it could only be applied directly to a large region if the velocity (u) was constant across the region (set equal to V). The energy equation was stated to apply to a streamline with velocity u. In many real fluid flow problems, u varies across a section (refer, for example, to pipe flow in Chapter 4). The velocity distribution across a section is not always known, and the mean velocity V (defined as discharge divided by area) must be used instead of u. Both the energy and momentum equations may still be used by introducing the energy and momentum coefficients, α and β, respectively. These are defined as follows:

$$\alpha = \int u^3 dA / V^3 A \qquad (2.6)$$

$$\beta = \int u^2 dA / V^2 A \qquad (2.7)$$

The derivation of the equation for the energy coefficient rests on the principle that the total kinetic energy possessed by a bundle of streamtubes is equal to α multiplied by the kinetic energy represented by the mean velocity V. Stated mathematically,

$$\sum \tfrac{1}{2} m u^2 = \alpha \tfrac{1}{2} M V^2$$

or for unit time

$$\sum \tfrac{1}{2} \rho \delta Q u^2 = \alpha \tfrac{1}{2} \rho Q V^2$$

hence, as $\delta Q = u\,\delta A$ and $Q = VA$,

$$\sum \tfrac{1}{2}\rho u^3\,\delta A = \alpha \tfrac{1}{2}\rho V^3 A$$

This leads to (2.6). The momentum coefficient β may be derived similarly by equating momentum for a bundle of streamtubes.

The Bernoulli energy equation may then be rewritten in terms of mean velocity V as

$$\frac{p}{\rho g} + \frac{\alpha V^2}{2g} + z = \text{constant} \tag{2.8}$$

and the momentum equation as

$$F_x = \rho Q \beta (V_{2x} - V_{1x}) \tag{2.9}$$

The values of α and β must be derived from the velocity distributions across a region. They always exceed unity, but usually by only a small margin, so they are frequently omitted. However, this is not always the case, and they should therefore not be forgotten.

Example calculations for these coefficients for open-channel flow are given in Chapter 5. In the case of turbulent pipe flow, $\alpha = 1.06$ and $\beta = 1.02$, and both may safely be ignored.

2.6 Application of the energy equation

Bernoulli's equation may be applied to any continuous flow system. The simplest system might be a single pipeline with a frictionless fluid discharging through it.

For such a system, (2.2) may be rewritten as

$$\frac{p_1}{\rho g} + \frac{u_1^2}{2g} + z_1 = \frac{p_2}{\rho g} + \frac{u_2^2}{2g} + z_2 = \text{constant}$$

where subscripts 1 and 2 refer to two points along any streamline. Hence Bernoulli's energy equation, in conjunction with the continuity equation, may be used to determine the variation of pressure and velocity along any streamline.

Example 2.1 Application of Bernoulli's equation

For the frictionless syphon shown in Fig. 2.4, determine the discharge and the pressure heads at A and B, given that the pipe diameter is 200 mm and the nozzle exit diameter is 150 mm.

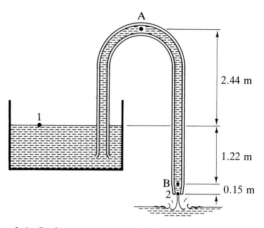

Figure 2.4 Syphon arrangement for Example 2.1.

Solution

To find the discharge, first apply Bernoulli's equation along the streamline between 1 and 2:

$$\frac{p_1}{\rho g} + \frac{u_1^2}{2g} + z_1 = \frac{p_2}{\rho g} + \frac{u_2^2}{2g} + z_2$$

At both 1 and 2, the pressure is atmospheric, and at 1 the velocity is negligible, hence

$$z_1 - z_2 = \frac{u_2^2}{2g}$$

From Figure 2.4,

$$z_1 - z_2 = 1.22 + 0.15 = 1.37 \text{ m}$$

hence

$$\frac{u_2^2}{2g} = 1.37$$

or

$$u_2 = 5.18 \text{ m/s}$$

Next apply the continuity equation to find the discharge:

$$Q = uA$$
$$= 5.18 \times \left(\frac{\pi \times 0.15^2}{4} \right)$$

hence

$$Q = 0.092 \text{ m}^3/\text{s}$$

To find the pressure head at A, again apply Bernoulli's equation along the streamline from 1 to A

$$\cancel{\frac{p_1}{\rho g}}^{0} + \cancel{\frac{u_1^2}{2g}}^{0} + z_1 = \frac{p_A}{\rho g} + \frac{u_A^2}{2g} + z_A$$

as $p_1 = 0$ and $u_1 = 0$ then

$$\frac{p_A}{\rho g} = (z_1 - z_A) - \frac{u_A^2}{2g}$$

$$= -2.44 - \frac{u_A^2}{2g}$$

and

$$u_A A_A = Q$$

$$u_A = 0.092 \bigg/ \left(\frac{\pi \times 0.2^2}{4}\right) = 2.93 \text{ m/s}$$

$$\frac{u_A^2}{2g} = 0.44 \text{ m}$$

hence

$$\frac{p_A}{\rho g} = -2.44 - 0.44 = -2.88 \text{ m}$$

The pressure head at B is found similarly:

$$\cancel{\frac{p_1}{\rho g}}^{0} + \cancel{\frac{u_1^2}{2g}}^{0} + z_1 = \frac{p_B}{\rho g} + \frac{u_B^2}{2g} + z_B \ (p_1, u_1 \rightarrow 0)$$

$$\frac{p_B}{\rho g} = (z_1 - z_B) - \frac{u_B^2}{2g}$$

$$(z_1 - z_B) = 1.22 \text{ m}$$

$$u_B = u_A = 2.93 \text{ m/s}$$

$$\frac{p_B}{\rho g} = 1.22 - 0.44 = 0.78 \text{ m}$$

It may be observed that the pressure head at A is negative. This is because atmospheric pressure was taken as zero (i.e. pressures are gauge pressures) and therefore the pressure head at A is still positive in terms of absolute pressure, because atmospheric pressure is about 10.1 m head of water.

Modifications to Bernoulli's equation

In practice, the total energy of a streamline does not remain constant. Energy is 'lost' through friction, and external energy may be added by means of a pump or extracted by a turbine. These energy gains and losses are explained in later chapters, but it is worth introducing them into Bernoulli's equation at this point so that it can be extended to cover a wider range of practical cases. Consider a streamline between two points 1 and 2. If the energy head lost through friction is denoted by h_f and the external energy head added (say by a pump) is E, then Bernoulli's equation may be rewritten as

$$H_1 + E = H_2 + h_f$$

or

$$\frac{p_1}{\rho g} + \frac{u_1^2}{2g} + z_1 + E = \frac{p_2}{\rho g} + \frac{u_2^2}{2g} + z_2 + h_f$$

This equation is really a restatement of the First Law of Thermodynamics for an incompressible fluid.

Example 2.2   Application of Bernoulli's equation with energy gains and losses

A pump delivers water from a lower to a higher reservoir. The difference in elevation between the reservoirs is 10 m. The pump provides an energy head of 11 m and the frictional head losses are 0.7 m. If the pipe diameter is 300 mm, calculate the discharge.

Solution

Apply the modified Bernoulli's equation from the lower to the higher reservoir:

$$H_1 + 11 = H_2 + 0.7$$

or

$$\frac{p_1}{\rho g} + \frac{u_1^2}{2g} + z_1 + 11 = \frac{p_2}{\rho g} + \frac{u_2^2}{2g} + z_2 + 0.7$$

Taking $p_1 = p_2 = 0$ and $u_1 = 0$,

$$\frac{u_2^2}{2g} = (z_1 - z_2) + 11 - 0.7$$

$$= -10 + 11 - 0.7$$

$$= 0.3 \text{ m}$$

$$u_2 = 2.426 \text{ m/s}$$

Applying the continuity equation,

$$Q = u_2 \left(\frac{\pi \times 0.3^2}{4} \right) = 0.171 \text{ m}^3/\text{s}$$

2.7   Application of the momentum equation

Range of applications

The momentum equation may be used directly to evaluate the force causing a change of momentum in a fluid. Such applications include determining forces on pipe bends and junctions, nozzles and hydraulic machines. Examples of some of these are given in this section.

In addition, the momentum equation is used to solve problems in which energy losses occur that cannot be evaluated directly, or when the flow is unsteady. Examples of such problems include local head losses in pipes, the hydraulic jump and unsteady flow in pipes and channels. These applications are discussed in later chapters.

Forces exerted on pipework

Whenever there is a change in geometry or direction, the fluid will exert a force on the pipework. These forces may be considerable, and must be resisted in order that the pipeline does not move. For underground pipes, the forces are normally resisted by thrust blocks which transfer the force to the surrounding earth. For exposed pipework, the forces are transmitted by supports at the pipe joints to the nearest structural member (e.g. a wall or beam). The calculation of these forces is now illustrated by three examples.

Example 2.3   Force exerted by a firehose

Calculate the force required to hold a firehose for a discharge of 5 l/s if the nozzle has an inlet diameter of 75 mm and an outlet diameter of 25 mm.

Solution

The nozzle, surrounded by a suitable control volume, is shown in Figure 2.5. The forces acting in the x-direction on the control volume are the pressure forces (F_P)

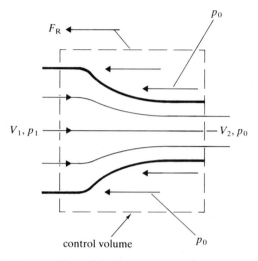

Figure 2.5 Forces on a nozzle.

and the reaction force (F_R) on the fluid. The sum of these must equal the momentum force (F_M). Hence (2.4) and (2.5) apply, i.e.

$$F_M = F_P + F_R$$

(where all terms are taken positive with x) and

$$F_M = \rho Q(V_2 - V_1)$$

V_1 and V_2 may be found using the continuity equation:

$$V_1 = Q/A_1 = 0.005 \bigg/ \left(\frac{\pi \times 0.075^2}{4}\right)$$

$$= 1.13 \ (\text{m/s})$$

and

$$V_2 = Q/A_2 = 0.005 \bigg/ \left(\frac{\pi \times 0.025^2}{4}\right)$$

$$= 10.19 \ (\text{m/s})$$

The pressure forces are

$$F_P = p_1 A_1 - p_0 A_2 - p_0(A_1 - A_2)$$

The pressure p_1 may be found using Bernoulli's equation:

$$\frac{p_1}{\rho g} + \frac{V_1^2}{2g} = \frac{p_0}{\rho g} + \frac{V_2^2}{2g}$$

Taking $p_0 = 0$ (Note: gauge pressures may be used because atmospheric pressure p_0 acts over the whole area A_1),

$$\frac{p_1}{\rho g} = \frac{V_2^2 - V_1^2}{2g}$$

or

$$p_1 = (\rho/2)(V_2^2 - V_1^2) = (1000^3/2)(10.19^2 - 1.13^2)$$

$$= 51.28 \text{ kN/m}^2$$

hence

$$F_P = 51.28 \times 10^3 \times \pi \times 0.075^2/4$$

$$= 0.226 \text{ kN}$$

The momentum force is

$$F_M = 10^3 \times 0.005(10.19 - 1.13)$$

$$= 0.0453 \text{ kN}$$

Hence, the reaction force is

$$F_R = F_M - F_P$$

$$= 0.0453 - 0.226$$

$$= -0.181 \text{ kN}$$

F_R is the force exerted by the nozzle on the fluid. The fireman must, of course, provide a force of equal magnitude but opposite direction.

Example 2.4   Force on a pipe bend

Calculate the magnitude and direction of the force exerted by the pipe bend shown in Figure 2.6 if the diameter is 600 mm, the discharge is 0.3 m^3/s and the upstream pressure head is 30 m.

Solution

In this case, the force exerted is due to the change of direction and the x- and y-components of the force must be calculated separately. Hence

$$F_{Mx} = F_{Px} + F_{Rx}$$

$$F_{My} = F_{Py} + F_{Ry}$$

and

$$F_{Mx} = \rho Q(V_{2x} - V_{1x})$$
$$F_{My} = \rho Q(V_{2y} - V_{1y})$$

As the pipe is of constant diameter

$$V_2 = V_1 = 0.3 \Big/ \left(\frac{\pi \times 0.6^2}{4}\right)$$

$$= 1.06 \text{ m/s}$$

Neglecting the small energy loss around the pipe bend,

$$p_2 = p_1 = 30\rho g = 294.3 \text{ kN}$$

Pressure forces

$$F_{Px} = p_1 A_1 - 0 = 294.3 \times \pi \times 0.6^2/4$$

$$= 83.21 \text{ kN}$$

and

$$F_{Py} = 0 - p_2 A_2 = -83.21 \text{ kN}$$

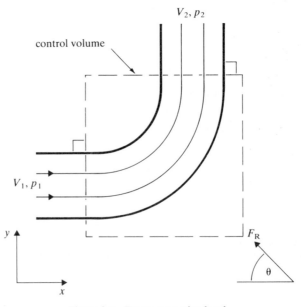

Figure 2.6 Forces on a pipe bend.

Momentum forces

$$F_{Mx} = \rho Q(0 - V_1)$$
$$= 10^3 \times 0.3(-1.06)$$
$$= -0.318 \text{ kN}$$

and

$$F_{My} = \rho Q(V_2 - 0)$$
$$= +0.318 \text{ kN}$$

Reaction forces

$$F_{Rx} = F_{Mx} - F_{Px}$$
$$= -0.318 - 83.21$$
$$= -83.528 \text{ kN}$$

and

$$F_{Ry} = F_{My} - F_{Py}$$
$$= 0.318 - (-83.21)$$
$$= 83.528 \text{ kN}$$

hence

$$F_R = \sqrt{F_{Rx}^2 + F_{Ry}^2}$$
$$= 118.1 \text{ kN}$$

and

$$\theta = \tan^{-1}(F_{Ry}/F_{Rx})$$
$$= 45° \text{ (from negative } x\text{-direction to positive } y\text{-direction)}$$

Example 2.5   Force on a T-junction

Calculate the magnitude and direction of the force exerted by the T-junction shown in Figure 2.7 if the discharges are $Q_1 = 0.3 \text{ m}^3/\text{s}$, $Q_2 = 0.15 \text{ m}^3/\text{s}$, $Q_3 = 0.15 \text{ m}^3/\text{s}$, the diameters are $D_1 = 450 \text{ mm}$, $D_2 = 300 \text{ mm}$, $D_3 = 200 \text{ mm}$ and the upstream pressure $p_1 = 500 \text{ kN/m}^2$.

Solution

In this case there are changes of direction and pressure and velocity. First find the three velocities by continuity, then apply Bernoulli's equation to find the pressures p_2 and p_3. Then apply the momentum equation.

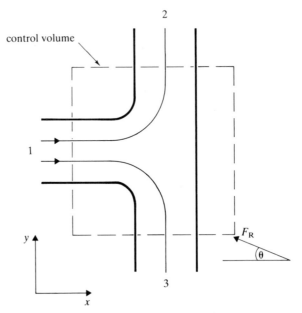

Figure 2.7   Forces on a T-junction.

Velocities

$$V_1 = Q_1/A_1 = 1.886 \text{ m/s}$$
$$V_2 = Q_2/A_2 = 2.122 \text{ m/s}$$
$$V_3 = Q_3/A_3 = 4.775 \text{ m/s}$$

Pressures

$$\frac{p_1}{\rho g} + \frac{V_1^2}{2g} = \frac{p_2}{\rho g} + \frac{V_2^2}{2g} = \frac{p_3}{\rho g} + \frac{V_3^2}{2g}$$

hence

$$p_2 = p_1 + \rho(V_1^2 - V_2^2)/2 = 499.53 \text{ kN/m}^2$$
$$p_3 = p_1 + \rho(V_1^2 - V_3^2)/2 = 490.38 \text{ kN/m}^3$$

Pressure forces

$$F_{Px} = p_1 A_1 - 0 = 79.52 \text{ kN}$$
$$F_{Py} = p_3 A_3 - p_2 A_2 = -19.96 \text{ kN}$$

Momentum forces

$$F_{Mx} = 0 - \rho Q_1 V_1 = -0.566 \text{ kN}$$
$$F_{My} = \rho Q_2 V_2 + (-\rho Q_3 V_3) - 0 = -0.40 \text{ kN}$$

Reaction forces

$$F_{Rx} = F_{Mx} - F_{Px} = -80.09 \text{ kN}$$

$$F_{Ry} = F_{My} - F_{Py} = +19.50 \text{ kN}$$

hence

$$F_R = \sqrt{F_{Rx}^2 + F_{Ry}^2} = 82.43 \text{ kN}$$

and

$$\theta = \tan^{-1}(F_{Ry}/F_{Rx}) = 13.7° \text{ (from negative } x\text{-direction to positive } y\text{-direction)}$$

2.8   Velocity and discharge measurement

Velocity measurement

The pitot tube, shown diagramatically in Figure 2.8a, is used to measure velocity. At the nose of the pitot tube, the fluid is brought to rest and the height of the fluid in the pitot tube therefore corresponds to $p/\rho g + u^2/2g$.

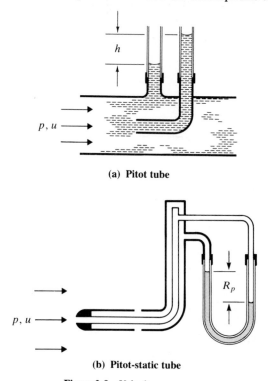

(a) **Pitot tube**

(b) **Pitot-static tube**

Figure 2.8   Velocity measurement.

This is known as the stagnation pressure. The pressure head ($p/\rho g$) is measured separately by a second tube, and hence

$$\frac{p}{\rho g} + \frac{u^2}{2g} = \frac{p}{\rho g} + h$$

or

$$u = \sqrt{2gh} \qquad (2.10)$$

The two pressure heads are normally measured by a single integrated instrument called the pitot-static tube, as shown in Figure 2.8b. This instrument, when connected to a suitable manometer, may be used to measure point velocities in pipes, channels and wind tunnels.

Discharge measurement in pipelines

The venturi meter, shown in Figure 2.9a, is an instrument which may be used to measure discharge in pipelines. It essentially consists of a narrowed section tapering out to the pipe diameter at each end. In the throat section, the velocity is increased, and consequently the pressure is decreased. By measuring the difference in pressure, an estimate of discharge may be made as follows.

Consider a streamline from the upstream position (1) to the throat position (2). Then

$$\frac{p_1}{\rho g} + \frac{V_1^2}{2g} = \frac{p_2}{\rho g} + \frac{V_2^2}{2g}$$

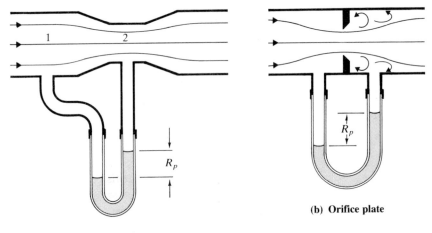

(a) Venturi meter

(b) Orifice plate

Figure 2.9 Discharge measurement.

(assuming no energy losses and using mean velocities), or

$$V_2^2 - V_1^2 = 2g\left(\frac{p_1}{\rho g} - \frac{p_2}{\rho g}\right)$$

The pressure head difference is indicated by the manometer reading, R_p, hence

$$V_2^2 - V_1^2 = 2gR_p\left(\frac{\rho_g}{\rho} - 1\right)$$

where ρ_g is the density of the gauge fluid. Also, by continuity,

$$Q = V_1 A_1 = V_2 A_2$$

Substituting for V_2 in terms of V_1 yields

$$V_1^2[(A_1/A_2)^2 - 1] = 2gh^*$$

where

$$h^* = R_p\left(\frac{\rho_g}{\rho} - 1\right)$$

Solving for V_1,

$$V_1 = \frac{1}{\sqrt{(A_1/A_2)^2 - 1}}\sqrt{2gh^*}$$

hence

$$Q_{ideal} = A_1 V_1 = \frac{A_1}{\sqrt{(A_1/A_2)^2 - 1}}\sqrt{2gh^*}$$

Taking $m = D_1/D_2$,

$$Q_{ideal} = \frac{\pi D_1^2}{4}\left(\frac{1}{\sqrt{m^4 - 1}}\right)\sqrt{2gh^*} \tag{2.11}$$

The actual discharge will be slightly less than this, due to energy losses in the converging section. These energy losses are accounted for by introducing a coefficient of discharge C_d such that $Q_{actual} = C_d Q_{ideal}$. Energy losses may be minimised by careful design, as evidenced in British Standard 1042, Part I Section 1.1 (1981) or the *Water measurement manual* (USBR 1967). Venturi meters designed in accordance with these standards have C_d values between 0.97 and 0.99.

It is useful to note that the indicated head (R_p), and consequently the discharge equation, is independent of the inclination of the meter. You may care to prove this for yourself, by deriving the expression for h^* for an inclined venturi meter.

A second device (shown in Fig. 2.9b) for measuring discharge in a pipeline is an orifice plate. Its function is similar to the venturi meter, in that a region of low pressure is created by a local constriction. The same discharge equation therefore applies. However, the major difference between the devices lies in the fact that, downstream of the orifice plate, the flow area expands instantaneously while the fluid is unable to expand at the same rate. This creates a 'separation zone' of turbulent eddies in which large energy losses occur. Consequently, the coefficient of discharge is considerably lower than that for the venturi meter (typically about 0.65). The advantages of the orifice plate is its lower cost and its compactness. So long as the increased energy losses are acceptable, it may be used instead of a venturi meter.

Discharge through a small orifice

Figure 2.10 shows a jet of water issuing from a large tank through a small orifice. At a small distance from the tank, the streamlines are straight and parallel, and the pressure is atmospheric. Application of Bernoulli's equation between this point and the water surface in the tank yields

$$\frac{p_0}{\rho g} + h = \frac{p_0}{\rho g} + \frac{u^2}{2g}$$

or

$$u = \sqrt{2gh}$$

This result is usually attributed to Torricelli, who demonstrated its validity experimentally in 1643. The discharge may be calculated by applying the continuity equation

$$Q = uA$$

or

$$Q_{ideal} = A\sqrt{2gh} \qquad (2.12)$$

where A is the area of the jet.

The area of the jet, however, is smaller than that of the orifice, due to the convergence of the streamlines, as is shown in Figure 2.10. The contraction of the jet is called the vena contracta. Experiments have shown that the jet area (A) and orifice area (A_0) are related by

$$A = C_c A_0$$

where C_c is the coefficient of contraction, and is normally in the range 0.61–0.66.

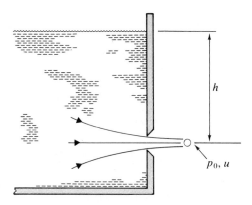

Figure 2.10 Discharge through a small orifice.

In addition, energy losses are incurred at the orifice, and therefore a second coefficient C_v, the velocity coefficient, is introduced to account for these. C_v is normally in the range 0.97–0.99. Hence, the true discharge may be written as

$$Q_{\text{actual}} = C_v C_c A_0 \sqrt{2gh}$$

or

$$Q_{\text{actual}} = C_d A_0 \sqrt{2gh} \qquad (2.13)$$

where C_d is the overall coefficient of discharge.

Discharge through a large orifice

If the orifice is large, then the small orifice equation will no longer be accurate. This is a result of the significant variation of h (and therefore u) across the orifice. In such cases, it is necessary to find the discharge as follows (see Fig. 2.11).

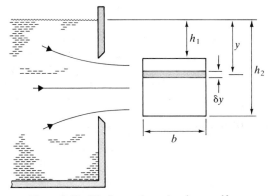

Figure 2.11 Discharge through a large orifice.

For any elemental strip of thickness δy, across a rectangular orifice,

$$\delta Q = b \, \delta y \sqrt{2gy}$$

and

$$Q = b\sqrt{2g} \int_{y=h_1}^{y=h_2} \sqrt{y} \, dy$$

$$= b\sqrt{2g} \left(\frac{2}{3} y^{3/2} \right)_{h_1}^{h_2}$$

$$= \tfrac{2}{3} b\sqrt{2g} (h_2^{3/2} - h_1^{3/2}) \tag{2.14}$$

If h_1 tends to zero, the upper edge of the orifice no longer influences the flow. This corresponds to a thin plate weir. Consequently, (2.14) forms the basis for a discharge formula of thin plate weirs. This is developed more fully in Chapter 13.

2.9 Potential flows

In the problems studied so far, each flow has been considered as a continuum. However, a great deal can be learned by investigating the behaviour of streamlines in a flow. The laws governing their behaviour were investigated by the school of pure mathematicians (the 'hydrodynamicists') that studied flow problems. In order to develop mathematical relationships, the hydrodynamicists developed the concept of an 'ideal' fluid, the characteristics of which were outlined at the beginning of the chapter.

It has already been pointed out that streamlines are everywhere tangential to the flow, so that no flow ever crosses a streamline. Flows which can be represented by streamlines are often known as 'potential flows' or 'ideal flows'.

If the flow vectors are known (or can be guessed) for a number of successive points in a flow, then the streamline may be drawn. Therefore, the streamline gives a graphical representation of the flow. A series of streamlines may be used to construct a 'map' of the flow pattern, though it will emerge that there are mathematical conditions which govern the validity of such a map. For certain flow patterns involving real fluids, these conditions are not fulfilled. It has already been stated that any solid boundary (e.g. a pipe wall) which encloses a flow may be regarded as a surface covered by a continuous series of adjacent streamlines (i.e. a 'stream surface' or streamtube).

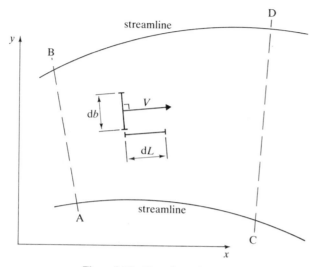

Figure 2.12 Two-dimensional flow.

Properties of streamlines

The stream function. Let two streamlines, 1 and 2 (Fig. 2.12), enclose a two-dimensional flow. The streamlines are not necessarily parallel or straight. It follows, from the definition of a streamline, that the discharge crossing boundary AB must equal the discharge crossing CD. Another approach to this statement is to argue that if point A is taken as an arbitrary 'zero', there is a certain 'increment' of discharge, δQ, between A and B. If point C is now taken as a 'zero' position, then the increment of discharge from C to D must be the same as that between A and B. Therefore any streamline may be assigned a numerical value corresponding to the increment of discharge between that streamline and some arbitrary 'zero' streamline. This number is called the stream function, ψ. The concept of a stream function may be expressed algebraically as follows.

Let the velocity of flow at any point be V (Fig. 2.12). Then for unit depth of flow (depth is assumed to lie in the z-direction in Cartesian co-ordinates) the increment of discharge between streamline 1 and streamline 2 is

$$\psi = \int_1^2 d\psi = \int_A^B V\, db = \int_C^D V\, db = \text{constant} \tag{2.15}$$

For convenience, in algebraic manipulations it is usual to use a sign convention as well as a magnitude to define ψ. The convention adopted here is that if an observer faces downstream (i.e. with the flow moving away from the

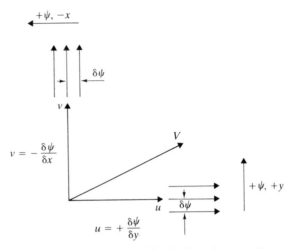

Figure 2.13 Components of flow in Cartesian co-ordinates.

observer in the direction of sight), ψ will increase positively to the left (see Fig. 2.13). Since ψ is an arbitrary function of position, the position of the reference or zero streamline is also arbitrary, so it is perfectly possible to have positive and negative values of ψ. The negative sign does not have any special significance (it does not indicate the existence of a 'negative flow', whatever that might be) and should be interpreted purely graphically.

The velocity potential function.   Another function can also be used to characterise a flow. This is known as the 'potential function' or 'velocity potential', which is defined by the equation $d\phi/dL = V$. Hence,

$$\phi = \int V \, dL \qquad (2.16)$$

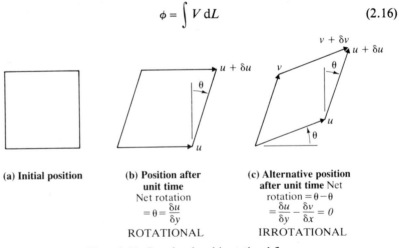

Figure 2.14 Rotational and irrotational flows.

Conventionally, ϕ increases in the direction of flow. Lines of constant ϕ are known as 'equipotential' lines, and are orthogonal to streamlines. A diagram which represents a flow in terms of its lines of ψ and ϕ is a 'flow net'. An accelerating flow (V increasing) is indicated by convergence of the streamlines and equipotential lines, whereas a decelerating flow will be represented by diverging streamlines and equipotential lines.

Conditions for the validity of a flow net

In order to preserve the condition of orthogonality of the lines representing the stream and potential functions, the flow must be 'irrotational', i.e. there must be no net rotation of a given element of fluid (Fig. 2.14). This may be expressed algebraically as follows.

The velocity, V, is resolved into its x- and y-components – u and v, respectively. If there is a velocity u at ordinate y, and a velocity $(u + \delta u)$ at $(y + \delta y)$, then the net rotation of an element (Fig. 2.14b) may be defined as $\delta u/\delta y$.

However, for an element with velocities u (at y), $(u + \delta u)$ (at $y + \delta y$) and v (at x), $v + \delta v$ (at $x + \delta x$), the net rotation is $\delta u/\delta y - \delta v/\delta x$ (see Fig. 2.14c). In an irrotational flow, and for the limit $\delta x \to 0$, $\delta y \to 0$,

$$\frac{\partial u}{\partial y} - \frac{\partial v}{\partial x} = 0 \tag{2.17}$$

Stream and potential functions in Cartesian co-ordinates

It is often useful to express the two flow functions in terms of their Cartesian co-ordinates, especially for some of the algebraic manipulations which arise (Fig. 2.13). The x-direction component of V is u, so from the definition of ψ

$$u = \frac{\partial \psi}{\partial y} \tag{2.18}$$

Similarly for the y-component,

$$v = -\frac{\partial \psi}{\partial x} \tag{2.19}$$

By a similar process of reasoning,

$$u = \frac{\partial \phi}{\partial x} \qquad v = \frac{\partial \phi}{\partial y} \tag{2.20}$$

If these equations are substituted into (2.17),

$$\frac{\partial u}{\partial y} = \frac{\partial \phi}{\partial x \partial y} \quad \text{and} \quad \frac{\partial v}{\partial x} = \frac{\partial \phi}{\partial x \partial y}$$

Therefore

$$\frac{\partial u}{\partial y} - \frac{\partial v}{\partial x} = \frac{\partial \phi}{\partial x \partial y} - \frac{\partial \phi}{\partial x \partial y} = 0$$

which demonstrates that a potential flow satisfies the condition of irrotationality. An alternative way of making the same statement is as follows:

$$u = \frac{\partial \psi}{\partial y} \qquad \text{so} \qquad \frac{\partial u}{\partial y} = \frac{\partial^2 \psi}{\partial y^2}$$

$$v = -\frac{\partial \psi}{\partial x} \qquad \text{so} \qquad \frac{\partial v}{\partial x} = -\frac{\partial^2 \psi}{\partial x^2}$$

Therefore

$$\frac{\partial u}{\partial y} - \frac{\partial v}{\partial x} = \frac{\partial^2 \psi}{\partial y^2} + \frac{\partial^2 \psi}{\partial x^2} = 0$$

This is the Laplace equation, which must be satisfied if the flow is to be a potential flow.

2.10   Some typical flow patterns

Uniform rectilinear flow

For the case of a positive uniform flow parallel to the x-axis (Fig. 2.15) (velocity $= u =$ constant), $u = \mathrm{d}\psi/\mathrm{d}y$ (from (2.18)). Therefore

$$\psi = \int \mathrm{d}\psi = \int u \, \mathrm{d}y = uy \tag{2.21}$$

(if $\psi = 0$ when $y = 0$, then the constant of integration $= 0$). Similarly, for a positive uniform flow parallel to the y-axis (from (2.19)),

$$\psi = \int -v \, \mathrm{d}x = -vx \tag{2.22}$$

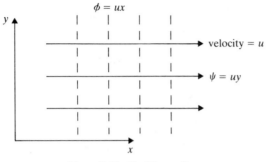

Figure 2.15   Rectilinear flow.

For a uniform flow with velocity V and parallel to neither axis,

$$V = \frac{\mathrm{d}\psi}{\mathrm{d}b}$$

therefore

$$\delta\psi = V\,\delta b$$

This may be expressed in terms of the x- and y-components:

$$\delta\psi = V\delta b = u\delta y - v\delta x$$

Therefore

$$\psi = uy - vx \tag{2.23}$$

Example 2.6   Drawing streamlines

Draw the streamlines for the following flows (up to a maximum ψ value of 20):

(a) a uniform flow of 5 m/s parallel to the x-axis positive direction;
(b) a uniform flow of 5 m/s parallel to the y-axis negative direction;
(c) the flow resulting from the combination of (a) and (b).

Solution

(a) For a flow of 5 m/s the stream function is

$$\psi_u = uy = +5y$$

The streamlines are drawn in Fig. 2.16.
(b) For a flow of 5 m/s in the negative y-direction the stream function is

$$\psi_v = -(-vx) = +5x$$

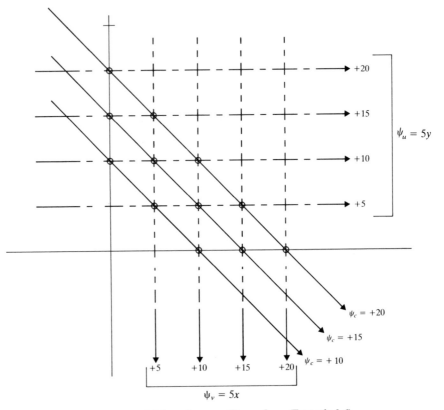

Figure 2.16 Addition of two rectilinear flows (Example 2.6).

These streamlines are also shown in Figure 2.16.

(c) The combined stream function is

$$\psi_c = 5y + 5x$$

This pattern of streamlines is obtained by simple graphical addition. Thus at the junction between the streamlines $\psi_u = 5$ and $\psi_v = 5$, the total is $\psi_c = 5 + 5 = +10$, and so on. Points having the same total are then joined to form the streamline pattern for the total flow.

Radial flows

Radial flows may be flows whose velocity vectors point away from the centre ('source') or towards the centre ('sink') (see Fig. 2.17). For unit depth of flow, the discharge between any pair of radial lines, subtending angle θ at the centre, must be equal to the product of radial velocity and cross-sectional area between those lines, i.e.

$$\delta Q = V_r r \, \delta\theta$$

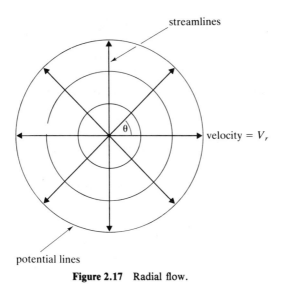

Figure 2.17 Radial flow.

for a source. Therefore, from the definition of ψ,

$$\psi = V_r r \theta \qquad (2.24)$$

If the total discharge from the source is $Q_s = V_r r 2\pi$, then

$$\psi = Q_s \theta / 2\pi \qquad (2.25)$$

Since a sink is simply a 'negative source', the stream function for a sink is

$$\psi = - Q_s \theta / 2\pi \qquad (2.26)$$

Since potential lines are orthogonal to streamlines, it follows that an equipotential is in the form of a circle. From (2.24) and (2.25), $V_r = Q_s/2\pi r$ for a source. Therefore

$$\phi = \int V dL = \int V_r \, dr = \int \frac{Q_s}{2\pi} \frac{dr}{r}$$

$$= \frac{Q_s}{2\pi} \ln\left(\frac{r}{c}\right) \qquad (2.27)$$

Flows in a curved path

A streamline may be curved in any arbitrary fashion, depending on the pattern of forces to which it is being subjected. Therefore, from Newton's Second Law, if a flow is following a curved path, then there must be some lateral force acting upon the fluid. An equation to express this is now developed.

Consider an element of fluid constrained to travel in a path which is curved as shown in Figure 2.18. The mean velocity of the element is $(V + \frac{1}{2}\delta V)$ and its mass is $\rho a\,\delta r$. The radial force exerted by the element on its surrounds is given by

$$\text{mass} \times \frac{(\text{velocity})^2}{\text{radius}} = \rho a\,\delta r\,\frac{(V + \frac{1}{2}\delta V)^2}{(r + \frac{1}{2}\delta r)}$$

For the element to be in equilibrium, the surrounding fluid must exert a reaction force on the element. This reaction is the result of an increase in pressure with radius. Therefore, equating forces,

$$\delta p a = \rho a\,\delta r\,\frac{(V + \frac{1}{2}\delta V)^2}{(r + \frac{1}{2}\delta r)}$$

So, if $\delta V \ll V$ and $\delta r \ll r$, we obtain

$$\frac{\delta p}{\delta r} = \frac{\rho V^2}{r} \tag{2.28}$$

The Bernoulli equation may be applied along any streamline. If the streamlines are assumed to lie in a horizontal plane, then the potential energy is constant so the 'z' term may be dropped, and the Bernoulli equation becomes

$$\frac{V^2}{2g} + \frac{p}{\rho g} = \text{constant} = H$$

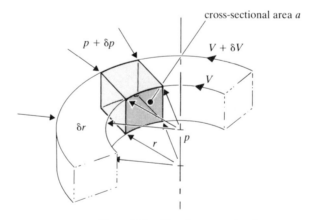

Figure 2.18　Flow in a curved path.

The rate of change of head between one streamline and another is therefore

$$\frac{\delta H}{\delta r} = \left(\frac{(V + \delta V)^2 - V^2}{2g\,\delta r} + \frac{(p + \delta p) - p}{\rho g\,\delta r} \right)$$

If products of small quantities are ignored,

$$\frac{\delta H}{\delta r} = \frac{2V\,\delta V}{2g\,\delta r} + \frac{\delta p}{\rho g\,\delta r}$$

Substituting for $\delta p/\delta r$ from (2.28),

$$\frac{\delta H}{\delta r} = \frac{V\,\delta V}{g\,\delta r} + \frac{V^2}{gr} = \frac{V}{g}\left(\frac{\delta V}{\delta r} + \frac{V}{r} \right)$$

or, in the limit ($\delta r \to 0$),

$$\frac{dH}{dr} = \frac{V}{g}\left(\frac{dV}{dr} + \frac{V}{r} \right) \tag{2.29}$$

Vortices

Equation (2.29) is a completely general equation for any flow with circumferential streamlines. Two cases of such flows are of particular interest. They are both members of the same 'family' of flows, which are given the general title of vortices. The two specific cases examined here are the free vortex and the forced vortex.

Forced vortex. A forced vortex is a circular motion approximating to the pattern generated by the action of a mechanical rotor on a fluid. The rotor 'forces' the fluid to rotate at uniform rotational speed ω rad/s, so $V = \omega r$. Therefore,

$$\frac{dV}{dr} = \frac{V}{r} = \omega$$

Substitution in (2.29) yields

$$\frac{dH}{dr} = \frac{\omega r}{g}(\omega + \omega) = \frac{2\omega^2 r}{g} \tag{2.30}$$

A forced vortex is a rotational flow, so it cannot be represented by a flow net.

Free vortex. A free vortex approximates to naturally occurring circular flows (e.g. the circumferential component of the flow down a drain hole or around a river bend) in which there is no external source of energy. It follows that there can be no difference of total head between one streamline and another, i.e.

$$\frac{dH}{dr} = 0 = \frac{V}{g}\left(\frac{dV}{dr} + \frac{V}{r}\right)$$

so

$$\frac{dV}{dr} = -\frac{V}{r}$$

This equation is satisfied by a flow such that

$$V = \frac{\text{constant}}{r} = \frac{K}{r}$$

since

$$\frac{dV}{dr} = -\frac{K}{r^2} = -\frac{V}{r}$$

A free vortex is irrotational. This may seem surprising, nevertheless the absence of a mechanical rotor does mean that for any fluid element at radius r the net rotation is zero, as defined by Figure 2.13. The element at the centre is an apparent exception to this (and is known as a 'singular point'), but as the statement $r \to 0$ also implies that $V \to \infty$ it is of no practical significance. The free vortex therefore conforms to the conditions for construction of a flow net. Since the flow is in a circumferential path, $d\psi/dr = V$. Therefore

$$\psi = \int V \, dr = \int \frac{K}{r} \, dr = K \ln\left(\frac{r}{c}\right) \tag{2.31}$$

As the potential lines are orthogonal to the streamlines, a line of constant potential must be a radial line. From the definition of ϕ,

$$\phi = \int d\phi = \int V \, dL = Vr\theta = K\theta \tag{2.32}$$

(The reader may like to compare the ψ and ϕ functions for a source or sink with those for the free vortex.)

Circulation and vorticity

The study of vortices forms a useful starting point for developing the concepts of circulation and vorticity (Fig. 2.19). Circulation is simply the

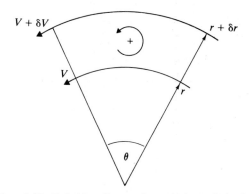

Figure 2.19 Definition diagram for vorticity and circulation.

product (velocity × length) around a circuit. For the element of fluid shown in the figure, the circulation Γ (assuming positive flows are anticlockwise) is

$$\Gamma = (V + \delta V)(r + \delta r)\theta - Vr\theta = V\,\delta r\theta + r\,\delta V\theta$$

The flow is assumed to be in a circular arc, so velocities in the radial direction are zero.

The vorticity ζ is simply the intensity of circulation, i.e.

$$\frac{\Gamma}{\text{area}} = \zeta = \frac{V\,\delta r\theta + r\,\delta V\theta}{r\theta\,\delta r} = \left(\frac{V}{r} + \frac{\delta V}{\delta r}\right) \tag{2.33}$$

which should be compared with (2.29).

Example 2.7   Vortex flow

A cylinder of diameter 350 mm is filled with water and is rotated at 130 rad/s. Assuming that there is zero gauge pressure on the centreline at the top of the cylinder, estimate:

(a) the pressure on the side wall of the cylinder;
(b) the loading on the upper end plate due to the pressure.

Solution

(a) As the cylinder is rotated mechanically, the flow induced in the fluid will be a forced vortex. Equation (2.30) is therefore appropriate:

$$\frac{\mathrm{d}H}{\mathrm{d}r} = \frac{2\omega^2 r}{g}$$

This is rearranged as

$$\int \mathrm{d}H = \int \frac{2\omega^2 r}{g}\,\mathrm{d}r$$

and hence

$$H = \frac{\omega^2 r^2}{g} + C$$

Given that pressure is zero at the centre, and knowing that velocity is also zero ($\omega r \to 0$ when $r \to 0$), the constant C must be zero.

Therefore, at $r = 0.175$ m,

$$H = \omega^2 r^2 / g$$

But

$$H = \frac{p}{\rho g} + \frac{V^2}{2g}$$

or, since $V = \omega r$,

$$H = \frac{p}{\rho g} + \frac{\omega^2 r^2}{2g}$$

Therefore

$$p = \tfrac{1}{2} \rho \omega^2 r^2 = \tfrac{1}{2} \times 1000 \times 130^2 \times 0.175^2$$

$$= 258.8 \text{ kN/m}^2$$

(b) At any radius r within the cylinder,

$$p = \tfrac{1}{2} \rho \omega^2 r^2$$

The force δF on a circular annulus of radius r and thickness δr is

$$\delta F = \tfrac{1}{2} \rho \omega^2 r^2 2 \pi r \, \delta r = \pi \rho \omega^2 r^3 \, \delta r$$

Therefore

$$F = \int_0^{0.175} dF = \int_0^{0.175} \pi \rho \omega^2 r^3 \, dr = \left[\frac{\pi \rho \omega^2 r^4}{4} \right]_0^{0.175}$$

$$= \frac{\pi \times 1000 \times 130^2 \times 0.175^4}{4} = 12.45 \text{ kN}$$

Note that hydrostatic forces have been ignored in this example, since they would be small compared with the forces due to the vortex.

Combinations of flow patterns

The real value and power of potential flow methods can only be appreciated when the various individual flow patterns are combined (or superposed). A large number of combined patterns have been developed, closely approximating to real flows. Some idea of the wide range of patterns available may be obtained by consulting Milne-Thompson (1968). There is space here only for a brief introduction.

It is possible to combine flows graphically or algebraically. An example of the first method has already been presented (Example 2.1) and another follows. For the algebraic approach it is sometimes best to apply polar, rather than Cartesian, co-ordinates.

Example 2.8 Streamlines

A linear flow, having a uniform negative velocity of 60 m/s parallel to the x-axis, is combined with a source with a discharge of 160 m^3/s. Draw the streamlines for the combined flow and locate the stagnation point.

Solution

The stream function for the linear flow is $-60y$, and that for the source is $+160\theta/2\pi$. The streamlines for the two flows are shown in Figure 2.20. The combined streamlines are obtained by using graphical addition, as outlined in Example 2.6.

A stagnation point (a point where the fluid comes to rest) occurs where the velocity due to the source is equal and opposite to the linear flow, i.e. where $V_r = +60$ m/s.

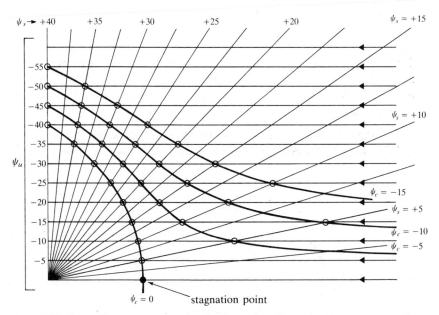

Figure 2.20 Streamline pattern – source, plus linear flow (Example 2.8). *Note.* Heavy lines represent ψ_c.

So, using (2.24) and (2.25),

$$V_r r \theta = \frac{Q_s \theta}{2\pi} = \psi$$

Hence $V_r = Q_s/2\pi r$, and for the stagnation point $60 = 160/2\pi r$, so $r = 0.424$ m. The velocities are equal and opposite only along the axis $y = 0$ (see Fig. 2.20).

The streamlines for the combined flow divide about the line $\psi_c = 0$. This line can be treated as the external surface of a solid body, all the streamlines with a negative ψ_c then represent the flow around such a shape.

Other important examples of combined flows are the following:

(a) Source plus sink plus linear flow. A source with discharge Q_s has its centre at $(-k, 0)$, and a sink with discharge $-Q_s$ has its centre at $(+k, 0)$. These are combined with a linear flow u parallel to the x-axis (Fig. 2.21). The streamline $\psi_c = 0$ is oval in shape, and the streamlines with negative ψ_c represent the flow around such a shape.

(b) If the combination of flows is as in (a) but $k \to 0$, the streamline $\psi_c = 0$ becomes a circle. (Note that there are mathematical conditions for this, e.g. as $k \to 0$, $Q_s \times 2k$ must remain finite, otherwise the source and sink simply cancel each other out – see Milne-Thompson, 1968.)

(c) For a combination of flows as in (b), but with the addition of a free vortex surrounding the circular streamline $\psi_c = 0$, the addition of the vortex displaces the position of the stagnation points. If the streamlines are plotted, it is found that the velocity pattern is symmetrical about the y-axis, but asymmetrical about the x-axis. This implies an asymmetrical pressure distribution, which in turn implies that there is a component of force perpendicular to the direction of the linear flow. This is known as a 'lift force'. This flow pattern leads to some important equations which are used in aerodynamics.

Applications of a number of the flow patterns mentioned in this section will be found scattered through the remainder of the book. Typical

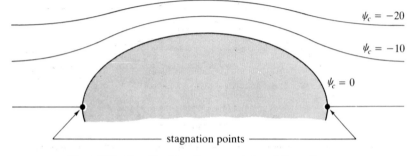

Figure 2.21 Combination of source, sink and linear flow.

examples are flows around bends in siphon spillways or rivers (which approximate to free vortex flows), and flows around streamlined bridge piers (which approximate to the combination source + sink + linear flow mentioned above).

References and further reading

British Standards Institution 1981. BS1042 Part 1 Sec 1.1 (1981). *Orifice plates, nozzles and venturi tubes inserted in circular cross section conduits running full.* BS1042, Part 1, Section 1.1. London: BSI.

Massey, B. S. 1983. *Mechanics of fluids*, 5th edn. Wokingham: Van Nostrand Reinhold.

Milne-Thompson, L. M 1968. *Theoretical hydrodynamics*, 5th edn. London: Macmillan.

US Bureau of Reclamation 1967. *Water measurement manual*, 2nd edn. Washington, DC: US Department of the Interior.

3

Behaviour of real fluids

3.1 Real and ideal fluids

In Chapter 2, equations describing fluid motion were developed. The equations were mathematically straightforward because the fluid was assumed to be ideal, i.e. it possessed the following characteristics:

(a) it was inviscid;
(b) it was incompressible;
(c) it had no surface tension;
(d) it always formed a continuum.

From a civil engineering standpoint, surface tension problems are encountered only under rather special circumstances (e.g. very low flows over weirs or in hydraulic models). Compressibility phenomena are particularly associated with the high speed gas flows which occur in chemical engineering or aerodynamics. The civil engineer may occasionally need to consider one particular case of compressibility associated with surge, and this will be investigated in Chapters 6 and 12.

In this chapter, it is the effect of viscosity which will dominate the discussion, since it is this characteristic which differentiates so many aspects

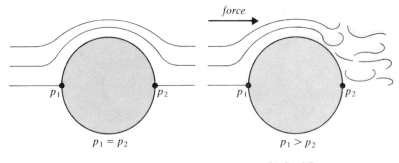

(a) **Ideal flow**   (b) **Real flow**

Figure 3.1 Ideal and real flow around a cylinder.

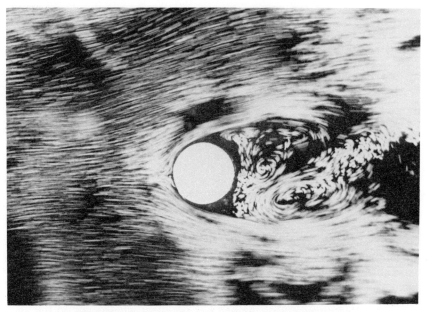

Photograph 1   Flow around a cylinder showing eddy formation.

of real flows from ideal flows. This difference is exemplified if the ideal and real flows around a bluff (i.e. non-streamlined) shape are compared (Fig. 3.1). The ideal fluid flows smoothly around the body with no loss of energy between the upstream and downstream sides of the body. The pressure distribution around the body is therefore symmetrical, and may be obtained by applying Bernoulli's equation. The real fluid flow approximates to the ideal flow only around the upstream portion of the body, where the streamlines are converging. Shortly after the streamlines start to diverge (as the fluid starts to negotiate the downstream part of the body), the streamlines fail to conform to the symmetrical pattern. A region of strongly eddying flow occurs (Photograph 1), in which there are substantial energy losses. Consequently, there is now an asymmetrical pressure distribution, which in turn implies that there is now a force acting on the body. An explanation for the behaviour of the real fluid will emerge as the nature of viscosity is considered.

3.2   Viscous flow

An approach to viscosity

Perhaps the best way to approach viscosity is to contrast the effect of shearing action on a real fluid with that on an ideal fluid. Imagine, for example, that the fluids are both lying sandwiched between a fixed solid surface

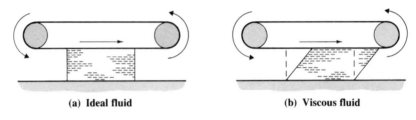

(a) Ideal fluid   (b) Viscous fluid

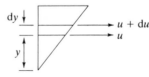

(c) Velocity variation for viscous flow

Figure 3.2   Effect of shear force on fluid.

on one side and a movable belt (initially stationary) on the other (Fig. 3.2). Turning first to the ideal fluid (Fig. 3.2a), if the belt is set in motion, experimental measurements will indicate:

(a) that the force required to move the belt is negligible;

(b) that the movement of the belt has no effect whatsoever on the ideal fluid, which therefore remains stationary.

By contrast, if the other belt is slowly started and comparable measurements are taken on the real fluid, it will be found that:

(a) A considerable force is required to maintain belt motion, even at slow speed.

(b) The whole body of fluid is deforming and continues to deform as long as belt motion continues. Closer investigation will reveal that the deformation pattern consists in the shearing, or sliding, of one layer of fluid over another. The layer of fluid immediately adjacent to the solid surface will adhere to that surface and, similarly, the layer adjacent to the belt adheres to the belt. Between the solid surface and the belt the fluid velocity is assumed to vary linearly as shown in Figure 3.2c. It is characteristic of a viscous fluid that it will deform continuously under a shear force. It is emphasised that this discussion is based on the assumption that the velocities are low, and that the flow is two-dimensional only.

A definition of viscosity

The pattern of events outlined in the previous section raises two questions. First, how must viscosity be defined? Secondly, based on this definition, can numerical values of viscosity be obtained for each fluid? Referring to the velocity diagram in Figure 3.2, it is evident that the shear force is being transmitted from one layer to another. If any pair of adjacent layers are taken, the lower one will have some velocity, u, whilst the upper one will be travelling with velocity $u + du$. The rate of shear strain is thus du/dy. Newton postulated that the shear force applied and the rate of fluid shear were related by the equation

$$F = A \times \text{constant} \times (du/dy)$$

where A is the area of the shear plane (i.e. the cross-sectional area of the fluid in the x–z plane). The equation may be rewritten as

$$\frac{F}{A} = \tau = \text{constant} \times \frac{du}{dy} \qquad (3.1)$$

(this is valid only for 'laminar' flows, see Section 3.3). The constant is known as the absolute coefficient of viscosity of a fluid, and is given the symbol μ. Experiments have shown that:

(a) μ is not a constant;
(b) for the group of fluids known as 'Newtonian fluids', μ remains constant only at constant temperature – if the temperature rises, viscosity falls, and if the temperature falls, viscosity rises (air and water are Newtonian fluids);
(c) there are some ('non-Newtonian') fluids in which μ is a function of both temperature and rate of shear.

Another form of the coefficient of viscosity is obtained if absolute viscosity is divided by fluid density to produce the coefficient of kinematic viscosity, ν:

$$\nu = \mu/\rho$$

Both coefficients have dimensions: μ is measured in N s/m^2 (or kg/m s) and ν is in m^2/s.

Example 3.1   Viscous flow

A moving belt system, such as that depicted in Figure 3.2, is to be used to transfer lubricant from a sump to the point of application. The working length of the belt is assumed to be straight and to be running parallel to a stationary metal surface. The belt speed is 200 mm/s, the perpendicular distance between the belt and the metal surface is 5 mm, and the belt is 500 mm wide. The lubricant is an oil, having an absolute viscosity of 0.007 N s/m^2. Estimate the force per unit length of belt, and the quantity of lubricant discharged per second.

Solution

$$\tau = \mu \frac{du}{dy} = 0.007 \times \frac{200 \times 10^{-3}}{5 \times 10^{-3}} = 0.28 \text{ N/m}^2$$

area A of a 1 m length of belt $= 1 \times 0.5 = 0.5$ m^2

force per unit length of belt $= 0.28 \times 0.5 = 0.14$ N

To estimate the discharge, note that the velocity is assumed to vary linearly with perpendicular distance from the belt. The discharge may therefore be obtained by taking the product of the mean velocity, V, and the area of the flow cross section, by:

$$Q = Vby = \frac{200 \times 10^{-3}}{2} \times 0.5 \times 0.005 = 2.5 \times 10^{-4} \text{ m}^3/\text{s}$$

3.3 The stability of laminar flows and the onset of turbulence

Introduction

The flows examined so far have involved only low speeds (of the belt and fluid). However, what happens if, say, the speed of the belt, and therefore the rate of shear, is increased? The answer is that the pattern of linearly sheared flow will continue to exist only up to a certain belt velocity. Above that velocity a dramatic transformation takes place in the flow pattern. But why?

Effect of a disturbance in a sheared flow

Consider first a slowly moving sheared flow. What might be the effect of a small disturbance (due perhaps to a small local vibration) on the otherwise rectilinear flow pattern? The pathlines might be slightly deflected (Fig. 3.3a), bunching together more closely at A and opening out correspondingly at B. This implies that the local velocity at A, u_A, increases slightly compared with the upstream velocity u, while at the same time u_B reduces. Now, from Bernoulli's equation,

$$\frac{p}{\rho g} + \frac{u^2}{2g} = \frac{p_A}{\rho g} + \frac{u_A^2}{2g} = \frac{p_B}{\rho g} + \frac{u_B^2}{2g}$$

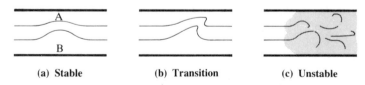

| (a) Stable | (b) Transition | (c) Unstable |

Figure 3.3   Effect of disturbance on a viscous flow.

Therefore $p_B > p_A$. It can thus be reasoned that the disturbance will produce a small transverse resultant force acting from B towards A. The lateral component of velocity will also produce a corresponding component of viscous shear force, which acts in the opposite sense to the resultant disturbing force. As long as the fluid is moving slowly, the resultant disturbing force tends to be outweighed by the viscous force. Disturbances are therefore damped out. As the rate of shear increases, the effect of the disturbance becomes more pronounced:

(a) The difference between u_A and u_B increases.
(b) The pressure difference $(p_A - p_B)$ increases with $(u_A^2 - u_B^2)$, so the deflection of the pathline becomes more pronounced.
(c) The greater shear results in a deformation of the crest of the pathline pattern (Fig. 3.3b). When the rate of shear is sufficiently great, the deformation is carried beyond the point at which the rectilinear pattern of pathlines can cohere (one can think of an analogy with a breaking wave). The flow pattern then disintegrates into a disorderly pattern of eddies in place of the orderly pattern of layers.

The above is (deliberately) a radical oversimplification of a complex physical pattern. It does, however, explain why the neat and mathematically tractable viscous flows only exist at low velocities. The other eddying flow is known as turbulent flow. The existence of these distinct patterns of flow was first investigated scientifically by Osborne Reynolds (1842–1912) at Manchester University.

Reynolds' experiment

Classical hydrodynamicists (most of them were primarily mathematicians) had long been puzzled by certain aspects of flow which did not conform to the known mathematical formulations. Towards the end of the 19th century, Reynolds designed an experiment in which a filament of dye was injected into a flow of water (Fig. 3.4). The discharge was carefully controlled, and passed through a glass tube so that observations could be made. Reynolds discovered that the dye filament would flow smoothly along the tube as long as the velocities remained very low. If the discharge was increased gradually, a point was reached at which the filament became wavy. A small further increase in discharge was then sufficient to trigger a vigorous eddying motion, and the dye mixed completely with the water. Thus three distinct patterns of flow were revealed:

'Viscous' or 'laminar' – in which the fluid may be considered to flow in discrete layers with no mixing.
'Transitional' – in which some degree of unsteadiness becomes apparent (the wavy filament). Modern experimentation has demonstrated that this

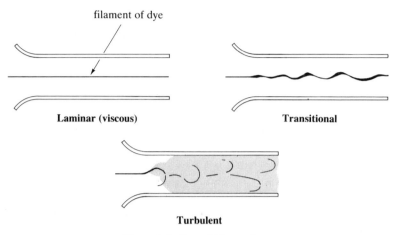

Figure 3.4 Reynolds' experiment.

type of flow may comprise short 'bursts' of turbulence embedded in a laminar flow.

'Turbulent' − in which the flow incorporates an eddying or mixing action. The motion of a fluid particle within a turbulent flow is complex and irregular, involving fluctuations in velocity and direction.

Most of the flows which are encountered by civil engineers are turbulent flows.

The Reynolds Number

An obvious question which now arises is 'can we predict whether a flow will be viscous or turbulent?'. Reynolds' experiments revealed that the onset of turbulence was a function of fluid velocity, viscosity and a typical dimension. This led to the formation of the dimensionless Reynolds Number (symbol Re):

$$Re = \frac{\rho u l}{\mu} = \frac{u l}{\nu} \tag{3.2}$$

It is possible to show that the Reynolds Number represents a ratio of forces (see Example 11.1):

$$Re = \frac{\text{inertia force}}{\text{viscous force}}$$

For this reason, any two flows may be 'compared' by reference to their respective Reynolds Numbers. The onset of turbulence therefore tends to occur within a predictable range of values of Re. For example, flows in commercial pipelines normally conform to the following pattern:

for Re < 2000, laminar flow exists;
for 2000 < Re < 4000, the flow is transitional;
for Re > 4000, the flow is turbulent.

These values of Re should be regarded only as a rough guide (in some experiments laminar flows have been detected for Re ≫ 4000).

3.4 Shearing action in turbulent flows

General description

Shearing in laminar flows may be visualised as a purely frictional action between adjacent fluid layers. By contrast, shearing in turbulent flows is both difficult to visualise and less amenable to mathematical treatment. As a consequence, the solutions of problems involving turbulent flows tend to invoke experimental data.

   In Section 3.3, a (rather crude) model of flow instability was proposed. This led to a description of the way in which a streamline might be broken down into an eddy formation. An individual eddy may be considered to possess a certain 'size' or scale. This size will obviously be related to the local rate of shearing and to the corresponding instability. The passage of a succession of eddies leads to a measurable fluctuation in the velocity at a given point (Fig. 3.5). The eddies are generally irregular in size and shape, so the fluctuation of velocity with time is correspondingly irregular. For convenience, this fluctuating velocity is broken down into two components:

(a) The time-averaged velocity of flow at a point, u. For the present this is assumed to be parallel to the x-axis.

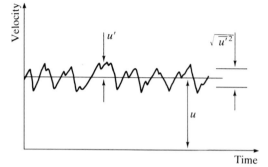

Figure 3.5 Variation of velocity with time in turbulent flow.

(b) Fluctuating components u' (in the x-direction), v' (y-direction) and w' (z-direction). The time average of u', v', or w' is obviously zero. The magnitudes of u', v' and w' and the rapidity with which they fluctuate give an indication of the structure of the eddy pattern.

A measure of the magnitude of the fluctuations may be based upon the root-mean-square of the quantity u' (i.e. $\sqrt{\overline{u'^2}}$). This measure is usually known as the 'intensity of turbulence', and is the ratio $\sqrt{\overline{u'^2}}/u$.

Simple models of turbulent flows

Turbulence implies that within a fluid flow individual particles of fluid are 'jostling' (migrating transversely) as they are carried along with the flow. An attempt to plot the pathlines of a series of particles on a diagram would result in a tangled and confused pattern of lines. Since particles are continuously interchanging, it follows that their properties will similarly be continuously interchanging (hence the rapid mixing of the dye in Reynolds' observations of turbulence). Due to the complexity of the motion, it is impossible to produce an equation or numerical model of turbulence which is simple, complete and accurate. Nevertheless, some useful insights may be gained by investigating two elementary models of turbulence, even though both are known to have serious shortcomings.

The 'Reynolds' Stress' model. It has been stated above that the fluctuating components of velocity are u', v' and w'. If we restrict our attention to a two-dimensional flow (x- and y-components), then only u' and v' are present (Fig. 3.6a). Hence, during a time interval dt, the mass of fluid flowing in the y-direction through a small horizontal element of area δA is

$$\rho v' \, \delta A \, \delta t$$

This mass has an instantaneous horizontal velocity of $u + u'$. Its momentum, δM, is therefore

$$\delta M = \rho v' \, \delta A \, \delta t(u + u')$$

Therefore, the rate of transport (or rate of interchange) of momentum during the particular instant is

$$\frac{\delta M}{\delta t} = \rho v' \, \delta A \, (u + u') = \rho \overline{v'} \, \delta A u + \rho \overline{v'u'} \, \delta A$$

The average rate of transport of momentum will be a function of the time-averaged velocities of the fluid particles. The magnitude of u remains constant. The averaged values of u' and v' are $\overline{u'}$ and $\overline{v'}$, but $\overline{u'}$ and $\overline{v'}$ must

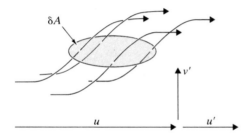

(a) Reynolds' eddy model

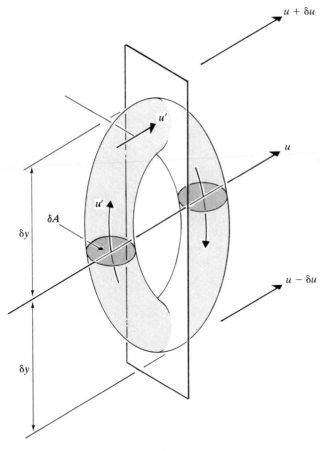

(b) Eddy for Prandtl model

Figure 3.6 Turbulent eddies.

both be zero (see Fig. 3.5 and the associated text, above). Despite this fact, the product $\overline{u'v'}$ may not be zero, so

$$\text{average } \frac{\delta M}{\delta t} = \rho \overline{u'v'} \, \delta A$$

The existence of a 'rate of interchange' of momentum necessarily implies the existence of a corresponding force within the fluid:

$$\delta F = \rho \overline{u'v'} \, \delta A$$

Alternatively, since force/area = stress, τ,

$$\tau = \frac{\delta F}{\delta A} = \rho \overline{u'v'} \tag{3.3}$$

This is termed a Reynolds' Stress.

Prandtl eddy model. A slightly more sophisticated approach is due to Prandtl, a German engineer, who pioneered much of the work on shear flows in the early years of the 20th century. Prandtl sought to develop a simple model of an eddy, from which an analogue to viscosity might be derived. Consider, then, a rectilinear flow. At a point (x, y) in the flow, the velocity of flow is u. Superimposed on that flow is an eddy (Fig. 3.6b) in the form of a ring of cross-sectional area δA. The eddy is rotating, rather like a wheel. The tangential velocity of the eddy is u', and thus the mass flow transferred at any one cross section will be $\rho \delta A u'$. The rate of interchange of mass is therefore $2\rho \delta A u'$ (since one mass of fluid migrates upwards as another equal mass migrates downwards).

Between the plane at y and the plane at $y + \delta y$, the velocity increases from u to $u + \delta u$, so it is reasonable to assume that

$$\delta u = u' = \delta y \frac{du}{dy}$$

so

$$\text{rate of interchange} = 2\rho \delta A \, \delta y \, du/dy$$

The rate of interchange of momentum will be the product of mass interchange and change in velocity. The change in velocity between the base of the eddy (at $y - \delta y$) and the top of the eddy (at $y + \delta y$) is $2u' = 2 \, \delta y \, du/dy$. Therefore, the rate of interchange of momentum is

$$\frac{dM}{dt} = 2\rho \delta A \left(\delta y \frac{du}{dy} \right) \left(2 \, \delta y \frac{du}{dy} \right)$$

Now, shear stress, $\tau = \text{force/area}$ of shear plane, and force $= dM/dt$, therefore

$$\tau = 4\rho \left(\delta y \frac{du}{dy} \right)^2 = \varepsilon \frac{du}{dy} \qquad (3.4)$$

where ε is the 'eddy viscosity' and is given by

$$\varepsilon = 4\rho\, \delta y^2 \left(\frac{du}{dy} \right)$$

$$= \rho (2\, \delta y)^2 \left(\frac{du}{dy} \right) \qquad (3.5)$$

ε is analogous to μ, but is a function of the rate of shear, and of eddy size $(2\delta y)$.

Eddy viscosity is therefore too complex to be reducible (for example) to a convenient viscosity diagram. Prandtl's treatment is now rather dated. Nevertheless, because of its links with the physics of turbulence, it is a convenient starting point for the student. Furthermore, Prandtl's model formed the basis of much of the progress in the analysis of turbulence which continues to the present.

Velocity distribution in turbulent shear flows

It is possible to develop a relationship between y and u on the basis of Prandtl's work. In order to achieve this, an assumption must be made regarding eddy size. The simplest guess might be that $(2\delta y) = \text{constant} \times y$, i.e. a linear relationship. If the constant is K, then (3.4) may be rewritten as

$$\tau = \rho (Ky)^2 \left(\frac{du}{dy} \right)^2$$

or

$$\frac{\tau}{\rho} = (Ky)^2 \left(\frac{du}{dy} \right)^2 \qquad (3.6)$$

Now the ratio $\sqrt{\tau/\rho}$ is known as the 'friction velocity', u_*. No such velocity actually exists in the flow; u_* is just a 'reference' value. Also, from (3.3), $u_* = \sqrt{u'v'}$.

Equation (3.6) may therefore be rewritten in terms of u_*:

$$u_* = Ky \frac{du}{dy} \qquad (3.6a)$$

Therefore

$$du = \frac{u_*}{K} \frac{dy}{y} \qquad (3.6b)$$

so

$$u = \frac{u_*}{K} \ln y + C$$

or

$$\frac{u}{u_*} = \frac{1}{K} \ln y + C \qquad (3.7)$$

No further progress can now be made on a purely mathematical basis; it is necessary to use experimental data to evaluate K and C. It was originally thought that K was a constant (equal to 0.4), but this is now known to be an oversimplification. C is a function of surface roughness, k_S, and must be determined for different materials.

3.5 The boundary layer

Description of a boundary layer

The concepts used in analysing shearing action in fluids have been introduced above. For illustrative purposes, it was assumed that the shearing action was occurring in a fluid sandwiched between a moving belt and a stationary solid surface. The fluid was thus bounded on two sides. It may have occurred to the reader that such a situation is not common in civil engineering. Some flows (e.g. the flow of air round a building) are bounded on one side only, while others (e.g. the flow through a pipe) are completely surrounded by a stationary solid surface. To develop the boundary layer concept, it is helpful to begin with a flow bounded on one side only. Consider, therefore, a rectilinear flow passing over a stationary flat plate which lies parallel to the flow (Fig. 3.7a). The incident flow (i.e. the flow just

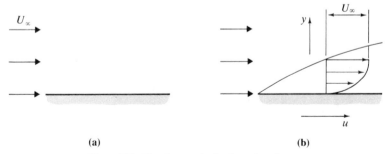

(a)   (b)

Figure 3.7 Development of a boundary layer.

upstream of the plate) has a uniform velocity, U_∞. As the flow comes into contact with the plate, the layer of fluid immediately adjacent to the plate decelerates (due to viscous friction) and comes to rest. This follows from the postulate that in viscous fluids a thin layer of fluid actually 'adheres' to a solid surface. There is then a considerable shearing action between the layer of fluid on the plate surface and the second layer of fluid. The second layer is therefore forced to decelerate (though it is not quite brought to rest), creating a shearing action with the third layer of fluid, and so on. As the fluid passes further along the plate, the zone in which shearing action occurs tends to spread further outwards (Fig 3.7b). This zone is known as a 'boundary layer'. Outside the boundary layer the flow remains effectively free of shear, so the fluid here is not subjected to viscosity-related forces. The fluid flow outside a boundary layer may therefore be assumed to act like an ideal fluid.

As with any other sheared flow, the flow within the boundary layer may be viscous or turbulent, depending on the value of the Reynolds Number. To evaluate Re we need a 'typical dimension' and in boundary layers this dimension is usually the distance in the x-plane from the leading edge of the solid boundary. The Reynolds Number thus becomes $Re_x = \rho U_\infty x / \mu$.

A moment's reflection should convince the reader that if the solid surface is sufficiently long, a point will be reached at which the magnitude of Re indicates the onset of turbulence. This hypothesis accords with experimental observations. To add further to the complications, the very low velocities in the flow close to the plate imply a low local Re, and the consequent possibility of laminar flow here. The structure of the boundary layer is therefore as is shown in Figure 3.8. A graph (or 'distribution') of velocity variation with y may be drawn. This reveals that:

(a) in the laminar zone there is a smooth velocity distribution to which a mathematical function can be fitted with good accuracy;
(b) in the turbulent zone the mixing or eddying action produces a steeply

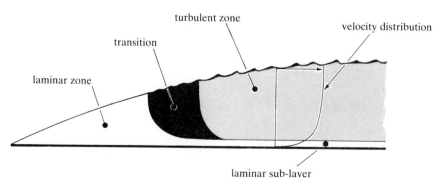

Figure 3.8 Structure of a boundary layer.

sheared profile near the surface of the plate, but a flatter, more uniform profile further out towards the boundary layer edge.

In both cases, the velocity distributions are assumed to be asymptotic to the free stream velocity, U_∞.

Boundary layer equations

Although the basic structure of a boundary layer is clear, the engineer usually needs a precise numerical description for each particular problem. The basic parameters and equations required will now be developed. In the interests of simplicity, this treatment will be restricted to a two-dimensional incompressible flow with constant pressure.

(a) The boundary layer thickness, δ, is the distance in the y-direction from the solid surface to the outer edge of the boundary layer. Since the velocity distribution in the boundary layer is asymptotic to U_∞, it is difficult to measure an exact value for δ. The usual convention is to assume that the edge of the boundary layer occurs where $u/U_\infty = 0.99$.

(b) The displacement thickness, δ_*, is the distance by which a streamline is displaced due to the boundary layer. Consider the velocity distribution at a section in the boundary layer (Fig. 3.9). Inside the boundary layer, the velocity is everywhere less than in the free stream. The discharge through this cross section is correspondingly less than the discharge through the same cross-sectional area in the free stream. This deficit in discharge can be quantified for unit width, and an equation may then be developed for δ_*:

$$\text{deficit of discharge through an element} = (U_\infty - u)\,\delta y$$

$$\text{deficit through whole boundary layer section} = \int_0^\delta (U_\infty - u)\,\mathrm{d}y$$

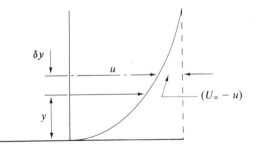

Figure 3.9   Velocity distribution in a boundary layer.

In the free stream an equivalent discharge would pass through a layer of depth δ_*, so

$$U_\infty \delta_* = \int_0^\delta (U_\infty - u) \, dy$$

Therefore

$$\delta_* = \int_0^\delta \left(1 - \frac{u}{U_\infty}\right) dy \tag{3.8}$$

(c) The momentum thickness, θ, is analogous to the displacement thickness. It may be defined as the depth of a layer in the free stream which would pass a momentum flux equivalent to the deficit due to the boundary layer:

$$\text{mass flow through element} = \rho u \, \delta y$$

$$\text{deficit of momentum flux} = \rho u \, \delta y (U_\infty - u)$$

$$\text{deficit through whole boundary layer section} = \int_0^\delta \rho u \, dy (U_\infty - u)$$

In the free stream, an equivalent momentum flux would pass through a layer of depth θ and unit width, so that

$$\rho U_\infty^2 \theta = \int_0^\delta \rho u (U_\infty - u) \, dy$$

$$\theta = \int_0^\delta \frac{u}{U_\infty} \left(1 - \frac{u}{U_\infty}\right) dy \tag{3.9}$$

(d) The definition of kinetic energy thickness δ_{**} follows the same pattern, leading to the equation

$$\delta_{**} = \int_0^\delta \frac{u}{U_\infty} \left(1 - \left(\frac{u}{U_\infty}\right)^2\right) dy \tag{3.10}$$

(e) The momentum integral equation is used to relate certain boundary layer parameters so that numerical estimates may be made. Consider the longitudinal section through a boundary layer (Fig. 3.10), the section is bounded on its outer side by a streamline, BC, and is 1 m wide. The discharge across CD is

$$Q_{CD} = \int_0^\delta u \, dy$$

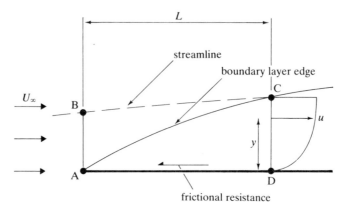

Figure 3.10 Longitudinal section through a boundary layer.

The momentum flux ($= \rho Q \times$ velocity) is therefore

$$\frac{\mathrm{d}M_{\mathrm{CD}}}{\mathrm{d}t} = \int_0^\delta \rho\, u^2\, \mathrm{d}y$$

As BC is a streamline, the discharge across AB must be the same as that across CD:

$$Q_{\mathrm{AB}} = \int_0^\delta u\, \mathrm{d}y$$

The incident velocity at AB is U_∞, so the momentum flux is

$$\frac{\mathrm{d}M_{\mathrm{AB}}}{\mathrm{d}t} = \int_0^\delta \rho U_\infty u\, \mathrm{d}y$$

Boundary layers are actually very thin, so it is reasonable to assume the velocities are in the x-direction. The loss of momentum flux is due to the frictional shear force (F_S) at the solid surface. Therefore

$$-F_S = \int_0^\delta \rho u^2\, \mathrm{d}y - \int_0^\delta \rho U_\infty u\, \mathrm{d}y$$

The negative sign follows from the fact that the frictional resistance acts in the opposite sense to the velocity. This equation may be rearranged to give

$$F_S = \int_0^\delta \rho (U_\infty u - u^2)\, \mathrm{d}y = \rho U_\infty^2 \int_0^\delta \frac{u}{U_\infty} \left(1 - \frac{u}{U_\infty}\right) \mathrm{d}y$$
$$= \rho U_\infty^2 \theta$$

The frictional shear at the solid surface is not a constant, but varies with x, due to the growth of the boundary layer. The shear force may therefore be expressed as

$$F_S = \int_0^L \tau_0 \, dx$$

where τ_0 is the shear stress between the fluid and the solid surface. The momentum integral equation is therefore

$$\int_0^L \tau_0 \, dx = \rho U_\infty^2 \theta \tag{3.11}$$

Solution of the momentum integral equation

In order to solve the momentum equation, some further information regarding θ and τ_0 is required. Both of these quantities are related to the velocity distribution in the boundary layer. Since velocity distribution depends on the nature of the flow, i.e. whether it is laminar or turbulent, the two cases are now considered separately.

Laminar flow.   The velocity distribution in a laminar boundary layer may be expressed in a number of forms. A typical equation is

$$\frac{u}{U_\infty} = \left[A \frac{y}{\delta} - B \left(\frac{y}{\delta} \right)^2 \right]$$

Therefore,

$$\theta = \int_0^\delta \frac{u}{U_\infty} \left(1 - \frac{u}{U_\infty} \right) dy$$

$$= \int_0^\delta \left[A \frac{y}{\delta} - B \left(\frac{y}{\delta} \right)^2 \right] \left[1 - \left[A \frac{y}{\delta} - B \left(\frac{y}{\delta} \right)^2 \right] \right] dy \tag{3.12}$$

If A and B are known, the above integral can be evaluated. Furthermore:

$$\tau_0 = \mu \frac{du}{dy} = \mu U_\infty \frac{d}{dy} \left[A \frac{y}{\delta} - B \left(\frac{y}{\delta} \right)^2 \right]_{y=0}$$

$$= \mu U_\infty \left[\frac{A}{\delta} - \frac{2By}{\delta} \right]_{y=0} = \frac{\mu U_\infty A}{\delta} \tag{3.13}$$

If (3.12) and (3.13) are substituted into (3.11), a solution can be obtained.

Turbulent flow. Experimental investigations have shown that the velocity distribution outside the laminar sublayer may be approximately represented by

$$\frac{u}{U_\infty} = \left(\frac{y}{\delta}\right)^{1/n}, \qquad \text{where } 6 < n < 11 \tag{3.14}$$

Typically (after Prandtl) a value $n = 7$ is used. This can be substituted into the equation for the momentum thickness (Equation (3.9)) and a solution obtained. This equation cannot, however, be differentiated to obtain τ_0, since

$$\frac{d}{dy}\left(\frac{y}{\delta}\right)^{1/7}_{y=0} = 0$$

It is again necessary to use experimental data to fill the gap in the mathematical procedure. This provides us with the alternative relationship

$$\tau_0 = \frac{0.023\rho U_\infty^2}{(\text{Re}_\delta)^{1/m}} \tag{3.15}$$

where

$$\text{Re}_\delta = \rho U_\infty \delta / \mu.$$

Example 3.2   Turbulent boundary layer

Water flows down a smooth wide concrete apron into a river. Assuming that a turbulent boundary layer forms, estimate the shear stress and the boundary layer thickness 50 m downstream of the entrance to the apron. Use the following data:

$$U_\infty = 7 \text{ m/s} \qquad \mu = 1.14 \times 10^{-3} \text{ kg/m s}$$

$$\frac{u}{U_\infty} = \left(\frac{y}{\delta}\right)^{1/7} \qquad \tau_0 = \frac{0.0225\rho U_\infty^2}{(\text{Re}_\delta)^{1/4}}$$

Solution

From (3.11),

$$\int_0^L \tau_0 \, dx = \rho U_\infty^2 \theta$$

This may be rewritten as

$$\tau_0 = \frac{d}{dx}(\rho U_\infty^2 \theta)$$

From (3.9),

$$\theta = \int_0^\delta \frac{u}{U_\infty} \left(1 - \frac{u}{U_\infty}\right) dy$$

Substituting $u/U_\infty = (y/\delta)^{1/7}$,

$$\theta = \int_0^\delta \left(\frac{y}{\delta}\right)^{1/7} \left[1 - \left(\frac{y}{\delta}\right)^{1/7}\right] dy = \frac{7}{72}\delta$$

Hence

$$\tau_0 = \frac{d}{dx}\left[\rho U_\infty^2 \left(\frac{7\delta}{72}\right)\right] = \frac{7}{72}\rho U_\infty^2 \frac{d\delta}{dx}$$

But it has been stated that $\tau_0 = 0.0225\rho U_\infty^2/(\mathrm{Re}_\delta)^{1/4}$. Therefore

$$\frac{0.0225\rho U_\infty^2}{(\mathrm{Re}_\delta)^{1/4}} = \frac{7}{72}\rho U_\infty^2 \frac{d\delta}{dx}$$

Substituting $\rho = 1000$ kg/m^3 and $U_\infty = 7$ m/s, and rearranging,

$$\frac{0.2314}{(\mathrm{Re}_\delta)^{1/4}} = \frac{d\delta}{dx}$$

$\mathrm{Re}_\delta = \rho U_\infty \delta/\mu$, therefore $(\mathrm{Re}_\delta)^{1/4} = 49.78\,\delta^{1/4}$, so, integrating,

$$x = \int \frac{49.78\delta^{1/4}}{0.2314}\, d\delta$$

$$= \frac{39.82\delta^{5/4}}{0.2314}$$

or

$$\delta = \left(\frac{0.2314x}{39.82}\right)^{4/5}$$

so if $x = 50$ m, then $\delta = 0.372$ m and

$$\tau_0 = \frac{0.0225\rho U_\infty^2}{(\mathrm{Re}_\delta)^{1/4}} = \frac{0.0225 \times 1000 \times 7^2}{\left(\dfrac{1000 \times 7 \times 0.372}{1.14 \times 10^{-3}}\right)^{1/4}}$$

$$= 28.36 \text{ N/m}^2$$

3.6   Some implications of the boundary layer concept

'Flow separation'

Returning to the problem of flow round a 'bluff' shape, which was considered at the beginning of the chapter, it is now possible to investigate the pattern of events in the light of our knowledge of boundary layer formation (see Fig. 3.11).

Around the upstream half of the body, the fluid is deflected outwards, the streamlines converge as the flow accelerates and a boundary layer grows progressively. After the fluid passes the Y–Y axis, the flow is decelerating. The fluid in the boundary layer is travelling at a lower speed than the fluid in the free stream, and a point is reached at which negative velocities arise at the inner part of the boundary layer. The line traced by the points of zero velocity downstream of the body divides the zones of positive and negative velocity, and indicates that flow separation has occurred. The development of the negative velocity zone further implies that the pressures within the zone are low compared with those in the free stream. Fluid from further out in the boundary layer is therefore drawn inwards to the low pressure zone. The effect of all this is that powerful eddies are generated, which are then drawn downstream by the flow, thus forming the 'wake' zone.

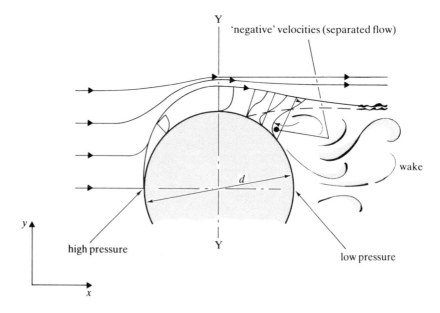

Figure 3.11   Flow separation.

Surface roughness and boundary layer development

A question may be posed: Different materials exhibit different degrees of roughness (e.g. plastic or concrete) – does this have any effect on the boundary layer? There are, broadly, three answers which could be given.

(a) In laminar flow, the friction is transmitted by pure shearing action. Consequently, the roughness of the solid surface has no effect, except to trap small 'pools' of stationary fluid in the interstices, and thus slightly increase the thickness of the stationary layer of fluid.
(b) In a turbulent flow, a laminar sub-layer forms close to the solid surface. If the average height of the surface roughness is smaller than the height of the laminar sub-layer, there will be little or no effect on the overall flow.
(c) Turbulent flow embodies a process of momentum transfer from layer to layer. Consequently, if the surface roughness protrudes through the laminar region into the turbulent region, then it will cause additional eddy formation and therefore greater energy loss in the turbulent flow. This implies that the apparent frictional shear will be increased.

'Drag' forces on a body

From the outline given in the two preceding paragraphs, it will be evident that a body immersed in an external flow may be subjected to two distinct force patterns:

(a) Due to the frictional shearing action between the body and the flow.
(b) Due to flow separation, if the body is 'bluff'. Under such conditions, the deflection of the incident flow around the upstream face of the body creates a region of locally increased pressure, whilst the separation zone downstream creates a region of locally low pressure. This pressure difference exerts a force on the body. The component of this force in the direction of the incident flow is known as the 'form drag'. If there is a component perpendicular to this direction, it is often referred to as 'lift'. The terminology originated in the aeronautical industry, which sponsored much of the research subsequent to Prandtl's pioneering work.

Measurements have been carried out on many body shapes, ranging from rectangular sections to aerofoil (wing) shapes. The measured forces are usually reduced to coefficient form, for example drag is expressed as 'drag coefficient' C_D, where

$$C_D = \frac{\text{drag force}}{\frac{1}{2}\rho A U_\infty^2}$$

A is a 'representative area', e.g. the cross-sectional area of a body perpendicular to the direction of the incident flow. Note that the drag force is usually the total force due to both friction and form components. C_D is therefore a function of the body shape and the Reynolds Number. Graphs of C_D versus the Reynolds Number for various shapes may be found in Schlichting (1979). Civil engineers encounter 'drag' problems in connection with wind flows around buildings, bridge piers in rivers, etc.

It is worth drawing attention to one further characteristic of separated flows. Under certain conditions it is possible for the eddy-forming mechanism to generate a rhythmic pattern in the wake. Eddies are shed alternately, first from one side of the body and then from the other. This eddy-shedding process sets up a transverse (lift) force on the body, whereas in a symmetrically generated wake the forces generated in this way are also symmetrical, and therefore effectively cancel. If the rhythmical pattern is set up it generates an alternating transverse force which can cause vibration problems. The frequency of the eddy-shedding is related to the Strouhal Number:

$$\text{St} = \text{frequency} \times d/U_\infty$$

Experiments have shown that, for circular cylinders,

$$\text{St} = 0.198(1 - 19.7/\text{Re}), \qquad \text{for } 250 < \text{Re} < 2 \times 10^5$$

Vibration problems of this type tend to occur when the natural frequency of a flexible structure (e.g. a suspension bridge or metal-clad chimney) coincides with the eddy frequency.

Bounded flows

Most civil engineering flows are 'bounded', either completely (as in pipe flows) or partly (as in open channels). Boundary layer growth will commence at the entry to the pipe or channel system, and continue downstream. Providing that the system is of sufficient length, a point will be reached at

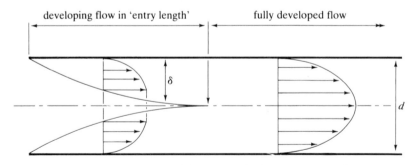

Figure 3.12 Development of flow pattern in the entrance to a pipe.

which the boundary layer thickness, δ, extends as far as the centreline (Fig. 3.12). From this point onwards it is clear that no boundary layer growth is possible. The flow has therefore become virtually uniform, and the boundary shear stress, τ_0, must, within close limits, be constant. This is known as the 'fully developed flow' condition. In most practical cases the 'entry length' is short compared with the overall length of the pipe or channel (for example, in laminar pipe flow the entry length is approximately $110d$, while for turbulent pipe flow it is approximately $50d$). It is therefore conventional and simplest to treat most problems as if the flow were fully developed throughout. In fully developed flows it is usual for the boundary shear stress to be expressed in terms of a 'coefficient of friction' (symbol C_f), which is defined by the equation

$$C_f = \tau_0/\tfrac{1}{2}\rho V^2 (= \lambda/4, \text{ see Ch. 4})$$

where V is the mean velocity of flow (i.e. Q/A).

In view of what was said previously, it is not surprising that the friction factor is:

(a) a function of the flow type (laminar or turbulent), and consequently a function of the Reynolds Number;
(b) a function of the boundary roughness (in turbulent flows), providing the roughness elements are sufficiently high – the evaluation and applications of friction factors will be developed more fully in Chapter 4.

3.7 Cavitation

At the beginning of this chapter, reference was made to the characteristics of an ideal fluid. One of those characteristics specified that an ideal fluid constituted a homogeneous continuum under all conditions of flow. This is also frequently true of real fluids – frequently, but not always. The principal exception to homogeneous flow will now be examined.

If the pressure of a liquid falls, the temperature at which boiling occurs also falls. It follows that a fall in pressure of sufficient magnitude will induce boiling, even at normal atmospheric temperatures, and this phenomenon is called cavitation. Cavitation arises under certain conditions, some typical examples being:

(a) In severely sheared separated flows of liquids. Such flows can occur on spillways or in control valves in pipelines.
(b) In fluid machines such as pumps or turbines. Fluid machines involve an energy interchange process (e.g. pumps may convert electrical power into fluid power). The interchange involves a moving element that is immersed in the fluid. Under some circumstances, very low pressures may occur at the interface between the element and the fluid.

The reason for regarding cavitation as a problem is that the boiling process involves the formation and collapse of vapour bubbles in the liquid. When these bubbles collapse, they cause a considerable local shock (or 'hammer blow'), i.e. a sharp rise and fall in the local pressure. The peak pressure during such a shock may be up to 400×10^6 N/m^2. Although one such event lasts only a few milliseconds, under cavitating conditions the event may be repeated in rapid succession. The potential for producing damage to concrete or metal surfaces is thus considerable. There have been a number of cases of major (and therefore expensive!) damage in large hydroelectric schemes. Theoretically, the onset of cavitation coincides with the vapour pressure of the liquid. However, the presence of dissolved gases or small solid particles in suspension frequently means that cavitation can occur at higher pressures. It is usually assumed that with water at 20 °C, for example, it is inadvisable to operate at pressure heads lower than 3 m absolute (i.e. 7 m vacuum).

3.8 Surface tension effects

Surface tension has been briefly discussed in the Introductory Notes. Surface tension, T_S, is usually very small compared with other forces in fluid flows (e.g. for a water surface exposed to air, $T_S \simeq 0.073$ N/m). In the vast majority of cases, engineers ignore surface tension. However, it may assume importance in low flows (for example, in hydraulic models or in weir flows under very low heads). Research into such situations led to the formation of a dimensionless parameter known as the Weber Number, We (named after the German naval architect M. Weber (1871–1951)):

$$\mathrm{We} = \sqrt{\frac{\text{inertia force}}{\text{surface tension}}} = \sqrt{\frac{\rho u^2 l}{T_S}}$$

For a given flow, the magnitude of We indicates whether T_S is likely to be significant.

3.9 Summary

The major topics covered in this chapter have related to turbulence and the boundary layer. Within this brief account, it is impossible to achieve more than an introductory coverage. More advanced studies may be pursued by consulting the work of Schlichting (1979) or Cebeci and Bradshaw (1977) on boundary layers. Also, Hinze (1975) and Tennekes and Lumley (1972) provide a more comprehensive outline of the physics and mathematics of turbulence.

References and further reading

Cebeci, T. and P. Bradshaw 1977. *Momentum transfer in boundary layers*. New York: McGraw-Hill.

Hinze, J. O. 1975. *Turbulence*, 2nd edn. New York: McGraw-Hill

Schlichting, H. 1979. *Boundary layer theory*, 7th edn. New York: McGraw-Hill

Tennekes, H. and J. L. Lumley 1972. *A first course in turbulence*. Cambridge, Massachusetts: MIT Press.

4

Flow in pipes and closed conduits

4.1 Introduction

The flow of water, oil and gas in pipes is of immense practical significance in civil engineering. Water is conveyed from its source, normally in pressure pipelines, to water treatment plants where it enters the distribution system and finally arrives at the consumer. Surface water drainage and sewerage is conveyed by closed conduits (Photograph 2), which do not usually operate under pressure, to sewage treatment plants, from where it is usually discharged to a river or the sea. Oil and gas are often transferred from their source by pressure pipelines to refineries (oil) or into a distribution network for supply (gas).

Surprising as it may seem, a comprehensive theory of the flow of fluids in pipes was not developed until the late 1930s, and practical design methods for the evaluation of discharges, pressures and head losses did not appear until 1958. Until these design tools were available, the efficient design of pipeline systems was not possible.

This chapter describes the theories of pipe flow, beginning with a review of the historical context and ending with the practical applications.

4.2 The historical context

Table 4.1 lists the names of the main contributors, and their contributions, to pipe flow theories in chronological order.

The Colebrook−White transition formula represents the culmination of all the previous work, and can be applied to any fluid in any pipe operating under turbulent flow conditions. The later contributions of Moody, Ackers and Barr are mainly concerned with the practical application of the Colebrook−White equation.

Photograph 2 Laying an urban drainage system.

There are three major concepts described in the table. These are:

(a) the distinction between laminar and turbulent flow;
(b) the distinction between rough and smooth pipes;
(c) the distinction between artificially roughened pipes and commercial pipes.

To understand these concepts, the best starting point is the contribution of Reynolds, followed by the laminar flow equations, before proceeding to the more complex turbulent flow equations.

Table 4.1 The chronological development of pipe flow theories.

Date	Name	Contribution
1839–1841	Hagen & Poiseuille	laminar flow equation
1850	Darcy & Weisbach	turbulent flow equation
1884	Reynolds	distinction between laminar and turbulent flow – Reynolds' Number
1913	Blasius	friction factor equation for smooth pipes
1914	Stanton & Pannell	experimental values of the friction factor for smooth pipes
1930	Nikuradse	experimental values of the friction factor for artificially rough pipes
1930s	Prandtl & von Kármán	equations for rough and smooth friction factors
1937–1939	Colebrook & White	experimental values of the friction factor for commercial pipes and the transition formula
1944	Moody	the Moody diagram for commercial pipes
1958	Ackers	the Hydraulics Research Station Charts and Tables for the design of pipes and channels
1975	Barr	direct solution of the Colebrook–White equation

Laminar and turbulent flow

Reynolds' experiments demonstrated that there were two kinds of flow – laminar and turbulent – as described in Chapter 3. He found that transition from laminar to turbulent flow occurred at a critical velocity for a given pipe and fluid. Expressing his results in terms of the dimensionless parameter $Re = \rho D V / \mu$, he found that for Re less than about 2000 the flow was always laminar, and that for Re greater than about 4000 the flow was always turbulent. For Re between 2000 and 4000, he found that the flow could be either laminar or turbulent, and termed this the transition region.

In a further set of experiments, he found that for laminar flow the frictional head loss in a pipe was proportional to the velocity, and that for turbulent flow the head loss was proportional to the square of the velocity.

These two results had been previously determined by Hagen and Poiseuille ($h_f \propto V$) and Darcy and Weisbach ($h_f \propto V^2$), but it was Reynolds who put these equations in the context of laminar and turbulent flow.

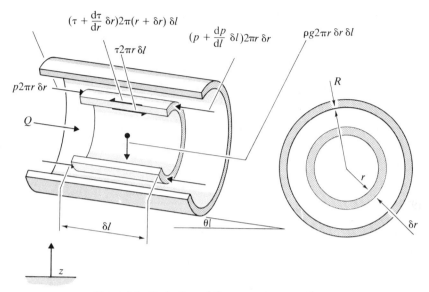

Figure 4.1 Derivation of the momentum equation.

4.3 Fundamental concepts of pipe flow

The momentum equation

Before proceeding to derive the laminar and turbulent flow equations, it is instructive to consider the momentum (or dynamic) equation of flow and the influence of the boundary layer.

Referring to Figure 4.1, showing an elemental annulus of fluid, thickness δr, length δl, in a pipe of radius R, the forces acting are the pressure forces, the shear forces and the weight of the fluid. The sum of the forces acting is equal to the change of momentum. In this case momentum change is zero, since the flow is steady and uniform. Hence

$$p2\pi r\,\delta r - \left(p + \frac{dp}{dl}\,\delta l\right)2\pi r\,\delta r + \tau 2\pi r\,\delta l - \left(\tau + \frac{d\tau}{dr}\,\delta r\right)2\pi(r + \delta r)\,\delta l + \rho g 2\pi r\,\delta l\,\delta r\sin\theta = 0$$

Setting $\sin\theta = -\,dz/dl$ and dividing by $2\pi r\,\delta r\,\delta l$ gives

$$-\frac{dp}{dl} - \frac{d\tau}{dr} - \frac{\tau}{r} - \rho g\frac{dz}{dl} = 0$$

(ignoring second-order terms), or

$$-\frac{dp^*}{dl} - \frac{\tau}{r} - \frac{d\tau}{dr} = 0$$

where p^* $(= p + \rho gz)$ is the piezometric pressure measured from the datum $z = 0$. As

$$\frac{1}{r}\frac{d}{dr}(\tau r) = \frac{1}{r}\left(r\frac{d\tau}{dr} + \tau\right) = \frac{d\tau}{dr} + \frac{\tau}{r}$$

then

$$-\frac{dp^*}{dl} - \frac{1}{r}\frac{d}{dr}(\tau r) = 0$$

Rearranging,

$$\frac{d}{dr}(\tau r) = -r\frac{dp^*}{dl}$$

Integrating both sides with respect to r,

$$\tau r = -\frac{dp^*}{dl}\frac{r^2}{2} + \text{constant}$$

At the centreline $r = 0$, and therefore constant $= 0$. Hence

$$\tau = -\frac{dp^*}{dl}\frac{r}{2} \tag{4.1}$$

Equation (4.1) is the momentum equation for steady uniform flow in a pipe. It is equally applicable to laminar or turbulent flow, and relates the shear stress τ at radius r to the rate of head loss with distance along the pipe. If an expression for the shear force can be found in terms of the velocity at radius r, then the momentum equation may be used to relate the velocity (and hence discharge) to head loss.

In the case of laminar flow, this is a simple matter. However, for the case of turbulent flow it is more complicated, as will be seen in the following sections.

The development of boundary layers

Figure 4.2a shows the development of laminar flow in a pipe. At entry to the pipe, a laminar boundary layer begins to grow. However, the growth of the boundary layer is halted when it reaches the pipe centreline, and thereafter the flow consists entirely of a boundary layer of thickness r. The resulting velocity distribution is as shown in Figure 4.2a.

For the case of turbulent flow shown in Figure 4.2b, the growth of the boundary layer is not suppressed until it becomes a turbulent boundary layer with the accompanying laminar sub-layer. The resulting velocity profile therefore differs considerably from the laminar case. The existence

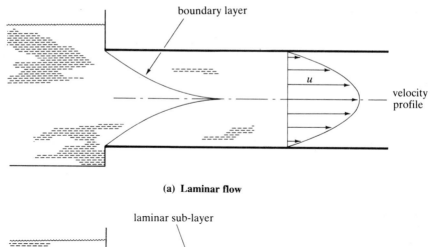

(a) Laminar flow

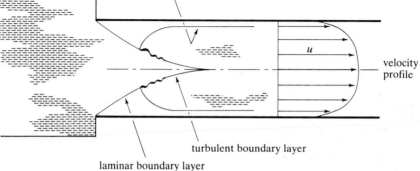

(b) Turbulent flow

Figure 4.2 Boundary layers and velocity distributions.

of the laminar sub-layer is of prime importance in explaining the difference between smooth and rough pipes.

Expressions relating shear stress to velocity have been developed in Chapter 3, and these will be used in explaining the pipe flow equations in the following sections.

4.4 Laminar flow

For the case of laminar flow, Newton's law of viscosity may be used to evaluate the shear stress (τ) in terms of velocity (u):

$$\tau = \mu \frac{\mathrm{d}u}{\mathrm{d}y} = -\mu \frac{\mathrm{d}u_r}{\mathrm{d}r}$$

Substituting into the momentum equation (4.1),

$$\tau = -\mu \frac{du_r}{dr} = -\frac{dp^*}{dl}\frac{r}{2}$$

or

$$\frac{du_r}{dr} = \frac{1}{2\mu}\frac{dp^*}{dl} r$$

Integrating,

$$u_r = \frac{1}{4\mu}\frac{dp^*}{dl} r^2 + \text{constant}$$

At the pipe boundary, $u_r = 0$ and $r = R$, hence

$$\text{constant} = -\frac{1}{4\mu}\frac{dp^*}{dl} R^2$$

and

$$u_r = -\frac{1}{4\mu}\frac{dp^*}{dl}(R^2 - r^2) \qquad (4.2)$$

Equation (4.2) represents a parabolic velocity distribution, as shown in Figure 4.2a. The discharge (Q) may be determined from (4.2). Returning to Figure 4.1 and considering the elemental discharge (δQ) through the annulus, then

$$\delta Q = 2\pi r\, \delta r\, u_r$$

Integrating

$$Q = 2\pi \int_0^R r u_r\, dr$$

and substituting for u_r from (4.2) gives

$$Q = -\frac{2\pi}{4\mu}\frac{dp^*}{dl} \int_0^R r(R^2 - r^2)\, dr$$

or

$$Q = -\frac{\pi}{8\mu}\frac{dp^*}{dl} R^4 \qquad (4.3)$$

Also the mean velocity (V) may be obtained directly from Q:

$$V = \frac{Q}{A} = -\frac{\pi}{8\mu} \frac{\mathrm{d}p^*}{\mathrm{d}l} R^4 \frac{1}{\pi R^2}$$

or

$$V = -\frac{1}{8\mu} \frac{\mathrm{d}p^*}{\mathrm{d}l} R^2 \qquad (4.4)$$

In practice, it is usual to express (4.4) in terms of frictional head loss by making the substitution

$$h_f = -\frac{\Delta p^*}{\rho g}$$

Equation (4.4) then becomes

$$V = \frac{1}{8\mu} \frac{h_f}{L} \rho g \frac{D^2}{4}$$

or

$$h_f = \frac{32\mu L V}{\rho g D^2} \qquad (4.5)$$

This is the Hagen–Poiseuille equation, named after the two people who first carried out (independently) the experimental work leading to it.

The wall shear stress (τ_0) may be related to the mean velocity (V) by eliminating $\mathrm{d}p^*/\mathrm{d}l$ from (4.1) and (4.4) to give

$$\tau = 4\mu V r / R^2 \qquad (4.6)$$

As $\tau = \tau_0$ when $r = R$, then

$$\tau_0 = 4\mu V / R \qquad (4.7)$$

Equation (4.6) shows that (for a given V) the shear stress is proportional to r, and is zero at the pipe centreline, with a maximum value (τ_0) at the pipe boundary.

Example 4.1   Laminar pipe flow

Oil flows through a 25 mm diameter pipe with a mean velocity of 0.3 m/s. Given that $\mu = 4.8 \times 10^{-2}$ kg/m s and $\rho = 800$ kg/m^3, calculate (a) the pressure drop in a 45 m length and (b) the maximum velocity, and the velocity 5 mm from the pipe wall.

Solution

First check that flow is laminar, i.e. Re < 2000.

$$\text{Re} = \rho D V / \mu = 800 \times 0.025 \times 0.3 / 4.8 \times 10^{-2}$$
$$= 125$$

(a) To find the pressure drop, apply (4.5):

$$h_f = 32 \mu L V / \rho_o g D^2$$
$$= (32 \times 4.8 \times 10^{-2} \times 45 \times 0.3) / (800 \times 9.81 \times 0.025^2)$$
$$= 4.228 \text{ m (of oil)}$$

or $\Delta p = -\rho_o g h_f = -33.18 \text{ kN/m}^2$. (Note: the negative sign indicates that pressure reduces in the direction of flow.)

(b) To find the velocities, apply (4.2):

$$u_r = -\frac{1}{4\mu} \frac{\mathrm{d}p^*}{\mathrm{d}l} (R^2 - r^2)$$

The maximum velocity (U_{max}) occurs at the pipe centreline, i.e. when $r = 0$, hence

$$U_{max} = -\frac{1}{4 \times 4.8 \times 10^{-2}} \times -\frac{33.18 \times 10^3}{45} (0.025/2)^2$$
$$= 0.6 \text{ m/s}$$

(Note: $U_{max} = 2 \times$ mean velocity (compare (4.2) and (4.4).))

To find the velocity 5 mm from the pipe wall (U_5), use (4.2) with $r = (0.025/2) - 0.005$, i.e. $r = 0.0075$:

$$u_5 = -\frac{1}{4 \times 4.8 \times 10^{-2}} \times -\frac{33.18 \times 10^3}{45} (0.0125^2 - 0.0075^2)$$
$$= 0.384 \text{ m/s}$$

4.5 Turbulent flow

For turbulent flow, Newton's viscosity law does not apply and, as described in Chapter 3, semi-empirical relationships for τ_0 were derived by Prandtl. Also, Reynolds' experiments, and the earlier ones of Darcy and Weisbach, indicated that head loss was proportional to mean velocity squared. Using the momentum equation (4.1), then

$$\tau_0 = -\frac{\mathrm{d}p^*}{\mathrm{d}l} \frac{R}{2}$$

and

$$-\frac{dp^*}{dl} = \frac{h_f \rho g}{L}$$

hence

$$\tau_0 = \frac{h_f}{L} \rho g \frac{R}{2}$$

Assuming $h_f = KV^2$, based on the experimental results cited above, then

$$\tau_0 = \frac{KV^2}{L} \rho g \frac{R}{2}$$

or

$$\tau_0 = K_1 V^2$$

(for $h_f = KV^2$).

Returning to the momentum equation and making the substitution $\tau_0 = K_1 V^2$, then

$$K_1 V^2 = -\frac{dp^*}{dl} \frac{R}{2}$$

hence

$$K_1 V^2 = \frac{h_f}{L} \rho g \frac{R}{2}$$

or

$$h_f = \frac{4K_1 L V^2}{\rho g D}$$

Making the substitution $\lambda = 8K_1/\rho$, then

$$h_f = \frac{\lambda L V^2}{2gD} \tag{4.8}$$

This is the Darcy–Weisbach equation, in which λ is called the pipe friction factor and is sometimes referred to as f (American practice) or $4f$ (early British practice). In current practice, λ is the normal usage and is found, for instance, in the Hydraulics Research Station charts and tables. It should be noted that λ is dimensionless, and may be used with any system of units.

The original investigators presumed that the friction factor was constant. This was subsequently found to be incorrect (as described in Section 3.6) Equations relating λ to both the Reynolds Number and the pipe roughness were developed later.

Smooth pipes and the Blasius equation

Experimental investigations by Blasius and others early in the 20th century led to the equation

$$\lambda = 0.316/\text{Re}^{0.25} \qquad (4.9)$$

The later experiments of Stanton and Pannel, using drawn brass tubes, confirmed the validity of the Blasius equation for Reynolds' Numbers up to 10^5. However, at higher values of Re the Blasius equation underestimated λ for these pipes. Before further progress could be made, the distinction between 'smooth' and 'rough' pipes had to be established.

Artificially rough pipes and Nikuradse's experimental results

Nikuradse made a major contribution to the theory of pipe flow by objectively differentiating between smooth and rought turbulence in pipes. He carried out a painstaking series of experiments to determine both the friction factor and the velocity distributions at various Reynolds' Numbers up to 3×10^6. In these experiments, pipes were artificially roughened by sticking uniform sand grains on to smooth pipes. He defined the relative roughness (k_S/D) as the ratio of the sand grain size to the pipe diameter. By using pipes of different diameter and sand grains of different size, he produced a set of experimental results of λ and Re for a range of relative roughness of 1/30 to 1/1014.

He plotted his results as log λ against log Re for each value of k_S/D, as shown in Figure 4.3. This figure shows that there are five regions of flow, as follows:

(a) **Laminar flow**. The region in which the relative roughness has no influence on the friction factor. This was assumed in deriving the Hagen–Poiseuille equation (4.5). Equating this to the Darcy–Weisbach equation (4.8) gives

$$\frac{32\mu VL}{\rho g D^2} = \frac{\lambda L V^2}{2gD}$$

or

$$\lambda = \frac{64\mu}{\rho D V} = \frac{64}{\text{Re}} \qquad (4.10)$$

Hence, the Darcy–Weisbach equation may also be used for laminar flow, provided that λ is evaluated by (4.10).

(b) **Transition from laminar to turbulent flow**. An unstable region between Re = 2000 and 4000. Fortunately, pipe flow normally lies outside this region.

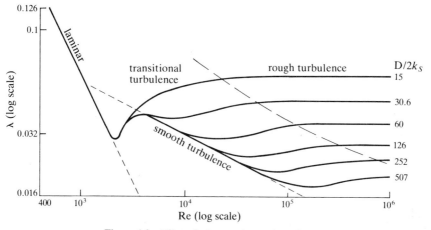

Figure 4.3 Nikuradse's experimental results.

(c) **Smooth turbulence**. The limiting line of turbulent flow, approached by all values of relative roughness as Re decreases.

(d) **Transitional turbulence**. The region in which λ varies with both Re and k_S/D. The limit of this region varies with k_S/D. In practice, most of pipe flow lies within this region.

(e) **Rough turbulence**. The region in which λ remains constant for a given k_S/D, and is independent of Re.

An explanation of why these five regions exist has already been given in Section 3.6. It may be summarised as follows:

Laminar flow. Surface roughness has no influence on shear stress transmission.

Transitional turbulence. The presence of the laminar sub-layer 'smooths' the effect of surface roughness.

Rough turbulence. The surface roughness is large enough to break up the laminar sub-layer giving turbulence right across the pipe.

The rough and smooth laws of von Kármán and Prandtl

The publication of Nikuradse's experimental results (particularly his velocity distribution measurements) were used by von Kármán and Prandtl to supplement their own work on turbulent boundary layers. By combining their theories of turbulent boundary layer flows with the experimental results, they derived the semi-empirical rough and smooth laws. These were:

for smooth pipes

$$\frac{1}{\sqrt{\lambda}} = 2 \log \frac{\text{Re}\sqrt{\lambda}}{2.51} \tag{4.11}$$

for rough pipes

$$\frac{1}{\sqrt{\lambda}} = 2 \log \frac{3.7D}{k_S} \qquad (4.12)$$

The smooth law is a better fit to the experimental data than the Blasius equation.

The Colebrook–White transition formula

The experimental work of Nikuradse and the theoretical work of von Kármán and Prandtl provided the framework for a theory of pipe friction. However, these results were not of direct use to engineers because they applied only to artificially roughened pipes. Commercial pipes have roughness which is uneven both in size and spacing, and do not, therefore, necessarily correspond to the pipes used in Nikuradse's experiments.

Colebrook and White made two major contributions to the development and application of pipe friction theory to engineering design. Initially, they carried out experiments to determine the effect of non-uniform roughness as found in commercial pipes. They discovered that in the turbulent transition region the λ–Re curves exhibited a gradual change from smooth to rough turbulence in contrast to Nikuradse's 'S'-shaped curves for uniform roughness, size and spacing. Colebrook then went on to determine the 'effective roughness' size of many commercial pipes. He achieved this by studying published results of frictional head loss and discharge for commercial pipes, ranging in size from 4 inches (101.6 mm) to 61 inches (1549.4 mm), and for materials, including drawn brass, galvanised, cast and wrought iron, bitumen-lined pipes and concrete-lined pipes. By comparing the friction factor of these pipes with Nikuradse's results for uniform roughness size in the rough turbulent zone, he was able to determine an 'effective roughness' size for the commercial pipes equivalent to Nikuradse's results. He was thus able to publish a list of k_S values applicable to commercial pipes.

A second contribution of Colebrook and White stemmed from their experimental results on non-uniform roughness. They combined the von Kármán–Prandtl rough and smooth laws in the form

$$\frac{1}{\sqrt{\lambda}} = -2 \log \left(\frac{k_S}{3.7D} + \frac{2.51}{\text{Re}\sqrt{\lambda}} \right) \qquad (4.13)$$

This gave predicted results very close to the observed transitional behaviour of commercial pipes, and is known as the Colebrook–White transition formula. It is applicable to the whole of the turbulent region for commercial pipes using an effective roughness value determined experimentally for each type of pipe.

The practical application of the Colebrook–White transition formula

Equation (4.13) was not at first used very widely by engineers, mainly because it was not expressed directly in terms of the standard engineering variables of diameter, discharge and hydraulic gradient. In addition, the equation is implicit and requires a trial-and-error solution. In the 1940s, slide-rules and logarithm tables were the main computational aids of the engineer, since pocket calculators and computers were not then available. So these objections to the use of the Colebrook–White equation were not unreasonable.

The first attempt to make engineering calculations easier was made by Moody. He produced a λ-Re plot based on (4.13) for commercial pipes, as shown in Figure 4.4 which is now known as the Moody diagram. He also presented an explicit formula for λ:

$$\lambda = 0.0055 \left[1 + \left(\frac{2000 \, k_S}{D} + \frac{10^6}{\mathrm{Re}} \right)^{\!\!1/3} \right] \tag{4.14}$$

which gives λ correct to ±5% for $4 \times 10^3 < \mathrm{Re} < 1 \times 10^7$ and for $k_S/D < 0.01$.

In a more recent publication, Barr (1975) gives another explicit formulation for λ:

$$\frac{1}{\sqrt{\lambda}} = -2 \log \left(\frac{k_S}{3.7D} + \frac{5.1286}{\mathrm{Re}^{0.89}} \right) \tag{4.15}$$

In this formula the smooth law component $(2.51/\mathrm{Re}\sqrt{\lambda})$ has been replaced by an approximation $(5.1286/\mathrm{Re}^{0.89})$. For $\mathrm{Re} > 10^5$ this provides a solution for $S_f \, (h_f/L)$ to an accuracy better than ±1%.

However, the basic engineering objections to the use of the Colebrook–White equation were not overcome until the publication of *Charts for the hydraulic design of channels and pipes* in 1958 by the Hydraulics Research Station. In this publication, the three dependent engineering variables (Q, D and S_f) were presented in the form of a series of charts for various k_S values, as shown in Figure 4.5. Additional information regarding suitable design values for k_S and other matters was also included. Table 4.2 lists typical values for various materials.

These charts are based on the combination of the Colebrook–White equation (4.13) with the Darcy–Weisbach formula (4.8), to give

$$V = -2\sqrt{2gDS_f} \, \log \left(\frac{k_S}{3.7D} + \frac{2.51\nu}{D\sqrt{2gDS_f}} \right) \tag{4.16}$$

where $S_f = h_f/L$, the hydraulic gradient. (Note: for further details concerning the hydraulic gradient refer to Ch. 12.) In this equation, the velocity

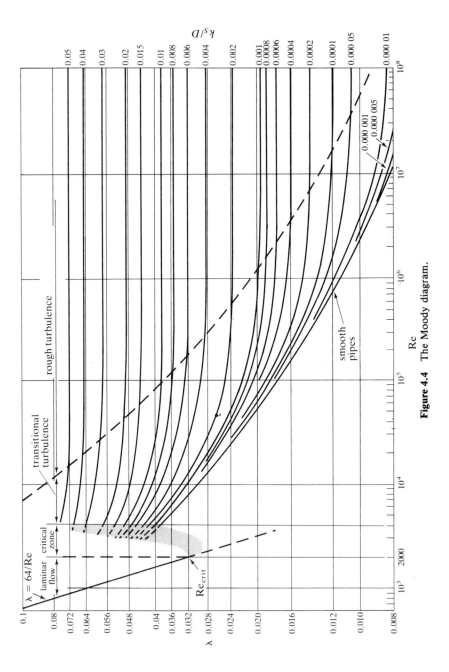

Figure 4.4 The Moody diagram.

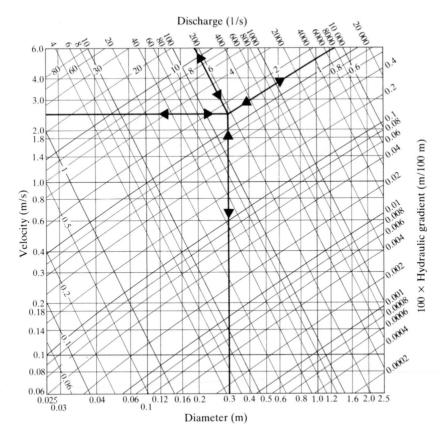

Figure 4.5 Hydraulics Research Station chart for $k_S = 0.03$ mm.

Table 4.2 Typical k_S values.

Pipe material	k_S(mm)
brass, copper, glass, Perspex	0.003
asbestos cement	0.03
wrought iron	0.06
galvanised iron	0.15
plastic	0.03
bitumen-lined ductile iron	0.03
spun concrete lined ductile iron	0.03
slimed concrete sewer	6.0

(and hence discharge) can be computed directly for a known diameter and frictional head loss.

More recently, the Hydraulics Research Station have also produced *Tables for the hydraulic design of pipes.*

In practice, any two of the three variables (Q, D and S_f) may be known, and therefore the most appropriate solution technique depends on circumstances. For instance, in the case of an existing pipeline, the diameter and available head are known and hence the discharge may be found directly from (4.16). For the case of a new installation, the available head and required discharge are known and the requisite diameter must be found. This will involve a trial-and-error procedure unless the HRS charts or tables are used. Finally, in the case of analysis of pipe networks, the required discharges and pipe diameters are known and the head loss must be computed. This problem may be most easily solved using an explicit formula for λ or the HRS charts.

Examples illustrating the application of the various methods to the solution of a simple pipe friction problem now follow.

Example 4.2   Estimation of discharge given diameter and head loss

A pipeline 10 km long, 300 mm in diameter and with roughness size 0.03 mm, conveys water from a reservoir (top water level 850 m above datum) to a water treatment plant (inlet water level 700 m above datum). Assuming that the reservoir remains full, estimate the discharge, using the following methods:

(a) the Colebrook–White formula;
(b) the Moody diagram;
(c) the HRS charts.

Note: Assume $\nu = 1.13 \times 10^{-6}\ \text{m}^2/\text{s}$.

Solution

(a) Using (4.16),

$$D = 0.3\ \text{m} \qquad k_S = 0.03\ \text{m}$$

$$S_f = (850 - 700)/10\,000 = 0.015$$

hence

$$V = -2\sqrt{2g \times 0.3 \times 0.015}\ \log \left(\frac{0.03 \times 10^{-3}}{3.7 \times 0.3} + \frac{2.51 \times 1.13 \times 10^{-6}}{0.3\sqrt{2g \times 0.3 \times 0.015}} \right)$$

$$= 2.514\ \text{m/s}$$

and

$$Q = VA = \frac{2.514 \times \pi \times 0.3^2}{4} = 0.178\ \text{m}^3/\text{s}$$

(b) The same solution should be obtainable using the Moody diagram; however, it is less accurate since it involves interpolation from a graph. The solution method is as follows:

(1) calculate k_S/D
(2) guess a value for V
(3) calculate Re
(4) estimate λ using the Moody diagram
(5) calculate h_f
(6) compare h_f with the available head (H)
(7) if $H \neq h_f$, then repeat from step 2

This is a tedious solution technique, but it shows why the HRS charts were produced!

(1) $k_S/D = 0.03 \times 10^{-3}/0.3 = 0.0001$.
(2) As the solution for V has already been found in part (a) take $V = 2.5$ m/s.

(3)
$$\text{Re} = \frac{DV}{\nu} = \frac{0.3 \times 2.5}{1.13 \times 10^{-6}}$$
$$= 0.664 \times 10^6$$

(4) Referring to Figure 4.4, Re $= 0.664 \times 10^6$ and $k_S/D = 0.0001$ confirms that the flow is in the transitional turbulent region. Following the k_S/D curve until it intersects with Re yields

$$\lambda \simeq 0.014$$

(Note: Interpolation is difficult due to the logarithmic scale.)
(5) Using (4.8),

$$h_f = \frac{\lambda L V^2}{2gD} = \frac{0.014 \times 10^4 \times 2.5^2}{2g \times 0.3}$$
$$= 148.7 \text{ m}$$

(6) $H = (850 - 700) = 150 \neq 148.7$.
(7) A better guess for V is obtained by increasing V slightly. This will not significantly alter λ, but will increase h_f. In this instance, convergence to the solution is rapid because the correct solution for V was assumed initially!

(c) If the HRS chart shown in Figure 4.5 is used, then the solution of the equation lies at the intersection of the hydraulic gradient line (sloping downwards right to left) with the diameter (vertical), reading off the corresponding discharge (line sloping downwards left to right).

$S_f = 0.015$ $100S_f = 1.5$
and $D = 300$ mm
giving $Q = 180$ 1/s $= 0.18$ m^3/s

Example 4.3　Estimation of pipe diameter given discharge and head

A discharge of 400 l/s is to be conveyed from a headworks at 1050 m above datum to a treatment plant at 1000 m above datum. The pipeline length is 5 km. Estimate the required diameter, assuming that $k_S = 0.03$ mm.

Solution

This requires an iterative solution if methods (a) or (b) of the previous example are used. However, a direct solution can be obtained using the HRS charts.

$$S_f = 50/5000 \qquad 100S_f = 1$$
$$\text{and} \qquad Q = 400 \text{ l/s}$$
$$\text{giving} \qquad D = 440 \text{ mm}$$

In practice, the nearest (larger) available diameter would be used (450 mm in this case).

Example 4.4　Estimation of head loss given discharge and diameter

The known outflow from a branch of a distribution system is 30 l/s. The pipe diameter is 150 mm, length 500 m and roughness coefficient estimated at 0.06 mm. Find the head loss in the pipe, using the explicit formulae of Moody and Barr.

Solution

Again, the HRS charts could be used directly. However, if the analysis is being carried out by computer, solution is more efficient using an equation.

$$Q = 0.03 \text{ m}^3/\text{s}, \quad D = 0.15 \text{ m}$$
$$V = 1.7 \text{ m/s}$$
$$\text{Re} = 0.15 \times 1.7/1.13 \times 10^{-6}$$
$$\text{Re} = 0.226 \times 10^6$$

Using the Moody formula (eqn (4.14))

$$\lambda = 0.0055 \left(1 + \left(\frac{2000 \times 0.06 \times 10^{-3}}{0.15} + \frac{1}{0.226}\right)^{\frac{1}{3}}\right)$$

$$\lambda = 0.0150$$

Using the Barr formula (eqn (4.15))

$$\frac{1}{\sqrt{\lambda}} = -2 \log \left(\frac{0.06 \times 10^{-3}}{3.7 \times 0.15} + \frac{5.1286}{(0.226 \times 10^6)^{0.89}}\right)$$

$$\lambda = 0.0182$$

The accuracy of these formulae may be compared by substituting in the Colebrook–White equation (4.13) as follows:

λ	$1/\sqrt{\lambda}$	$-2 \log\left(\dfrac{k_S}{3.7D} + \dfrac{2.51}{Re\sqrt{\lambda}}\right) = \dfrac{1}{\sqrt{\lambda}}$
Moody 0.0150	8.165	7.403
Barr 0.0182	7.415	7.441

This confirms that the Barr formula is the more accurate in this case.

The head loss may now be computed using the Darcy–Weisbach formula (4.8):

$$h_f = \frac{0.0182 \times 500 \times 1.7^2}{2g \times 0.15}$$

$$= 8.94 \text{ m}$$

The Hazen–Williams formula

The emphasis here has been placed on the development and use of the Colebrook–White transition formula. Using the charts or tables it is simple to apply to single pipelines. However, for pipes in series or parallel or for the more general case of pipe networks it rapidly becomes impossible to use for hand-calculations. For this reason, simpler empirical formulae are still in common use. Perhaps the most notable is the Hazen–Williams formula, which takes the form

$$V = 0.355CD^{0.63}(h_f/L)^{0.54}$$

or, alternatively,

$$h_f = \frac{6.78L}{D^{1.165}}\left(\frac{V}{C}\right)^{1.85}$$

where C is a coefficient. The value of C varies from about 70 to 150, depending on pipe diameter, material and age.

This formula gives reasonably accurate results over the range of Re commonly found in water distribution systems, and because the value of C is assumed to be constant, it can be easily used for hand-calculation. In reality, C should change with Re, and caution should be exercised in its use. An interesting problem is to compare the predicted discharges as calculated by the Colebrook–White equation and by the Hazen–Williams formula over a large range of Re for a given pipe. The use of a microcomputer is recommended for this exercise.

4.6   Local head losses

Head losses, in addition to those due to friction, are always incurred at pipe bends, junctions and valves, etc. These additional losses are due to eddy formation generated in the fluid at the fitting, and, for completeness, they must be taken into account. In the case of long pipelines (e.g. several kilometres) the local losses may be negligible, but for short pipelines, they may be greater than the frictional losses.

A general theoretical treatment for local head losses is not available. It is usual to assume rough turbulence since this leads to the simple equation

$$h_L = k_L V^2/2g \qquad (4.17)$$

where h_L is the local head loss and k_L is a constant for a particular fitting.

For the particular case of a sudden enlargement (for instance, exit from a pipe to a tank) an expression may be derived for k_L in terms of the area of the pipe. This result may be extended to the case of a sudden contraction (for instance, entry to a pipe from a tank). For all other cases (e.g. bends, valves, junctions, bellmouths, etc.) values for k_L must be derived experimentally.

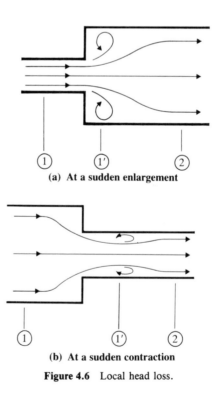

(a) At a sudden enlargement

(b) At a sudden contraction

Figure 4.6   Local head loss.

Figure 4.6a shows the case of a sudden enlargement. From position (1) to (2) the velocity decreases and therefore the pressure increases. At position (1') turbulent eddies are formed, which gives rise to a local energy loss. As the pressure cannot change instantaneously at the sudden enlargement, it is usually assumed that at position (1') the pressure is the same as at position (1). Applying the momentum equation between (1) and (2),

$$p_1 A_2 - p_2 A_2 = \rho Q (V_2 - V_1)$$

The continuity equation ($Q = A_2 V_2$) is now used to eliminate Q, so, with some rearrangement,

$$\frac{p_2 - p_1}{\rho g} = \frac{V_2}{g} (V_1 - V_2) \tag{a}$$

The local head loss may now be found by applying the energy equation from (1) to (2):

$$\frac{p_1}{\rho g} + \frac{V_1^2}{2g} = \frac{p_2}{\rho g} + \frac{V_2^2}{2g} + h_L$$

or

$$h_L = \frac{(V_1^2 - V_2^2)}{2g} - \frac{(p_2 - p_1)}{\rho g} \tag{b}$$

If (a) and (b) are combined and rearranged,

$$h_L = (V_1 - V_2)^2 / 2g$$

The continuity equation may now be used again to express the result in terms of the two areas. Hence, substituting $V_1 A_1 / A_2$ for V_2

$$h_L = \left(V_1 - V_1 \frac{A_1}{A_2} \right)^2 \bigg/ 2g$$

or

$$h_L = \left(1 - \frac{A_1}{A_2} \right)^2 \frac{V_1^2}{2g} \tag{4.18}$$

Equation (4.18) relates h_L to the areas and the upstream velocity. Comparing this equation with (4.17) yields

$$k_L = \left(1 - \frac{A_1}{A_2} \right)^2$$

Table 4.3 Local head loss coefficients.

	k_L value		
		Design	
Item	Theoretical	practice	Comments
bellmouth entrance	0.05	0.10	V = velocity in pipe
exit	0.2	0.5	
$90°$ bend	0.4	0.5	
$90°$ tees			
in-line flow	0.35	0.4	(for equal diameters)
branch to line	1.20	1.5	(for equal diameters)
gate valve (open)	0.12	0.25	

For the case of a pipe discharging into a tank, A_2 is much greater than A_1, and hence $k_L = 1$. In other words, for a sudden large expansion, the head loss equals the velocity head before expansion.

Figure 4.6b shows the case of a sudden contraction. From position (1) to (1′) the flow contracts, forming a vena contracta. Experiments indicate that the contraction of the flow area is generally about 40%. If the energy loss from (1) to (1′) is assumed to be negligible, then the remaining head loss occurs in the expansion from (1′) to (2). Since an expansion loss gave rise to (4.18), that equation may now be applied here. As

$$A_{1'} \simeq 0.6A_2$$

then

$$h_L = \left(1 - \frac{0.6A_2}{A_2}\right)^2 \frac{(V_2/0.6)^2}{2g}$$

or

$$h_L = 0.44 \, V_2^2/2g \qquad (4.19)$$

i.e. $k_L = 0.44$.

Typical k_L values for other important local losses (bends, tees, bellmouths and valves) are given in Table 4.3.

Example 4.5 Discharge calculation for a simple pipe system including local losses

Solve Example 4.2 allowing for local head losses incurred by the following items:

20	$90°$ bends
2	gate valves
1	bellmouth entry
1	bellmouth exit

Solution

The available static head (150 m) is dissipated both by friction and local head losses. Hence

$$H = h_f + h_L$$

Using Table 4.3,

$$h_L = [(20 \times 0.5) + (2 \times 0.25) + 0.1 + 0.5] \, V^2/2g$$
$$= 11.1 V^2/2g$$

Using the Colebrook–White formula (as in Example 4.2) now requires an iterative solution, since h is initially unknown. A solution procedure is as follows:

(1) assume $h_f \simeq H$ (i.e. ignore h_L)
(2) calculate V
(3) calculate h_L using V
(4) calculate $h_f + h_L$
(5) if $h_f + h_L \neq H$, set $h_f = H - h_L$ and return to (2)

Using example 4.2, an initial solution for V has already been found, i.e.

$$V = 2.514 \text{ m/s}$$

Hence,

$$h_L = 11.1 \times 2.514^2/2g = 3.58 \text{ m}$$

Adjust h ,

$$h_f = 150 - 3.58 = 146.42 \text{ m}$$

$$\text{Hence, } S_f = 146.42/10\,000 = 0.01464$$

Substitute in (4.16),

$$V = -2\sqrt{2g \times 0.3 \times 0.01464} \, \log \left(\frac{0.03 \times 10^{-3}}{3.7 \times 0.15} + \frac{2.51 \times 1.13 \times 10^{-6}}{0.3\sqrt{2g \times 0.3 \times 0.01464}} \right)$$

$$= 2.386 \text{ m/s}$$

Recalculate h_L,

$$h_L = 11.1 \times 2.386^2/2g = 3.22$$

Check

$$h_L + h_f = 146.42 + 3.22 = 149.64 \simeq 150$$

This is sufficiently accurate to be acceptable.

Hence,

$$Q = 2.386 \times \pi \times 0.3^2/4 = 0.17 \text{ m}^3/\text{s}$$

Note. Ignoring h_L gives $Q = 0.18 \text{ m}^3/\text{s}$.

4.7 Partially full pipes

Pipe systems for surface water drainage and sewerage are normally designed to flow full, but not under pressure. This contrasts with water mains, which are normally full and under pressure. The Colebrook–White equation may be used for drainage pipes by noting that, because the pipe flow is not pressurised, the water surface is parallel to the pipe invert, so the hydraulic gradient equals the pipe gradient:

$$h_f/L = S_0$$

where S_0 is the pipe gradient.

Additionally, an estimate of the discharge and velocity for the partially full condition is required. This enables the engineer to check if self-cleansing velocities are maintained at the minimum discharge. Self-cleansing velocities are of crucial importance in the design of surface water drainage and sewerage networks, where the flow may contain a considerable suspended solids load.

A free surface flow has one more variable than full pipe flow, namely the height of the free surface. This can introduce considerable complexity (refer to Ch. 5). However, for the case of circular conduits, the Colebrook–White equation may be modified to provide a solution.

Starting from the assumption that the friction factor for the partially full condition behaves similarly to that for the full condition, it remains to find a parameter for the partially full pipe which is equivalent to the diameter for the full pipe case. The hydraulic radius R is such a parameter:

$$R = A/P$$

where A is the water cross-sectional area and P is the wetted perimeter. For a pipe flowing full,

$$A/P = \pi D^2/4\pi D = D/4$$

or

$$4R = D$$

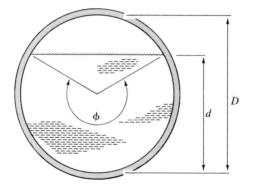

Figure 4.7 Pipe running partially full.

Hence the Colebrook–White transition law applied to partially full pipes becomes

$$\frac{1}{\sqrt{\lambda}} = -2 \log \left(\frac{k_S}{3.7 \times 4R} + \frac{2.51}{Re\sqrt{\lambda}} \right) \qquad (4.20)$$

where $Re = 4RV/\nu$.

Figure 4.7 shows a pipe with partially full flow (at a depth d). Starting from the Darcy–Weisbach equation (4.8) and replacing h_f/L by S_0 gives

$$V^2 = 2gS_0D/\lambda$$

Hence, for a given pipe with partially full flow,

$$V = (2gS_0 4R/\lambda)^{\frac{1}{2}}$$

or

$$V = \text{constant} \cdot R^{\frac{1}{2}}/\lambda^{\frac{1}{2}}$$

Forming the ratio $V_d/V_D = V_p$ gives

$$V_p = \lambda_D^{\frac{1}{2}} R_p^{\frac{1}{2}}/\lambda_d^{\frac{1}{2}} \qquad (4.21)$$

where the subscripts p, D and d refer, respectively, to the proportional value, the full depth (D) and the partially full depth (d). Similarly,

$$Q_p = \lambda_D^{\frac{1}{2}} A_p R_p^{\frac{1}{2}}/\lambda_d^{\frac{1}{2}} \qquad (4.22)$$

For a circular pipe,

$$A_d = \left(\frac{\phi - \sin \phi}{8} \right) D^2$$

$$P_d = \phi D/2$$

$$R_d = \left(1 - \frac{\sin \phi}{\phi}\right)\frac{D}{4}$$

and hence

$$A_p = \left(\frac{\phi - \sin \phi}{2\pi}\right) \tag{4.23}$$

$$R_p = \left(\frac{1 - \sin \phi}{\phi}\right) \tag{4.24}$$

Substitution of (4.23) and (4.24) into (4.21) and (4.22) allows calculation of the proportional velocity and discharge for any proportional depth (d/D). The expression for λ (Equation (4.20)) is, however, rather awkward to manipulate. Consider first the case of rough turbulence. Then,

$$\frac{1}{\sqrt{\lambda}} = 2 \log \left(\frac{3.7D}{k_S}\right)$$

Hence,

$$\frac{\sqrt{\lambda_D}}{\sqrt{\lambda_d}} = \frac{2 \log (3.7 \times 4R_d/k_S)}{2 \log (3.7D/k_S)}$$

This may be expressed by its equivalent:

$$\frac{\sqrt{\lambda_D}}{\sqrt{\lambda_d}} = 1 + \frac{\log R_p}{\log (3.7D/k_S)} \tag{4.25}$$

as

$$1 + \frac{\log R_p}{\log(3.7D/k_S)} = \frac{\log (3.7D/k_S) + \log R_p}{\log(3.7D/k_S)}$$

$$= \frac{\log [(3.7D/k_S)(R_d/R_D)]}{\log(3.7D/k_S)}$$

$$= \frac{\log [3.7 \times 4 R_d/k_S]}{\log(3.7D/k_S)}$$

Equation (4.25) may be substituted into (4.21) and (4.22) to yield

$$V_p = \left(1 + \frac{\log R_p}{\log(3.7D/k_S)}\right) R_p^{\frac{1}{2}}$$

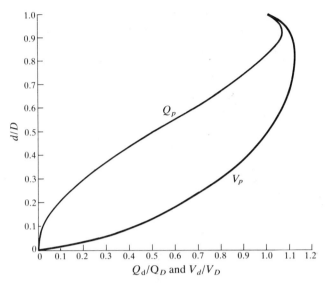

Figure 4.8 Proportional discharge and velocity for pipes flowing partially full (with $\theta = 1000$).

and

$$Q_p = \left(1 + \frac{\log R_p}{\log(3.7D/k_s)}\right) A_p R_p^{1/2}$$

The equivalent expressions for the transition region (as derived in Hydraulics Research Paper No. 2, published in 1959) are

$$V_p = \left(1 + \frac{\log R_p}{\log 3.7\theta}\right) R_p^{1/2} \tag{4.26}$$

and

$$Q_p = \left(1 + \frac{\log R_p}{\log 3.7\theta}\right) A_p R_p^{1/2} \tag{4.27}$$

where

$$\theta \simeq \left(\frac{k_s}{D} + \frac{1}{3600DS_0^{1/3}}\right)^{-1} \tag{4.28}$$

These results for $\theta = 1000$ are plotted in Figure 4.8. Tabulated values for various θ may be found in Hydraulics Research Ltd (1983a). Neither V_p nor Q_p are very sensitive to θ.

Figure 4.8 shows that the discharge in a partially full pipe may be greater than the discharge for a full pipe. This is because the wetted perimeter reduces rapidly immediately the pipe ceases to be full whereas the area does not, with a consequent increase in velocity. However, this condition is usually ignored for design purposes because, if the pipe runs full at any section (e.g. due to wave action or unsteady conditions), then the discharge will rapidly reduce to the full pipe condition and cause a 'backing up' of the flow upstream.

Example 4.5 Hydraulic design of a sewer

A sewerage pipe is to be laid at a gradient of 1 in 300. The design maximum discharge is 75 l/s and the design minimum flow is estimated to be 10 l/s. Determine the required pipe diameter to both carry the maximum discharge and maintain a self-cleansing velocity of 0.75 m/s at the minimum discharge.

Solution

The easiest way to solve this problem is to use the HRS design charts or tables. For a sewer, $k_S = 6.00$ mm (Table 4.2). However, to illustrate the solution, Figure 4.5 is used (for which $k_S = 0.03$ mm):

$$Q = 75 \text{ l/s}$$

$$100h_f/L = 100/300 = 0.333$$

Using Figure 4.5

$$D = 300 \text{ mm and } V = 1.06 \text{ m/s}$$

Next check the velocity for $Q = 10$ l/s

$$Q_p = 10/75 = 0.133$$

Using Figure 4.8 (neglecting the effect of θ),

$$d/D = 0.25 \text{ for } Q_p = 0.133$$

Hence $V_p = 0.72$ and

$$V_d = 0.72 \times 1.06 = 0.76 \text{ m/s}$$

This value exceeds the self-cleansing velocity, and hence the solution is $D = 300$ mm. In cases where the self-cleansing velocity is not maintained, it is necessary to increase the diameter or the pipe gradient.

Note: The solution using $k_S = 6$ mm and accounting for θ gives the following values:

$$D = 375 \text{ mm for } Q = 81 \text{ l/s and } V = 0.73 \text{ m/s}$$

$$\theta = 45$$

$Q_p = 10/81 = 0.123$

$d/D = 0.024$

$V_p = 0.67$ m/s

$V_d = 0.49$ m/s

Hence it would be necessary to increase D or S_0. In this case, increasing S_0 would be preferable.

References and further reading

Ackers, P. 1958. *Resistance of fluids flowing in channels and pipes*. Hydraulics Research Paper No. 1. London: HMSO.

Barr, D. I. H. 1975. Two additional methods of direct solution of the Colebrook–White function. *Proc. Instn Civ. Engrs* **59**, 827.

Colebrook, C. F. 1939. Turbulent flows in pipes, with particular reference to the transition region between the smooth and rough pipe laws. *J. Instn Civ. Engrs* **11**, 133.

Colebrook, C. F. and White C. M. 1937. Experiments with fluid friction in roughened pipes. *Proc R. Soc. A* **161**, 367.

Hydraulics Research Limited 1983a, *Tables for the hydraulic design of pipes* 4th edn. London: Thomas Telford.

Hydraulics Research Limited 1983b *Charts for the hydraulic design of channels and pipes* 5th edn. London: Thomas Telford.

Moody, L. F. 1944. Friction factors for pipe flows. *Trans. Am. Soc. Mech. Engrs* **66**, 671.

Webber, N. B. 1971. *Fluid mechanics for civil engineers*. London: Chapman and Hall.

5

Open channel flow

5.1 Flow with a free surface

Open channel flow is characterised by the existence of a free surface (the water surface). In contrast to pipe flow, this constitutes a boundary at which the pressure is atmospheric, and across which the shear forces are negligible. The longitudinal profile of the free surface defines the hydraulic gradient and determines the cross-sectional area of flow, as is shown in Figure 5.1. It also necessitates the introduction of an extra variable – the stage (see Fig. 5.1) – to define the position of the free surface at any point in the channel.

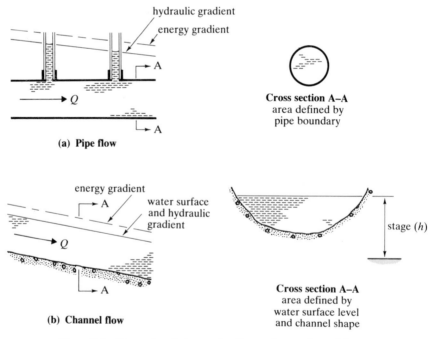

Figure 5.1 Comparison between pipe flow and open channel flow.

In consequence, problems in open channel flow are more complex than those of pipe flow, and the solutions are more varied, making the study of such problems both interesting and challenging. In this chapter the basic concepts are introduced and a variety of common engineering applications are discussed.

5.2 Flow classification

The fundamental types of flow have already been discussed in Chapter 2. However, it is appropriate here to expand those descriptions as applied to open channels. Recalling that flow may be steady or unsteady and uniform or non-uniform, the major classifications applied to open channels are as follows:

Steady uniform flow, in which the depth is constant, both with time and distance. This constitutes the fundamental type of flow in an open channel

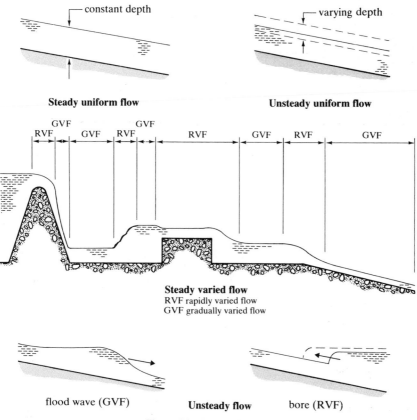

Figure 5.2   Types of flow.

in which the gravity forces are in equilibrium with the resistance forces. It is considered in Section 5.6.

Steady non-uniform flow, in which the depth varies with distance, but not with time. The flow may be either (a) gradually varied or (b) rapidly varied. Type (a) requires the joint application of energy and frictional resistance equations, and is considered in Section 5.10. Type (b) requires the application of energy and momentum principles, and is considered in Sections 5.7 and 5.8.

Unsteady flow, in which the depth varies with both time and distance (unsteady uniform flow is very rare). This is the most complex flow type, requiring the solution of energy, momentum and friction equations through time. It is considered in Section 5.11.

The various flow types are all shown in Figure 5.2.

5.3 Natural and artificial channels and their properties

Artificial channels comprise all man-made channels, including irrigation and navigation canals, spillway channels, sewers, culverts and drainage ditches. They are normally of regular cross-sectional shape and bed slope, and as such are termed prismatic channels. Their construction materials are varied, but commonly used materials include concrete, steel and earth.

Table 5.1 Geometric properties of some common prismatic channels.

	Rectangle	Trapezoid	Circle
area, A	by	$(b+xy)y$	$\frac{1}{8}(\phi - \sin\phi)D^2$
wetted perimeter, P	$b+2y$	$b+2y\sqrt{1+x^2}$	$\frac{1}{2}\phi D$
top width, B	b	$b+2xy$	$\left(\sin\dfrac{\phi}{2}\right)D$
hydraulic radius, R	$\dfrac{by}{b+2y}$	$\dfrac{(b+xy)y}{b+2y\sqrt{1+x^2}}$	$\dfrac{1}{4}\left(1-\dfrac{\sin\phi}{\phi}\right)D$
hydraulic mean depth, D_m	y	$\dfrac{(b+xy)y}{b+2xy}$	$\dfrac{1}{8}\left(\dfrac{\phi-\sin\phi}{\sin(1/2\ \phi)}\right)D$

The surface roughness characteristics of these materials are normally well defined within engineering tolerances. In consequence, the application of hydraulic theories to flow in artificial channels will normally yield reasonably accurate results.

In contrast, natural channels are normally very irregular in shape, and their materials are diverse. The surface roughness of natural channels changes with time, distance and water surface elevation. Therefore, it is more difficult to apply hydraulic theory to natural channels and obtain satisfactory results. Many applications involve man-made alterations to natural channels (e.g. river control structures and flood alleviation measures). Such applications require an understanding not only of hydraulic theory, but also of the associated disciplines of sediment transport, hydrology and river morphology (refer to Chs 9, 10 & 14).

Various geometric properties of natural and artificial channels need to be determined for hydraulic purposes. In the case of artificial channels, these may all be expressed algebraically in terms of the depth (y), as is shown in Table 5.1. This is not possible for natural channels, so graphs or tables relating them to stage (h) must be used.

The commonly used geometric properties are shown in Figure 5.3 and defined as follows:

Depth (y) – the vertical distance of the lowest point of a channel section from the free surface;
Stage (h) – the vertical distance of the free surface from an arbitrary datum;
Area (A) – the cross-sectional area of flow normal to the direction of flow;
Wetted perimeter (P) – the length of the wetted surface measured normal to the direction of flow;
Surface width (B) – the width of the channel section at the free surface;
Hydraulic radius (R) – the ratio of area to wetted perimeter (A/P);
Hydraulic mean depth (D$_m$) – the ratio of area to surface width (A/B).

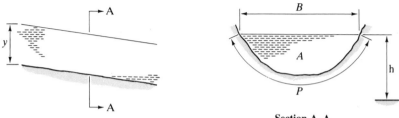

Figure 5.3 Definition sketch of geometric channel properties.

5.4 Velocity distributions, energy and momentum coefficients

The point velocity in an open channel varies continuously across the cross section because of friction along the boundary. However, the velocity distribution is not axisymmetric (as in pipe flow) due to the presence of the free surface. One might expect to find the maximum velocity at the free surface, where the shear stress is negligible, but the maximum velocity normally occurs below the free surface. Typical velocity distributions are shown in Figure 5.4 for various channel shapes.

The depression of the point of maximum velocity below the free surface may be explained by the presence of secondary currents which circulate from the boundaries towards the channel centre. Detailed experiments on velocity distributions have demonstrated the existence of such secondary currents, but no complete explanation has been found for the existence of these secondary currents in straight channels.

The energy and momentum coefficients (α and β) defined in Chapter 2 can only be evaluated for a channel if the velocity distribution has been measured. This contrasts with pipe flow, where theoretical velocity distributions for laminar and turbulent flow have been derived, which enable direct integration of the defining equations to be made.

For turbulent flow in regular channels, α rarely exceeds 1.15 and β rarely exceeds 1.05. In consequence, these coefficients are normally assumed to be unity. However, in irregular channels where the flow may divide into distinct regions, α may exceed 2 and should therefore be included in flow computations. Referring to Figure 5.5, which shows a natural channel with two flood banks, the flow may be divided into three regions. By making the assumption that $\alpha = 1$ for each region, the value of α for the whole channel may be found as follows

$$\alpha = \frac{\int u^3 \, \mathrm{d}A}{\bar{V}^3 A} = \frac{V_1^3 A_1 + V_2^3 A_2 + V_3^3 A_3}{\bar{V}^3 (A_1 + A_2 + A_3)} \tag{5.1}$$

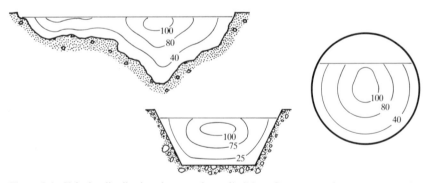

Figure 5.4 Velocity distributions in open channels. (*Note.* Contour numbers are expressed as a percentage of the maximum velocity.)

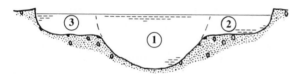

Figure 5.5 Division of a channel into a main channel and flood banks.

where

$$\bar{V} = \frac{Q}{A} = \frac{V_1 A_1 + V_2 A_2 + V_3 A_3}{(A_1 + A_2 + A_3)}$$

5.5 Laminar and turbulent flow

In Section 5.2, the state of flow was not discussed. It may be either laminar or turbulent, as in pipe flow. The criterion for determining whether the flow is laminar or turbulent is the Reynolds Number (Re), which was introduced in Chapters 3 and 4:

for pipe flow $\qquad$ $Re = \rho D V / \mu$ $\qquad\qquad$ (3.2)

for laminar pipe flow $\qquad$ $Re < 2000$

for turbulent pipe flow $\qquad$ $Re > 4000$

These results can be applied to open channel flow if a suitable form of the Reynolds Number can be found. This requires that the characteristic length dimension, the diameter (for pipes), be replaced by an equivalent characteristic length for channels. The one adopted is termed the hydraulic radius (R) as defined in Section 5.3.

Hence, the Reynolds Number for channels may be written as

$$Re_{(channel)} = \rho R V / \mu \qquad\qquad (5.2)$$

For a pipe flowing full, $R = D/4$, so

$$Re_{(channel)} = Re_{(pipe)}/4$$

and for laminar channel flow

$$Re_{(channel)} < 500$$

and for turbulent channel flow

$$Re_{(channel)} > 1000$$

In practice, the upper limit of Re is not so well defined for channels as it is for pipes, and is normally taken to be 2000.

In Chapter 4, the Darcy–Weisbach formula for pipe friction was introduced, and the relationship between laminar, transitional and turbulent flow was depicted on the Moody diagram. A similar diagram for channels has been developed. Starting from the Darcy–Weisbach formula,

$$h_f = \lambda L V^2/2gD \qquad (4.8)$$

and making the substitutions $R = D/4$ and $h_f/L = S_0$ (where S_0 = bed slope), then for uniform flow in an open channel

$$S_0 = \frac{\lambda V^2}{2g4R}$$

or

$$\lambda = 8gRS_0/V^2 \qquad (5.3)$$

The λ–Re relationship for pipes is given by the Colebrook–White transition law, and by substituting $R = D/4$ the equivalent formula for channels is

$$\frac{1}{\sqrt{\lambda}} = -2 \log \left(\frac{k_S}{14.8R} + \frac{0.6275}{Re\sqrt{\lambda}} \right) \qquad (5.4a)$$

Also combining (5.4a) with (5.3) yields

$$V = -2\sqrt{8gRS_0} \log \left(\frac{k_S}{14.8R} + \frac{0.6275\nu}{R\sqrt{8gRS_0}} \right) \qquad (5.4b)$$

A λ-Re diagram for channels may be derived using (5.4a) and channel velocities may be found directly from (5.4b). However, its application of these equations to a particular channel is more complex than the pipe case, due to the extra variables involved (i.e. for channels, R changes with depth and channel shape). In addition, the validity of this approach is questionable because the presence of the free surface has a considerable effect on the velocity distributions, as previously discussed. Hence, the frictional resistance is non-uniformly distributed around the boundary, in contrast to pressurised pipe flow, where the frictional resistance is uniformly distributed around the pipe wall.

In practice, the flow in open channels is normally in the rough turbulent zone, and consequently it is possible to use simpler formulae to relate frictional losses to velocity and channel shape, as is discussed in the next section.

5.6 Uniform flow

The development of friction formulae

Historically, the development of uniform flow resistance equations preceded the detailed investigations of pipe flow resistance. These developments are outlined below, and comparisons are made with pipe flow theory.

For uniform flow to occur, the gravity forces must exactly balance the frictional resistance forces which constitute the boundary shear force. Figure 5.6 shows a small longitudinal section in which uniform flow exists.

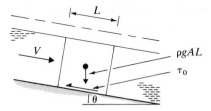

Figure 5.6 Derivation of uniform flow equations.

The gravity force resolved in the direction of flow $= \rho g A L \sin \theta$ and the shear force resolved in the direction of flow $= \tau_0 P L$, where τ_0 is the mean boundary shear stress. Hence,

$$\tau_0 P L = \rho g A L \sin \theta$$

Considering channels of small slope only, then

$$\sin \theta \simeq \tan \theta = S_0$$

hence,

$$\tau_0 = \rho g A S_0 / P$$

or

$$\tau_0 = \rho g R S_0 \tag{5.5}$$

The Chézy equation

To interpret (5.5), an estimate of the magnitude of τ_0 is required. Assuming a state of rough turbulent flow, then

$$\tau_0 \propto V^2 \quad \text{or} \quad \tau_0 = K V^2$$

Substituting into (5.5) for τ_0,

$$V = \sqrt{\frac{\rho g}{K}} \, RS_0$$

which may be written as

$$V = C\sqrt{RS_0} \tag{5.6}$$

This is known as the Chézy equation. It is named after the French engineer who developed the formula when designing a canal for the Paris water supply in 1768. The Chézy coefficient C is not, in fact, constant but depends on the Reynolds Number and boundary roughness, as can be readily appreciated from the previous discussions of the λ–Re diagram. A direct comparison between C and λ can be found by substituting (5.6) into (5.3) to yield

$$C = \sqrt{8g/\lambda}$$

In 1869, an elaborate formula for Chézy's C was published by two Swiss engineers: Ganguillet and Kutter. It was based on actual discharge data from the River Mississippi and a wide range of natural and artificial channels in Europe. The formula (in metric units) is

$$C = 0.552\left(\frac{41.6 + 1.811/n + 0.00281/S_0}{1 + [41.65 + (0.000281/S_0)]\,n/\sqrt{R}}\right) \tag{5.7}$$

where n is a coefficient known as Kutter's n, and is dependent solely on the boundary roughness.

The Manning equation

In 1889, the Irish engineer Robert Manning presented another formula (at a meeting of the Institution of Civil Engineers of Ireland) for the evaluation of the Chézy coefficient, which was later simplified to

$$C = R^{\frac{1}{6}}/n \tag{5.8}$$

This formula was developed from seven different formulae, and was further verified by 170 observations. Other research workers in the field derived similar formulae independently of Manning, including Hagen in 1876, Gauckler in 1868 and Strickler in 1923. In consequence, there is some confusion as to whom the equation should be attributed to, but it is generally known as the Manning equation.

Substitution of (5.8) into (5.6) yields

$$V = (1/n) R^{\frac{2}{3}} S_0^{\frac{1}{2}}$$

and the equivalent formula for discharge is

$$Q = \frac{1}{n} \frac{A^{\frac{5}{3}}}{P^{\frac{2}{3}}} S_0^{\frac{1}{2}} \tag{5.9}$$

where n is a constant known as Manning's n (it is numerically equivalent to Kutter's n).

Manning's formula has the twin attributes of simplicity and accuracy. It provides reasonably accurate results for a large range of natural and artificial channels, given that the flow is in the rough turbulent zone and that an accurate assessment of Manning's n has been made. It has been widely adopted for use by engineers throughout the world.

Evaluation of Manning's n

The value of the roughness coefficient n determines the frictional resistance of a given channel. It can be evaluated directly by discharge and stage measurements for a known cross section and slope. However, for design purposes, this information is rarely available, and it is necessary to rely on documented values obtained from similar channels.

For the case of artificially lined channels, n may be estimated with reasonable accuracy. For natural channels, the estimates are likely to be rather less accurate. In addition, the value of n may change with stage (particularly with flood flows over flood banks) and with time (due to changes in bed material as a result of sediment transport) or season (due to presence of vegetation). In such cases, a suitably conservative design value is normally adopted. Table 5.2 lists typical values of n for various materials and channel conditions. More detailed guidance is given by Chow (1959).

Table 5.2 Typical values of Manning's n.

Channel type	Surface material and alignment	$n(s/m^{\frac{1}{3}})$
river	earth, straight	0.02–0.025
	earth, meandering	0.03–0.05
	gravel (75–150 mm), straight	0.03–0.04
	gravel (75–150 mm), winding	
	or braided	0.04–0.08
unlined canals	earth, straight	0.018–0.025
	rock, straight	0.025–0.045
lined canals	concrete	0.012–0.017
models	mortar	0.011–0.013
	Perspex	0.009

Uniform flow computations

Manning's formula may be used to determine steady uniform flow. There are two types of commonly occurring problems to solve. The first is to determine the discharge given the depth, and the second is to determine the depth given the discharge. The depth is referred to as the normal depth, which is synonymous with steady uniform flow. As uniform flow can only occur in a channel of constant cross section, natural channels should be excluded. However, in solving the equations of gradually varied flow applicable to natural channels, it is still necessary to solve Manning's equation. Therefore it is useful to consider the application of Manning's equation to irregular channels in this section. The following examples illustrate the application of the relevant principles.

Example 5.1 Discharge from depth for a trapezoidal channel

The normal depth of flow in a trapezoidal concrete lined channel is 2 m. The channel base width is 5 m and has side slopes of 1 : 2. Manning's n is 0.015 and the bed slope, S_0, is 0.001. Determine the discharge (Q), mean velocity (V) and the Reynolds Number (Re).

Solution

Using Table 5.1,

$$A = (5 + 2y)y \qquad P = 5 + 2y\sqrt{1 + 2^2}$$

Hence, applying (5.9) for $y = 2$ m

$$Q = \frac{1}{0.015} \times \frac{[(5 + 4)2]^{5/3}}{[5 + (2 \times 2\sqrt{5})]^{2/3}} \times 0.001^{1/2}$$

$$= 45 \text{ m}^3/\text{s}$$

To find the mean velocity, simply apply the continuity equation:

$$V = \frac{Q}{A} = \frac{45}{(5 + 4)2} = 2.5 \text{ m/s}$$

The Reynolds Number is given by

$$\text{Re} = \rho R V / \mu$$

where $R = A/P$. In this case,

$$R = \frac{(5 + 4)2}{[5 + (2 \times 2\sqrt{5})]} = 1.29 \text{ m}$$

and

$$Re = \frac{10^3 \times 1.29 \times 2.5}{1.14 \times 10^{-3}} = 2.83 \times 10^6$$

Note. Re is very high, and corresponds to the rough turbulent zone. Therefore, Manning's equation is applicable. The interested reader may care to check the validity of this statement by applying the Colebrook–White equation (5.4b). Firstly calculate a k_S value equivalent to $n = 0.015$ for $y = 2$ m [$k_S = 2.225$ mm]. Then using these values of k_S and n compare the discharges as calculated using the Manning and Colebrook–White equations for a range of depths. Provided the channel is operating in the rough turbulent zone, the results are very similar.

Example 5.2   Depth from discharge for a trapezoidal channel

If the discharge in the channel given in Example 5.1 were 30 m³/s, find the normal depth of flow.

Solution

From Example 5.1

$$Q = \frac{1}{0.015} \times \frac{[(5 + 2y)y]^{5/3}}{[5 + 2\sqrt{5}y]^{2/3}} \times 0.001^{1/2}$$

or

$$Q = 2.108 \times \frac{[(5 + 2y)y]^{5/3}}{[5 + 2\sqrt{5}y]^{2/3}}$$

At first sight this may appear to be an intractable equation, and it will also be different for different channel shapes. The simplest method of solution is to adopt a trial-and-error procedure. Various values of y are tried, and the resultant Q is compared with that required. Iteration ceases when reasonable agreement is found. In this case $y < 2$ as $Q < 45$, so an initial value of 1.7 is tried.

Trial depths (m)	Resultant discharge (m³/s)
1.7	32.7
1.6	29.1
1.63	30.1

Hence the solution is $y = 1.63$ m for $Q = 30$ m³/s.

Example 5.3 Compound channels

During large floods, the water level in the channel given in Example 5.1 exceeds the bankfull level of 2.5 m. The flood banks are 10 m wide and are grassed with side slopes of 1 : 3. The estimated Manning's n for these flood banks is 0.035. Estimate the discharge for a maximum flood level of 4 m and the value of the velocity coefficient α.

Solution

In this case, it is necessary to split the section into subsections (1), (2), (3) as shown in Figure 5.7. Manning's formula may be applied to each one in turn, and the discharges can be summed. The division of the section into sub-sections is a little arbitrary. Since the shear stress across the arbitrary divisions will be small compared with the bed shear stresses, it may be ignored.

For section (1)

$$A_1 = \left(\frac{5 + 15}{2}\right)2.5 + (15 \times 1.5) = 47.5 \text{ m}^2$$

and

$$P_1 = 5 + (2\sqrt{5} \times 2.5) = 16.18 \text{ m}$$

Sections (2) and (3) have the same dimensions, hence

$$A_2 = A_3 = \left(\frac{10 + 14.5}{2}\right) \times 1.5 = 18.38 \text{ m}^2$$

$$P_2 = P_3 = 10 + (\sqrt{10} \times 1.5) = 14.74 \text{ m}$$

$$Q_1 = \frac{1}{0.015} \times \frac{47.5^{5/3}}{16.18^{2/3}} \times 0.001^{1/2}$$
$$= 205.3 \text{ m}^3/\text{s}$$

$$Q_2 = Q_3 = \frac{1}{0.035} \times \frac{18.38^{5/3}}{14.74^{2/3}} \times 0.001^{1/2}$$
$$= 19.2 \text{ m}^3/\text{s}$$

$$Q = Q_1 + Q_2 + Q_3 = 243.7 \text{ m}^3/\text{s}$$

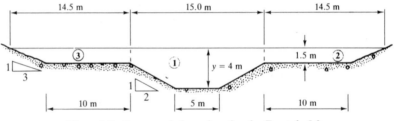

Figure 5.7 Compound channel section for Example 5.3.

The velocity coefficient may be found directly from (5.1):

$$\alpha = \frac{V_1^3 A_1 + V_2^3 A_2 + V_3^3 A_3}{\overline{V}^3 A}$$

$$V_1 = \frac{Q_1}{A_1} = \frac{205.3}{47.5} = 4.32 \text{ m/s}$$

$$V_2 = V_3 = \frac{Q_2}{A_2} = \frac{19.2}{18.38} = 1.04 \text{ m/s}$$

$$\overline{V} = \frac{Q}{A} = \frac{243.7}{47.5 + 18.38 + 18.38} = 2.89 \text{ m/s}$$

Hence

$$\alpha = \frac{4.32^3 \times 47.5 + [2(1.04^3 \times 18.38)]}{2.89^3 \times 84.26}$$

$$= 1.9$$

Note. This example illustrates a case in which the velocity coefficient should not be ignored.

5.7 Rapidly varied flow: the use of energy principles

Applications and methods of solution

Rapidly varied flow occurs whenever there is a sudden change in the geometry of the channel or in the regime of the flow. Typical examples of the first type include flow over sharp-crested weirs and flow through regions of rapidly varied cross section (e.g. venturi flumes and broad-crested weirs). The second type is normally associated with the hydraulic jump phenomenon in which flow with high velocity and small depth is rapidly changed to flow with low velocity and large depth. The regime of flow is defined by the Froude Number, a concept which is explained later in this section.

In regions of rapidly varied flow, the water surface profile changes suddenly and therefore has pronounced curvature. The pressure distribution under these circumstances departs considerably from the hydrostatic distribution. The assumptions of parallel streamlines and hydrostatic pressure distribution which are used for uniform and gradually varied flow do not apply. Solutions to rapidly varied flow problems have been found using the concepts of ideal fluid flow coupled with the use of finite-element techniques. However, such solutions are complex and do not include the boundary layer effects in real fluids.

Many rapidly varied flow problems may be solved approximately using energy and momentum concepts, and for engineering purposes this is often sufficiently accurate. This section describes and explains the use of these concepts.

The energy equation in open channels

Bernoulli's equation, derived in Chapter 2, may be applied to any streamline. If the streamlines are parallel, then the pressure distribution is hydrostatic.

Referring to Figure 5.8, which shows uniform flow in a steep channel, consider point A on a streamline. The pressure force at point A balances the component of weight normal to the bed, i.e.

$$p_A \, \Delta S = \rho g d \, \Delta S \cos \theta$$

$$p_A = \rho g d \cos \theta = \rho g y_2$$

It is more convenient to express this pressure force in terms of y_1 (the vertical distance from the streamline to the free surface). Since

$$d = \frac{y_2}{\cos \theta} \text{ and } d = y_1 \cos \theta$$

then

$$y_2 = y_1 \cos^2 \theta$$

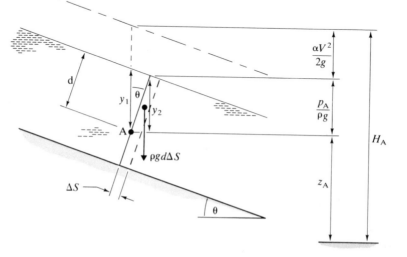

Figure 5.8 Application of Bernoulli's equation to uniform flow in open channels.

and hence

$$p_A = \rho g y_1 \cos^2 \theta$$

or

$$p_A/\rho g = y_1 \cos^2 \theta$$

Most channels have very small bed slopes (e.g. less than 1 : 100, corresponding to θ less than $0.57°$ or $\cos^2 \theta$ greater than 0.9999) and therefore for such
channels

$$\cos^2 \theta \simeq 1$$

and

$$p_A/\rho g \simeq y_1 \qquad\qquad (5.10)$$

Hence, Bernoulli's equation becomes

$$H = y + \frac{\alpha V^2}{2g} + z \qquad\qquad (5.11)$$

Application of the energy equation

Consider the problem shown in Figure 5.9. Steady uniform flow is interrupted by the presence of a hump in the streambed. The upstream depth and
the discharge are known, and it simply remains to find the depth of flow
at position (2).

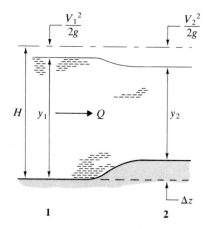

Figure 5.9 Flow transition problem.

Applying the energy equation (5.11) and assuming that frictional energy losses between (1) and (2) are negligible, then

$$y_1 + \frac{V_1^2}{2g} = y_2 + \frac{V_2^2}{2g} + \Delta z \qquad (5.12)$$

(taking $\alpha = 1$).

There are two principal unknown quantities – V_2 and y_2 – and hence to solve this a second equation is required: the continuity equation:

$$V_1 y_1 = V_2 y_2 = q$$

where q is the discharge per unit width. Combining these equations,

$$y_1 + \frac{q^2}{2gy_1^2} = y_2 + \frac{q^2}{2gy_2^2} + \Delta z$$

and rearranging,

$$2gy_2^3 + y_2^2 \left(2g\,\Delta z - 2gy_1 - \frac{q^2}{y_1^2}\right) + q^2 = 0$$

This is a cubic equation, in which y_2 is the only unknown, and to which there are three possible mathematical solutions. As far as the fluid flow is concerned, however, only one solution is possible. To determine which of the solutions for y_2 is correct requires a more detailed knowledge of the flow.

Specific energy

To solve the above problem, the concept of specific energy was introduced by Bakhmeteff in 1912. Specific energy (E_S) is defined as the energy of the flow referred to the channel bed as datum:

$$E_S = y + \alpha V^2 / 2g \qquad (5.13)$$

Application of the specific energy equation provides the solution to many rapidly varied flow problems.

For steady flow, (5.13) may be rewritten as

$$E_S = y + \alpha \frac{(Q/A)^2}{2g}$$

Now consider a rectangular channel:

$$\frac{Q}{A} = \frac{bq}{by} = \frac{q}{y}$$

where b is the width of the channel. Hence

$$E_S = y + \alpha q^2/2gy^2$$

As q is constant,

$$(E_S - y)y^2 = \alpha q^2/2g = \text{constant}$$

or

$$(E_S - y) = \text{constant}/y^2$$

Again, this is a cubic equation for the depth y for a given E_S. Considering only positive solutions, then the equation is a curve with two asymptotes:

$$\text{as } y \to 0, \quad E_S \to \infty$$

$$\text{as } y \to \infty, \quad E_S \to y$$

This curve is now used to solve the problem given in Figure 5.9. In Figure 5.10, the problem has been redrawn alongside a graph of depth versus specific energy.

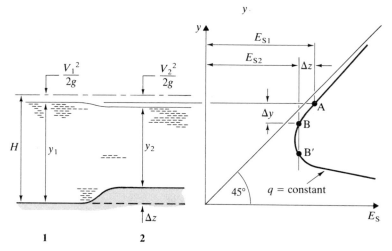

Figure 5.10 The use of specific energy to solve flow transition problems.

Equation (5.12) may be written as

$$E_{S1} = E_{S2} + \Delta z \qquad (5.14)$$

This result is plotted on the specific energy diagram in Figure 5.10. Point A on the curve corresponds to conditions at point (1) in the channel. Point (2) in the channel must be at either point B or B' on the specific energy curve (from (5.14)). All points between (1) and (2) must lie on the specific energy curve between A and B or B'. To arrive at point B' would imply that at some intermediate point $E_{S1} - E_{S2} > \Delta z$, which is physically impossible. Hence, the flow depth at (2) must correspond to point B on the specific energy curve. This is a very significant result, so an example of such a flow transition follows.

Example 5.4 Flow transition

The discharge in a rectangular channel of width 5 m and maximum depth 2 m is 10 m³/s. The normal depth of flow is 1.25 m. Determine the depth of flow downstream of a section in which the bed rises by 0.2 m over a distance of 1 m.

Solution

The solution is shown graphically in Figure 5.10. Assuming frictional losses are negligible, then (5.14) applies, i.e.

$$E_{S1} = E_{S2} + \Delta z$$

In this case,

$$E_{S1} = y_1 + V_1^2/2g = 1.25 + [10/(5 \times 1.25)]^2/2g = 1.38 \text{ m}$$

$$E_{S2} = y_2 + [10/(5y_2)]^2/2g = y_2 + 2^2/2gy_2^2$$

$$\Delta z = 0.2$$

hence

$$1.38 = y_2 + (2^2/2gy_2^2) + 0.2$$

or

$$1.18 = y_2 + 2/gy_2^2$$

This is a cubic equation for y_2, but the correct solution is that given by point B in Figure 5.10, which in this case is about 0.9 m. This is used as the initial estimate in a trial-and-error solution, as follows:

y_2 (m)	$E_{S2} = y_2 + 2/gy_2^2$ (m)
0.9	1.15
1.0	1.2
0.96	1.18

Hence the solution is $y_2 = 0.96$ m.

This result is often a source of puzzlement, and a simple physical explanation is called for. The answer lies in realising that the fluid is accelerating. Consider, initially, that the water surface remains at the same level over the bed rise. As the water depth is less (over the bed rise) and the discharge constant, then the velocity must have increased; i.e. the fluid must have accelerated. However, to accelerate the fluid, a force is required. This is provided by a fall in the water surface elevation, which implies that the hydrostatic force acting in the downstream direction is greater than that acting in the upstream direction.

Subcritical, critical and supercritical flow

In Figure 5.11, the specific energy curve (for constant discharge) has been redrawn alongside a second curve of depth against discharge for constant E_S. The figure is now used to illustrate several important principles of rapidly varied flow.

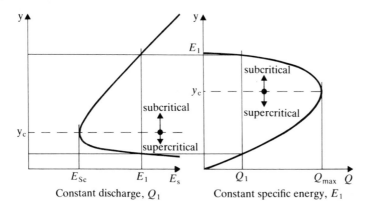

Figure 5.11 Variation of specific energy and discharge with depth.

(a) For a given constant discharge:
  (i) The specific energy curve has a minimum value E_{Sc} at point c with a corresponding depth y_c – known as the critical depth;
  (ii) for any other value of E_S there are two possible depths of flow known at alternate depths, one of which is termed subcritical and the other supercritical;

$$\text{for supercritical flow, } y < y_c$$

$$\text{for subcritical flow, } y > y_c$$

(b) For a given constant specific energy:
  (i) The depth–discharge curve shows that discharge is a maximum at the critical depth.
  (ii) For all other discharges there are two possible depths of flow (sub- and supercritical) for any particular value of E_S.

The general equation of critical flow

Referring to Figure 5.11, the general equation for critical flow may be derived by determining E_{Sc} and Q_{max}, independently, from the specific energy equation. It will be seen that the two methods both result in the same solution.

For Q = constant	For E_S = constant
$$E_S = y + \alpha V^2/2g$$	$$E_S = y + \alpha V^2/2g$$
or	or
$$E_S = y + \frac{\alpha}{2g}\frac{Q^2}{A^2} \qquad \text{(a)}$$	$$Q = \sqrt{\frac{2g}{\alpha}}\, A(E_S - y)^{\frac{1}{2}} \qquad \text{(a)}$$
For a minimum value (i.e. E_{Sc}),	For Q_{max} with $E_S = E_0$ = constant,
$$\frac{dE_S}{dy} = 0$$	$$\frac{dQ}{dy} = \sqrt{\frac{2g}{\alpha}}\left(\frac{A(E_0 - y)^{-\frac{1}{2}}}{-2} + \right.$$
$$= 1 + \frac{\alpha Q^2}{2g}\frac{d}{dA}\left(\frac{1}{A^2}\right)\frac{dA}{dy} \qquad \text{(b)}$$	$$\left.\frac{dA}{dy}(E_0 - y)^{\frac{1}{2}}\right) = 0 \qquad \text{(b)}$$
Since $\delta A = B\,\delta y$,	Since $\delta A = B\,\delta y$
then in the limit	then in the limit
$$\frac{dA}{dy} = B$$	$$\frac{dA}{dy} = B$$

substituting, into (b) | substituting into (b)

$$0 = 1 - \frac{\alpha Q^2}{2g} B_c \, 2A_c^{-3}$$

$$E_0 = \frac{A}{2B} + y$$

or

and substituting into (a)

$$\frac{\alpha Q^2 B_c}{g A_c^3} = 1$$

$$\frac{\alpha Q_{max}^2 \, B}{g A^3} = 1$$

Comparing,

$$\frac{\alpha Q_{max}^2 B_c}{g A_c^3} = 1 \qquad\qquad (5.15)$$

In other words, at the critical depth the discharge is a maximum and the specific energy is a minimum.

Critical depth and critical velocity (for a rectangular channel)

Determination of the critical depth (y_c) in a channel is necessary for both rapidly and gradually varied flow problems. The associated critical velocity (V_c) will be used in the explanation of the significance of the Froude Number. Both of these parameters may be derived directly from (5.15) for any shape of channel. For illustrative purposes, the simplest artificial channel shape is used here, i.e. rectangular. For critical flow,

$$\frac{\alpha Q^2 B}{g A^3} = 1$$

For a rectangular channel, $Q = qb$, $B = b$ and $A = by$. Substituting in (5.15) and taking $\alpha = 1$, then

$$y_c = (q^2/g)^{\frac{1}{3}} \qquad\qquad (5.16)$$

and as

$$V_c y_c = q$$

then

$$V_c = \sqrt{g y_c} \qquad\qquad (5.17)$$

Also, as

$$E_{Sc} = y_c + V_c^2/2g$$

then

$$E_{Sc} = y_c + y_c/2$$

or

$$y_c = \frac{2}{3} E_{Sc} \qquad\qquad (5.18)$$

An application of the critical depth line

The application of critical depth is illustrated in Figure 5.12, which shows the effect of a local bed rise on the free surface elevation for various initial depths of flow. It should be noted from (5.16) that the critical depth is dependent only on the discharge. Hence, for a given discharge the critical depth line may be drawn on the longitudinal channel section as shown in Figure 5.12. If the upstream flow is initially subcritical, then a depression of the water surface will occur above the rise in bed level. Conversely, if the upstream water level is initially supercritical, then an increase in the water surface level will occur. As the upstream water level approaches the critical depth, then the change in elevation of the free surface becomes more marked. All of these results may easily be verified by reference to the specific energy diagram drawn alongside in Figure 5.12. A practical example

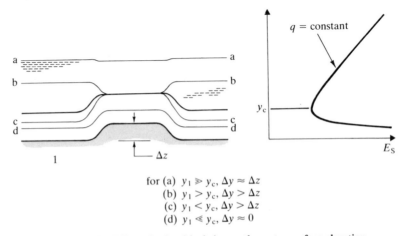

for (a) $y_1 \gg y_c$, $\Delta y \approx \Delta z$
(b) $y_1 > y_c$, $\Delta y > \Delta z$
(c) $y_1 < y_c$, $\Delta y > \Delta z$
(d) $y_1 \ll y_c$, $\Delta y \approx 0$

Figure 5.12 Effect of a local bed rise on the water surface elevation.

of this phenomenon often occurs in rivers under flood conditions. The normal depth of flow may be nearly critical, and any undulations in the bed result in large standing wave formations.

A further result may be gleaned from this example. If the local bed rise is sufficiently large, then critical flow will occur. It has already been shown that discharge is a maximum for critical flow and hence, under these circumstances, the local bed rise is acting as a 'choke' on the flow, so called because it limits the discharge. It is left to the reader to decide what happens if Δz is so large that E_{S2} is less than any point on the E_S curve. The question will be discussed again later in Section 5.9 and in Example 5.10.

The Froude Number

The Froude Number is defined as

$$Fr = V/\sqrt{gL}$$

where L is a characteristic dimension. It is attributed to William Froude (1810–97), who used such a relationship in model studies for ships. If L is replaced by D_m, the hydraulic mean depth, then the resulting dimensionless parameter

$$Fr = V/\sqrt{gD_m} \tag{5.19}$$

is applicable to open channel flow. This is extremely useful, as it defines the regime of flow, and as many of the energy and momentum equations may be written in terms of the Froude Number.

The physical significance of Fr may be understood in two different ways. First, from dimensional analysis,

$$Fr^2 = \frac{\text{inertial force}}{\text{gravitational force}} \equiv \frac{\rho L^2 V^2}{\rho g L^3} = \frac{V^2}{gL}$$

Secondly, by consideration of the speed of propagation, c, of a wave of low amplitude and long wavelength. Such waves may be generated in a channel as oscillatory waves or surge waves. Oscillatory waves (e.g. ocean waves) are considered in Chapter 8, and surge waves in Section 5.11. Both types of wave lead to the result that $c = \sqrt{gy}$. For a rectangular channel $D_m = y$, and hence

$$Fr = \frac{V}{\sqrt{gy}} = \frac{\text{water velocity}}{\text{wave velocity}}$$

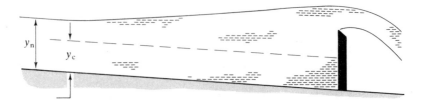

(a) Caused by the weir is transmitted upstream as $\text{Fr} < 1$ and $y_n > y_c$

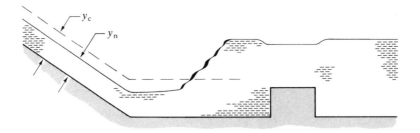

(b) Caused by a change of slope is not transmitted upstream as $\text{Fr} > 1$ and $y_n < y_c$

Figure 5.13   Flow disturbance.

Also, for a rectangular channel $V_c = \sqrt{gy_c}$, and hence, for critical flow,

$$\text{Fr} = \frac{V}{\sqrt{gy}} = \frac{\sqrt{gy_c}}{\sqrt{gy_c}} = 1$$

This is a general result for all channels (as it can be shown that for non-rectangular channels $V_c = \sqrt{gD_m}$ and $c = \sqrt{gD_m}$).

For subcritical flow,

$$V < V_c \text{ and } \text{Fr} < 1$$

For supercritical flow

$$V > V_c \text{ and } \text{Fr} > 1$$

The Froude Number therefore defines the regime of flow. There is a second consequence of major significance. Flow disturbances are propagated at a velocity of $c = \sqrt{gy}$. Hence, if the flow is supercritical, any flow disturbance can only travel downstream as the water velocity exceeds the wave velocity. The converse is true for subcritical flow. Such flow disturbances are introduced by all channel controls and local features. For example, if a pebble is dropped into a still lake, then small waves are generated in all directions at equal speeds, resulting in concentric wave fronts. Now imagine dropping the pebble into a river. The wave fronts will move upstream slower than they do downstream, due to the current. If the water velocity exceeds the wave velocity, the wave will not move upstream at all.

An example of how channel controls transmit disturbances upstream and downstream is shown in Figure 5.13. To summarise:

for Fr > 1 – supercritical flow
             – water velocity > wave velocity
             – disturbances travel downstream
             – upstream water levels are unaffected by downstream control

for Fr < 1 – subcritical flow
             – water velocity < wave velocity
             – disturbances travel upstream and downstream
             – upstream water levels are affected by downstream control

5.8 Rapidly varied flow: the use of momentum principles

The hydraulic jump

The hydraulic jump phenomenon is an important type of rapidly varied flow. It is also an example of a stationary surge wave. An hydraulic jump occurs when a supercritical flow meets a subcritical flow. The resulting flow transition is rapid, and involves a large energy loss due to turbulence. Under these circumstances, a solution to the hydraulic jump problem cannot be found using a specific energy diagram. Instead, the momentum equation is used. Figures 5.14a and b depict an hydraulic jump and the associated specific energy and force–momentum diagrams. Initially, ΔE_S is unknown, as only the discharge and upstream depth are given. By using the momentum principle, the sequent depth (y_2) may be found in terms of the initial depth (y_1) and the upstream Froude Number (Fr$_1$).

The momentum equation

The momentum equation derived in Chapter 2 may be written as

$$F_m = \Delta M$$

In this case, the forces acting are the hydrostatic pressure forces upstream (F_1) and downstream (F_2) of the jump. Their directions of action are as shown in Figure 5.14b. This is a point which often causes difficulty. It may be understood most easily by considering the jump to be inside a control

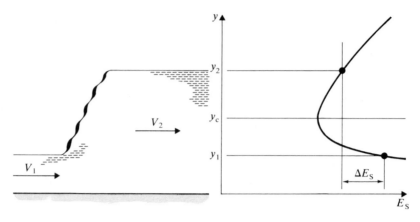

(a) With associated specific energy diagram

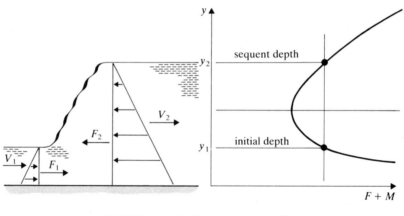

(b) With associated force–momentum diagram

Figure 5.14 The hydraulic jump.

volume, and to consider the external forces acting on the control volume (as described in Chapter 2).

Ignoring boundary friction, and for small channel slopes:

$$\text{net force in } x\text{-direction} = F_1 - F_2$$

$$\text{momentum change} = M_2 - M_1$$

hence

$$F_1 - F_2 = M_2 - M_1$$

or

$$F_1 + M_1 = F_2 + M_2 = \text{constant (for constant discharge)}$$

If depth (y) is plotted against force + momentum $(F + M)$ for a constant discharge, as in Figure 5.14b, then for a stable hydraulic jump

$$F + M = \text{constant.} \qquad (5.20)$$

Therefore, for any given initial depth, the sequent depth is the corresponding depth on the force–momentum diagram.

Solution of the momentum equation for a rectangular channel

For a rectangular channel, (5.20) may be evaluated as follows:

$$F_1 = \rho g(y_1/2)\, y_1 b \qquad F_2 = \rho g(y_2/2)\, y_2 b$$

$$M_1 = \rho Q V_1 \qquad M_2 = \rho Q V_2$$

$$= \rho Q \frac{Q}{y_1 b} \qquad = \rho Q \frac{Q}{y_2 b}$$

Substituting in (5.20) and rearranging,

$$\frac{\rho g b}{2}(y_1^2 - y_2^2) = \frac{\rho Q^2}{b}\left(\frac{1}{y_2} - \frac{1}{y_1}\right)$$

Substituting $q = Q/b$ and simplifying,

$$\frac{1}{2}(y_1^2 - y_2^2) = \frac{q^2}{g}\left(\frac{1}{y_2} - \frac{1}{y_1}\right)$$

$$\frac{1}{2}(y_1 + y_2)(y_1 - y_2) = \frac{q^2}{g}\frac{(y_1 - y_2)}{y_2 y_1}$$

$$\frac{1}{2}(y_1 + y_2) = \frac{q^2}{g}\frac{1}{y_2 y_1}$$

Substituting $q = V_1 y_1$ and dividing by y_1^2,

$$\frac{1}{2}\frac{y_2}{y_1}\frac{(1 + y_2)}{y_1} = \frac{V_1^2}{g y_1} = \text{Fr}_1^2$$

This is a quadratic equation in y_2/y_1, whose solution is

$$y_2 = (y_1/2) \, (\sqrt{1 + 8\mathrm{Fr}_1^2} - 1) \tag{5.21}$$

Energy dissipation in an hydraulic jump

Using (5.21), y_2 may be evaluated in terms of y_1 and hence the energy loss through the jump determined:

$$\Delta E = E_1 - E_2 = \left(y_1 + \frac{V_1^2}{2g} \right) - \left(y_2 + \frac{V_2^2}{2g} \right)$$

Substituting $q = Vy$,

$$\Delta E = y_1 - y_2 + \frac{q^2}{2g} \left(\frac{1}{y_1^2} - \frac{1}{y_2^2} \right) \tag{5.22}$$

Equation (5.21) may be rewritten as

$$\mathrm{Fr}_1^2 = \frac{1}{2} \frac{y_2}{y_1} \left(\frac{y_2}{y_1} + 1 \right) \tag{5.23}$$

The upstream Froude Number is related to q as follows:

$$\mathrm{Fr}_1 = V_1/\sqrt{gy_1} = (q/y_1)/\sqrt{gy_1}$$

$$\mathrm{Fr}_1^2 = q^2/gy_1^3 \tag{5.24}$$

Substituting (5.23) and (5.24) into (5.22), the solution is

$$\Delta E = (y_2 - y_1)^3/4 \, y_1 y_2 \tag{5.25}$$

It should be noted that the energy loss increases very sharply with the relative height of the jump.

Significance of the hydraulic jump equations

In the preceding section, the hydraulic jump equation was deliberately formulated in terms of the upstream Froude Number. The usefulness of this form of the equation is now demonstrated. For critical flow $\mathrm{Fr} = 1$, and substitution into (5.21) gives $y_2 = y_1$. This is an unstable condition resulting in an undular standing wave formation. For supercritical flow upstream, $\mathrm{Fr} > 1$ and hence $y_2 > y_1$. This is the required condition for the formation of a hydraulic jump. For subcritical flow upstream, $Fr_1 < 1$ and hence $y_2 < y_1$. This is not physically possible without the presence of some obstruction in the flow path. Such an obstruction would introduce an associated resistance force (F_R) and the momentum equation would then become $F_1 - F_2 - F_R = M_2 - M_1$. Under these circumstances an hydraulic jump would not form, but the momentum equation could be used to evaluate the resistance force (F_R). To summarise, for an hydraulic jump to form, the upstream flow must be supercritical, and the higher the upstream Froude Number, the greater the height of the jump and the greater the loss of energy through the jump.

Stability of the hydraulic jump

The hydraulic jump equation contains three independent variables: y_1, y_2 and Fr_1. A stable (i.e. stationary) jump will form only if these three independent variables conform to the relationship given in (5.21). To illustrate this, Figure 5.15 shows an hydraulic jump formed between two sluice gates. The upstream depth (y_1) and Froude Number (Fr_1) are controlled by the upstream sluice gate for a given discharge. The downstream depth (y_2) is controlled by the downstream sluice gate, and not by the hydraulic jump. Denoting the sequent depth by y_2', then the following observations may be made:

if $y_2 = y_2'$, then a stable jump forms;
if $y_2 > y_2'$, then the downstream force + momentum is greater than the upstream force + momentum, and the jump moves upstream;
if $y_2 < y_2'$, then the downstream force + momentum is less than the upstream force + momentum and the jump moves downstream.

Even if a stable jump forms, it may occur anywhere between the two sluice gates.

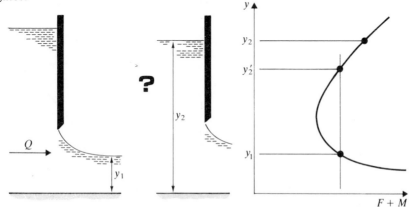

Figure 5.15 Stability of the hydraulic jump.

The occurrence and uses of an hydraulic jump

The hydraulic jump phenomenon has important applications in hydraulic engineering. Jumps may form downstream of hydraulic structures such as spillways, sluice gates and venturi flumes. They also sometimes occur downstream of bridge piers (normally under flood flow conditions).

Jumps may be deliberately induced to act as an energy dissipation device (e.g. in stilling bays) and to localise bed scour. These uses are discussed in Chapter 13.

5.9 Critical depth meters

The effect of a local bed rise on the flow regime has previously been discussed. It was concluded that critical flow over the bed rise would occur if the change in specific energy was such that it was reduced to the minimum specific energy. This situation is shown in Figure 5.16. This condition of critical depth over the bed rise, at first sight, may be thought only to occur for one particular step height (Δz_1) and upstream depth (y_1).

Consider what would happen if Δz_1 was further increased to Δz_2. The change in energy from (1) to (2) would be such that no position on the specific energy curve corresponding to (2) could be found. This apparent impossibility may be explained by reconsidering the specific energy curve. It was drawn for a particular fixed value of discharge (Q_1). In Figure 5.16, a second (thinner) curve is drawn for a smaller discharge (Q_2) for which the specific energy at position (2) is a minimum. Hence, if the step height was Δz_2, then the discharge would initially be limited to Q_2. However, if the discharge in the channel was Q_1, then the difference in the discharges ($Q_1 - Q_2$) would have to be stored in the channel upstream, resulting in an increase in the depth y_1. After a short time, the new upstream depth (y_1') would be sufficient to pass the required discharge Q_1 over the bed rise. In

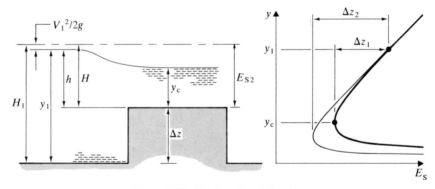

Figure 5.16 The broadcrested weir.

other words, the upstream energy will adjust itself such that the given discharge (Q_1) can pass over the bed rise with the minimum specific energy obtained at the critical depth. Of course, this will only apply provided that Δz is sufficiently large. Under these conditions, the local bed rise is acting as a control on the discharge by providing a choke. The upstream water depth is controlled by the bed rise, not by the channel. Such a bed rise is known as a broadcrested weir. Its usefulness lies in its ability to control the discharge and hence it may be used as a discharge measuring device.

Derivation of the discharge equation for broadcrested weirs

Referring to Figure 5.16, and assuming that the depth is critical at position (2), then from (5.17)

$$V_2 = V_c = \sqrt{gy_c}$$

and

$$Q = VA = \sqrt{gy_c} \, by_c$$

where b is the width of the channel at position (2). Also, from (5.18)

$$y_c = \tfrac{2}{3} E_{S2}$$

Assuming no energy losses between (1) and (2), then

$$E_{S2} = h + V_1^2/2g = H$$

$$y_c = \tfrac{2}{3} H$$

Substituting for y_c

$$Q = g^{1/2} \, b(\tfrac{2}{3})^{3/2} H^{3/2}$$

or

$$Q = \frac{2}{3} \sqrt{\frac{2g}{3}} \, bH^{3/2} \tag{5.26}$$

In practice, there are energy losses and it is the upstream depth rather than energy that is measured. Equation (5.26) is modified by the inclusion of the coefficient of discharge (C_d) to account for energy losses and the coefficient of velocity (C_v) to account for the upstream velocity head, giving

$$Q = C_d \, C_v \frac{2}{3} \sqrt{\frac{2g}{3}} \, bh^{3/2} \qquad (5.27)$$

Broadcrested weirs are discussed in more detail in Chapter 13.

Venturi flumes

This is a second example of a critical depth meter. In this case, the channel width is contracted to choke the flow, as shown in Figure 5.17. To demonstrate that critical flow is produced, it is necessary to draw two specific energy curves. Although the discharge (Q) is constant, the discharge per unit width (q) is greater in the throat of the flume than it is upstream. To force critical depth to occur in the throat, the specific energy in the throat must be a minimum. Neglecting energy losses, then the upstream specific energy (E_{S1}) will have the same value. Providing that the upstream energy (E_n) for uniform flow is less than E_{S1}, then the venturi flume, not the channel, will control the discharge. The equation for the discharge is the same as for the broadcrested weir (i.e. Equation (5.27)), remembering that b is the width at the section at which flow is critical, i.e. the throat.

Downstream of the throat the width expands and the flow, rather than returning to subcritical flow, continues to accelerate and becomes supercritical. As channel flow is normally subcritical, then at some point

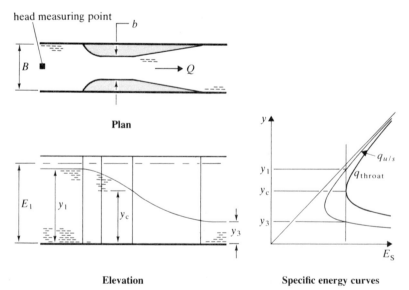

Figure 5.17   The venturi flume.

downstream an hydraulic jump will form. If the hydraulic jump moves upstream into the throat, then critical flow will no longer be induced and the flume will operate in the submerged condition. Under these circumstances, the discharge equation (5.27) no longer applies, and venturi flumes are therefore designed to prevent this happening for as wide a range of conditions as possible.

Venturi flumes are discussed in more detail in Chapter 13.

5.10 Gradually varied flow

The significance of bed slope and channel friction

In the discussions of rapidly varied flow, the influence of bed slope and channel friction was not mentioned. It was assumed that frictional effects may be ignored in a region of rapidly varied flow. This is a reasonable assumption, since the changes take place over a very short distance. However, bed slope and channel friction are very important because they determine the flow regime under gradually varied flow conditions.

The discussion of the specific energy curve and the criteria for maximum discharge indicated that for a given specific energy or discharge there are two possible flow depths at any point in a channel. The solution of Manning's equation results in only one possible flow depth (the normal depth). This apparent paradox is resolved by noting the influence of the bed slope and channel friction. These parameters determine which of the two possible flow depths will occur at any given point for uniform flow. In other words, for uniform flow, the bed slope and channel friction determine whether the flow regime is sub- or supercritical.

For a given channel and discharge, the normal depth of flow may be found using Manning's equation and the critical depth using (5.16). The normal depth of flow may be less than, equal to or greater than the critical depth. For a given channel shape and roughness, only one value of slope will produce the critical depth, and this is known as the critical slope (S_c). If the slope is steeper than S_c, then the flow will be supercritical and the slope is termed a steep slope. Conversely, if the slope is less steep than S_c, then the flow will be subcritical and the slope is termed a mild slope.

Critical bed slope in a wide rectangular channel

To illustrate this concept, the equation for the critical bed slope is now derived for the case of a wide rectangular channel.

For uniform flow,

$$Q = \frac{1}{n} \frac{A^{5/3}}{P^{2/3}} S_0^{1/2} \qquad (5.9)$$

and for critical flow

$$\frac{Q^2 B}{g A^3} = 1$$

(taking $\alpha = 1$ in (5.15)).

For a wide rectangular channel of width b, $B = b$, $A = by$ and $P \simeq b$. Hence, combining (5.9) and (5.15) to eliminate Q,

$$\frac{1}{n} \frac{A^{5/3}}{P^{2/3}} S_c^{1/2} = \sqrt{\frac{g A^3}{B}}$$

where S_c is the critical bed slope.

Making the above substitutions for a wide rectangular channel,

$$\frac{1}{n} \frac{(by_c)^{5/3}}{b^{2/3}} S_c^{1/2} = \frac{g^{1/2} (by_c)^{3/2}}{b^{1/2}}$$

where y_c is the critical depth. Hence

$$S_c = g n^2 / y_c^{1/3} \tag{5.28}$$

The implications of this equation are best seen from a numerical example.

Example 5.5 Determination of critical bed slope

Given a wide rectangular channel of width 20 m, determine the critical bed slope and discharge for critical depths of 0.2, 0.5 and 1.0 m. Assume that $n = 0.035$.

Solution

Using

$$S_c = g n^2 / y_c^{1/3} \tag{5.28}$$

and

$$Q = \frac{1}{n} \frac{A^{5/3}}{P^{2/3}} S_0^{1/2} \tag{5.9}$$

For a particular y_c, substitute into (5.28) to find S_c, and then substitute into (5.9) to find Q, hence

y(m)	S_c(m/m)	Q(m³/s)
0.2	0.02	5.6
0.5	0.016	22.0
1.0	0.013	62.6

The results of Example 5.5 demonstrate that the critical bed slope is dependent on discharge. In other words, for a given channel with a given slope, it is the discharge which determines whether that slope is mild or steep.

Example 5.6 Critical depth and slope in a natural channel

The data given below were derived from the measured cross section of a natural stream channel. Using the data, determine the critical depth and associated critical bed slope for a discharge of 60 m^3/s assuming $n = 0.04$.

Depth (m)	Area (m^2)	Perimeter (m)	Surface width (m)
0.5	3.5	9.5	9.0
1.0	9.0	13.9	13.0
1.5	16.0	16.7	15.0
2.0	24.0	19.5	17.0

Solution

In this case, (5.16) and (5.28) for critical depth and slope cannot be used, as these were both derived assuming a rectangular channel. Instead, (5.15) – the general equation for critical flow – must be used, i.e.

$$\frac{\alpha Q^2 B_c}{g A_c^3} = 1 \qquad (5.15)$$

Both A and B are functions of depth (y), as given in the above table. In this case, the best method of solution is a graphical one. Rearranging (5.15),

$$\frac{\alpha Q^2}{g} = \frac{A_c^3}{B_c}$$

Hence, if y is plotted against A^3/B, then for $y = y_c$, $A_c^3/B_c = \alpha Q^2/g$, and y_c may be read directly from the graph for any value of $\alpha Q^2/g$ on the A^3/B axis. From the given data, values of A^3/B for various y may be calculated, i.e.

y (m)	A^3/B (m^5)
0.5	4.8
1.0	56.1
1.5	273.1
2.0	813.2

and for critical flow $\alpha Q^2/g = 60^2/g = 367$ (assuming that $\alpha = 1$).

By inspection, the critical depth must be between 1.5 and 2.0. Using linear interpolation,

$$\frac{(y_c - 1.5)}{(2.0 - 1.5)} = \frac{(367 - 273.1)}{(813.2 - 273.1)}$$

$$y_c = 1.59 \text{ m}$$

To find the critical bed slope, apply Manning's equation with $y = y_c = 1.59$, $Q = 60$:

$$Q = \frac{1}{n} \frac{A_c^{5/3}}{P_c^{2/3}} S_c^{1/2}$$

or, in this case,

$$S_c = \left(\frac{Q n P_c^{2/3}}{A_c^{5/3}} \right)^2$$

For $y_c = 1.59$, again using linear interpolation,

$$P_c = 16.7 + \frac{0.09}{0.5} (19.5 - 16.7) = 17.2$$

$$A_c = 16 + \frac{0.09}{0.5} (24 - 16) = 17.44$$

hence

$$S_c = \left(\frac{60 \times 0.04 \times 17.2^{2/3}}{17.44^{5/3}} \right)^2$$

$$= 0.0186$$

Flow transitions

Having developed the idea of mild and steep slopes, the critical depth line and subcritical and supercritical flow, the concept of flow transitions is now introduced. Figure 5.18 shows two types of transition, due to changes of bed slope. In Figure 5.18a it is presupposed that the channel is of mild slope upstream and steep slope downstream. The critical depth (for a given discharge) is constant. Upstream, the flow is subcritical and the depth is greater than the critical depth. Downstream, the converse is true. In the vicinity of the intersection of the mild and steep slopes, gradually varied flow is taking place and the flow regime is in transition from sub- to supercritical. At the intersection the flow is critical.

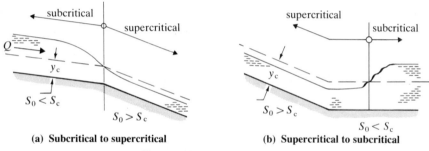

Figure 5.18   Flow transitions.

In Figure 5.18b, the slopes have been reversed and the resulting flow transition is both more spectacular and more complex. Upstream, the flow is supercritical, and downstream the flow is subcritical. This type of transition is only possible through the mechanism of the hydraulic jump. Gradually varied flow takes place between the intersection of the slopes and the upstream end of the jump.

An explanation as to why these two types of transition exist may be found in terms of the Froude Number. Consider what would happen if a flow disturbance was introduced in the transition region shown in Figure 5.18a. On the upstream (mild) slope Fr < 1, and the disturbance would propagate both upstream and downstream. On the downstream (steep) slope the disturbance would propagate downstream. The net result is that all flow disturbances are swept away from the transition region, resulting in the smooth flow transition shown.

Conversely, for the transition shown in Figure 5.18b, flow disturbances introduced upstream (on the steep slope) propagate downstream only. Those introduced downstream propagate both upstream and downstream. The net result in this case is that the disturbances are concentrated into a small region, which is the hydraulic jump.

To determine the flow profile through a region of gradually varied flow, due to changes of slope or cross section, the general equation of gradually varied flow must first be derived.

The general equation of gradually varied flow

The equation is derived by assuming that for gradually varied flow the change in energy with distance is equal to the frictional losses. Hence

$$\frac{dH}{dx} = \frac{d}{dx}\left(y + \frac{\alpha V^2}{2g} + z\right) = -S_f \tag{5.29}$$

where S_f is the friction slope. Rewriting,

$$\frac{d}{dx}\left(y + \frac{\alpha V^2}{2g}\right) = -\frac{dz}{dx} - S_f$$

or

$$\frac{dE_S}{dx} = S_0 - S_f \qquad (5.30)$$

where S_0 is the bed slope. From the section (5.7),

$$\frac{dE_S}{dy} = 1 - \frac{Q^2 B}{gA^3} \quad \text{(taking } \alpha = 1\text{)},$$

and as

$$Fr^2 = \frac{V^2}{gA/B} = \frac{Q^2 B}{gA^3}$$

then

$$\frac{dE_S}{dy} = 1 - Fr^2$$

Combining this with (5.30) gives

$$\frac{dy}{dx} = \frac{S_0 - S_f}{1 - Fr^2} \qquad (5.31)$$

which is the general equation of gradually varied flow.

S_f represents the slope of the total energy line (dH/dx). Since the bed slope (S_0) and the friction slope (S_f) are coincident for uniform flow, the friction slope (S_f) may be evaluated using Manning's equation or the Colebrook–White equation (5.4b).

Equations (5.30) and (5.31) are differential equations relating depth to distance. There is no general explicit solution (although particular solutions are available for prismatic channels). Numerical methods of solution are normally used in practice. These methods are considered in a later section.

Classification of flow profiles

Before examining methods of solution of (5.30) and (5.31), a deeper understanding of this type of flow may be gained by taking a general over-view of (5.31).

For a given discharge, S_f and Fr^2 are functions of depth (y), e.g.

$$\left. \begin{array}{l} S_f = n^2 Q^2 P^{4/3}/A^{10/3} \\ Fr^2 = Q^2 B/gA^3 \end{array} \right\} \begin{array}{l} \text{both decrease with increasing } A \\ \text{and hence increasing } y \end{array}$$

Hence it follows that as

$$S_f = S_0 \text{ when } y = y_n \text{ (uniform flow)}$$

then

$$S_f > S_0 \text{ when } y < y_n$$

$$S_f < S_0 \text{ when } y > y_n$$

and

$$\text{Fr}^2 > 1 \text{ when } y < y_c$$

$$\text{Fr}^2 < 1 \text{ when } y > y_c$$

These inequalities may now be used to find the sign of dy/dx in (5.31) for any condition.

Figure 5.19 shows a channel of mild slope with the critical and normal depths of flow marked. For gradually varied flow, the surface profile may occupy the three regions shown, and the sign of dy/dx can be found for each region:

Region 1
$y > y_n > y_c$, $S_f < S_0$ and $\text{Fr}^2 < 1$, hence dy/dx is positive.

Region 2
$y_n > y > y_c$, $S_f > S_0$ and $\text{Fr}^2 < 1$, hence dy/dx is negative.

Region 3
$y_n > y_c > y$, $S_f > S_0$ and $\text{Fr}^2 > 1$, hence dy/dx is positive.

The boundary conditions for each region may be determined similarly.

Region 1. As $y \to \infty$, S_f and $\text{Fr} \to 0$ and $dy/dx \to S_0$, hence the water surface is asymptotic to a horizontal line (as y is referred to the channel bed).

For $y \to y_n$, $S_f \to S_0$ and $dy/dx \to 0$, hence the water surface is asymptotic to the line $y = y_n$.

This water surface profile is termed an M1 profile. It is the type of profile which would form upstream of a weir or reservoir, and is known as a backwater curve.

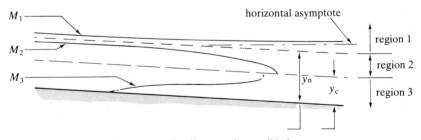

Figure 5.19 Profile types for a mild slope.

Regions 2 and 3. The profiles may be derived in a similar manner. They are shown in Figure 5.19. However, there are two anomalous results for Regions 2 and 3. First, for the M2 profile, $dy/dx \to \infty$ as $y \to y_c$. This is physically impossible, and may be explained by the fact that as $y \to y_c$ the fluid enters a region of rapidly varied flow, and hence (5.30) and (5.31) are no longer valid. The M2 profile is known as a drawdown curve, and would occur at a free overfall.

Secondly, for the M3 profile, $dy/dx \to \infty$ as $y \to y_c$. Again, this is impossible, and in practice an hydraulic jump will form before $y = y_c$.

So far, the discussion of surface profiles has been restricted to channels of mild slope. For completeness, channels of critical, steep, horizontal and adverse slopes must be considered. The resulting profiles can all be derived by similar reasoning, and are shown in Figure 5.20.

Outlining surface profiles and determining control points

An understanding of these flow profiles and how to apply them is an essential prerequisite for numerical solution of the equations. Before particular types of flow profile can be determined for any given situation, two things must be ascertained:

(a) Whether the channel slope is mild, critical or steep. To determine its category, the critical and normal depth of flow must be found for the particular design discharge.
(b) The position of the control point or points must be established. A control point is defined as any point where there is a known relationship between head and discharge. Typical examples are weirs, flumes and gates or, alternatively, any point in a channel where critical depth occurs (e.g. at the brink of a free overfall), or the normal depth of flow at a suitably remote distance from the point of interest.

Having established the slope category and the position of any control points, the flow profile(s) may then be sketched. For subcritical flow, the profiles are controlled from a point downstream. For supercritical flow, the profiles are controlled from upstream.

Figure 5.21 shows three typical flow profiles. Figure 5.21a shows the influence of a broadcrested weir on upstream water levels. Numerical solution of this problem proceeds upstream from the weir (refer to Example 5.7).

Figure 5.21b shows the influence of bridge piers under flood conditions. Many old masonry bridges (with several bridge piers) act in a similar manner to venturi flumes in choking the flow (particularly at high discharges). Flow through the bridge is rapidly varied and, on exit from the bridge piers, the flow is supercritical. However, supercritical flow cannot exist for long, as the downstream slope is mild and the downstream flow uniform (assuming that it is unaffected by downstream control). An

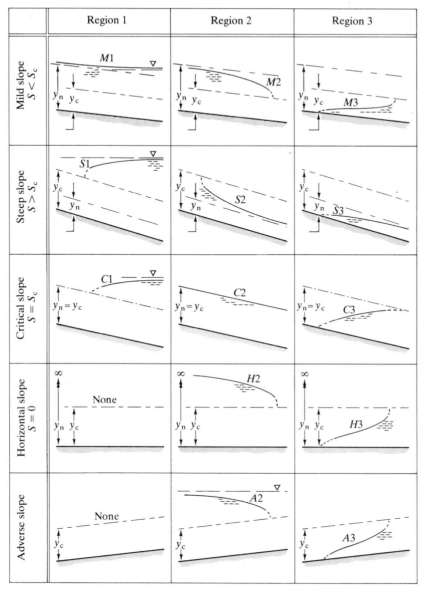

Figure 5.20 Classification of gradually varied flow profiles.

hydraulic jump must form to return the flow to the subcritical condition. The position and height of the jump are determined as shown in Figure 5.21b. An example of this type of problem is given in Example 5.8.

Figure 5.21c shows the flow profile for a side channel spillway and stilling basin. In this example, it is assumed that the side channel is not so steep as to invalidate the equations of gradually varied flow, and that aeration does not occur (refer to Chapter 13 for more details).

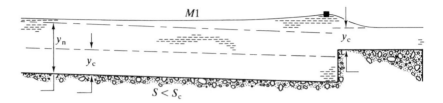

(a) Influence of broadcrested weir on river flow

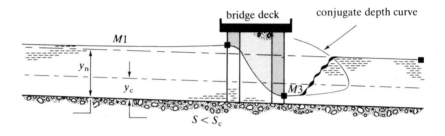

(b) Influence of bridge piers on river flow under flood conditions

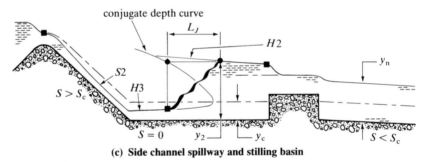

(c) Side channel spillway and stilling basin

Figure 5.21 Examples of typical surface profiles. (*Note.* ■ denotes a control point.)

Methods of solution of the gradually varied flow equation

The three forms of the gradually varied flow equation are:

$$\frac{\mathrm{d}H}{\mathrm{d}x} = -S_f \tag{5.29}$$

$$\frac{\mathrm{d}E_\mathrm{S}}{\mathrm{d}x} = S_0 - S_f \tag{5.30}$$

$$\frac{\mathrm{d}y}{\mathrm{d}x} = \frac{S_0 - S_f}{1 - \mathrm{Fr}^2} \tag{5.31}$$

There are three types of solution to the above equations:

(a) direct integration;
(b) graphical integration;
(c) numerical integration.

These three methods of solution are discussed, and examples are given where appropriate.

Direct integration. Equation (5.31) may be solved directly only for regular channels. Integration methods have been developed by various workers, starting with Dupuit in 1848, who found a solution for wide rectangular channels. He used the Chézy resistance equation and ignored changes of kinetic energy. In 1860, Bresse found a solution including changes of kinetic energy (again for wide rectangular channels using the Chézy equation). Bakhmeteff, starting in 1912, extended the work to trapezoidal channels. Using Manning's resistance equation, various other workers have extended Bakhmeteff's method.

   Full details of these methods may be found in Henderson (1966) or Chow (1959). They are not discussed here as they have been superseded by numerical integration methods which may be used for both regular and irregular channels.

Graphical integration. Equation (5.31) may be rewritten as

$$\frac{dx}{dy} = \frac{1 - Fr^2}{S_0 - S_f}$$

Hence,

$$\int_0^X dx = \int_{y_1}^{y_2} \frac{1 - Fr^2}{S_0 - S_f} dy$$

or

$$X = \int_{y_1}^{y_2} f(y)\, dy$$

where

$$f(y) = \frac{1 - Fr^2}{S_0 - S_f}$$

If a graph of y against $f(y)$ is plotted, then the area under the curve is equivalent to X. The value of the function $f(y)$ may be found by substitution of A, P, S_0 and S_f for various y for a given Q. Hence, the distance X between given depths (y_1 and y_2) may be found graphically.

This method was quite popular until the widespread use of computers facilitated the use of the more versatile numerical methods.

Numerical integration. Using simple numerical techniques, all types of gradually varied flow problems may be quickly and easily solved using only a microcomputer. A single program may be written which will solve most problems.

However, as an aid to understanding, the numerical methods are discussed under three headings:

(a) the direct step method (distance from depth for regular channels);
(b) the standard step method, regular channels (depth from distance for regular channels) and
(c) the standard step method, natural channels (depth from distance for natural channels).

The direct step method. Equation (5.31) may be rewritten in finite difference form as

$$\Delta x = \Delta y \left(\frac{1 - \mathrm{Fr}^2}{S_0 - S_f} \right)_{\text{mean}} \tag{5.32}$$

where 'mean' refers to the mean value for the interval (Δx). This form of the equation may be used to determine, directly, the distance between given differences of depth for any trapezoidal channel. The method is best illustrated by an example.

Example 5.7   Determining a backwater profile by the direct step method

Using Figure 5.21a, showing a backwater curve, determine the profile for the following (flood) conditions:

$$Q = 600 \text{ m}^3/\text{s} \qquad S_0 = 2 \text{ m/km} \qquad n = 0.04$$

$$\text{channel: rectangular, width 50 m}$$
$$\text{weir:} \quad C_d = 0.88$$

$$\text{sill height: } P_S = 2.5 \text{ m}$$

Solution

First, establish a control point as follows.

(a) Find normal depth (y_n) from Manning's equation:

$$Q = \frac{1}{n} \frac{A^{5/3}}{P^{2/3}} S_0^{1/2}$$

In this case,

$$600 = \frac{1}{0.04} \times \frac{(50y_n)^{5/3}}{(50 + 2y_n)^{2/3}} \times 0.002^{1/2}$$

Solution by trial and error yields

$$y_n = 4.443 \text{ m}$$

(b) Find the critical depth (y_c):

$$y_c = \left(\frac{q^2}{g}\right)^{1/3} \quad \text{(for a rectangular channel).}$$

In this case,

$$y_c = \left(\frac{(600/50)^2}{g}\right)^{1/3} = 2.448 \text{ m}$$

(c) Find the depth over the weir (y_w):

$$Q = 1.705 \, C_d B h^{3/2}$$

In this case,

$$600 = 1.705 \times 0.88 \times 50 \times h^{3/2}$$

so

$$h = 4 \text{ m}$$

As

$$y_w = h + P_s$$

$$y_w = 6.5 \text{ m}$$

Hence $y_w > y_n > y_c$ (i.e. Region 1), and $S < S_c$ (i.e. a mild slope).
This confirms (a) that the control is at the weir and (b) that there is an M1 type profile.

The profile is found by taking $y_w = 6.5$ m as the initial depth and $y = 4.5$ m (slightly greater than y_n) as the final depth, and proceeding upstream at small intervals of depth Δy. Fr, S_0 and S_f are evaluated at each intermediate depth, and a solution is found for Δx using the finite difference equation. A tabular solution is shown in Table 5.3, and the resulting profile is shown in Figure 5.22. A computer program for the direct step method is given in Appendix B.

Table 5.3 Computer solution of Example 5.7.

Discharge (cumecs)?600
Channel width (m)?50
Mannings's n?.04
Number of intervals(max 99)?10
Slope (in decimals)?.002
SLOPE +OR−?1
Normal Depth 4.4333
Critical Depth 2.4483

Initial depth?6.5
Final depth?4.5

y(m)	A(m^2)	P(m)	Fr	$(1 - Fr^2)$ mean	S_f	$(S_0 - S_f)$ mean	x(m)
6.5000	325.0000	63.0000	0.2312		0.0006		0.0000
				0.9439		0.0014	
6.3000	315.0000	62.6000	0.2423		0.0007		− 139.1034
				0.9383		0.0013	
6.1000	305.0000	62.2000	0.2543		0.0007		− 284.4225
				0.9319		0.0012	
5.9000	295.0000	61.8000	0.2673		0.0008		− 437.6747
				0.9246		0.0011	
5.7000	285.0000	61.4000	0.2815		0.0009		− 601.3459
				0.9163		0.0010	
5.5000	275.0000	61.0000	0.2970		0.0010		− 779.2213
				0.9066		0.0009	
5.3000	265.0000	60.6000	0.3140		0.0011		− 977.4688
				0.8954		0.00008	
5.1000	255.0000	60.2000	0.3327		0.0013		− 1207.1478
				0.8823		0.0006	
4.9000	245.0000	59.8000	0.3532		0.0015		− 1491.1912
				0.8669		0.0004	
4.7000	235.0000	59.4000	0.3760		0.0017		− 1890.7515
				0.8488		0.0002	
4.5000	225.0000	59.0000	0.4014		0.0019		− 2695.5699

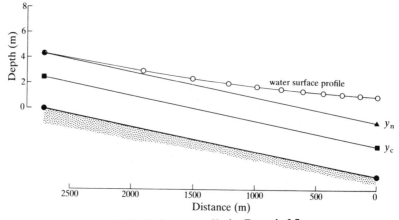

Figure 5.22 Backwater profile for Example 5.7.

Table 5.3 is self-explanatory, but the following points should be kept in mind:

(a) *Signs*:

   x is positive in the direction of flow
   S_0 is positive (except for adverse slopes)
   S_f is positive by definition
   Δy is positive if final depth > initial depth
   Δy is negative if initial depth > final depth

In Example 5.7, Δy is negative and S_0 is positive, which makes Δx negative.

(b) *Accuracy*: Normal depth is always approached asymptotically, so for depths approaching normal depth $S_0 - S_f$ is very small and must be calculated accurately. Of course, it is not possible to calculate the distance of normal depth from the control point as this is theoretically infinite. In Example 5.7, a depth slightly larger than normal depth was chosen for this reason.

(c) *Choice of step interval*: Numerical solutions always involve approximations, and here the choice of step interval affects the solution. The smaller the step interval, the greater the accuracy. If the calculations are carried out by computer, then successively smaller steps may be chosen until the solution converges to the desired degree of accuracy. In Example 5.7, such an exercise yielded the following results:

No. of steps	Δy	Length of reach	Percentage change
5	− 0.4	2587	
			4.0
10	− 0.2	2696	
			2.2
20	− 0.1	2757	
			0.6
30	− 0.067	2775	

The results of this exercise suggest that 20 steps would be more appropriate than the 10 steps originally selected.

In regions of large curvature (i.e. approaching critical depth), the step interval will need to be much smaller than in regions of small curvature. For example, drawdown curves require smaller steps than backwater curves.

(d) *Validity of solution*: It has been suggested that the identification of a control point is of paramount importance in determining gradually varied flow profiles. This is reasonable if a broad understanding of the flow pattern is the aim. However, the equations may be solved between any two depths provided they are within the same region of flow. Thus, in the case of

Example 5.7, the solution may proceed upstream or downstream, provided that both the initial and final depths are greater than the normal depth. If this condition is not met, then the sign of Δx will change at some intermediate point, demonstrating that the solution has passed through an asymptote and the solution is no longer valid.

Composite profiles. Channels may have more than one control point, as already shown in Figure 5.21b and c. In the case of Figure 5.21b, an hydraulic jump effects the transition from supercritical to subcritical flow, but its height and position are determined by the upstream and downstream control points. An example of how to solve such problems follows.

Example 5.8 Composite profiles

For the situation shown in Figure 5.21b, determine the distance of the jump from the bridge for the following conditions:

$$Q = 600 \text{ m}^3/\text{s} \qquad S_0 = 3 \text{ m/km} \qquad n = 0.04$$

$$\text{Channel: rectangular, width 50 m}$$

$$\text{Depth at exit from the bridge} = 1.2 \text{ m}$$

Solution

First, normal and critical depth must be found using the same methods as in Example 5.7. These are:

$$y_n = 3.897\text{m} \qquad y_c = 2.448\text{m}$$

In this case, the sequent depth of the jump equals the normal depth of flow. Utilising the hydraulic jump equation (5.21) in its alternative form, i.e.

$$y_1 = (y_2/2)(\sqrt{1 + 8\text{Fr}_2^2} - 1)$$

gives $y_1 = 1.41$ m for $y_2 = y_n = 3.897$ m.

Hence, to find the distance of the jump from the bridge an M3 profile must be calculated starting from the bridge (at $y = 1.2$) and ending at the initial depth of the jump ($y = 1.41$). This is most easily achieved by using the direct step method. Table 5.4 shows the calculations using a step length of 0.05 m. The position of $y = 1.41$ m is found by interpolation:

$$y = 1.4 \qquad x = 11.368$$

$$y = 1.45 \qquad x = 14.151$$

$$\frac{(1.41 - 1.4)}{(X - 11.368)} = \frac{(1.45 - 1.4)}{(14.151 - 11.368)}$$

$$X = 11.925 \text{ m}$$

Table 5.4   Computer solution of Example 5.8.

Discharge (cumecs)?600
Channel width (m)?50
Mannings's n?.04
Number of intervals(max 99)?5
Slope (in decimals)?.003
SLOPE + OR − ?1
Normal Depth 3.8967
Critical Depth 2.4483

Initial depth?1.2
Final depth?1.45

y(m)	A(m^2)	P(m)	Fr	$(1 - \text{Fr}^2)$mean	S_f	$(S_0 - S_f)$ mean	x(m)
1.2000	60.0000	52.4000	2.9146		0.1336		0.0000
				− 7.0052		− 0.1222	
1.2500	62.5000	52.5000	2.7415		0.1169		2.8657
				− 6.0985		− 0.1068	
1.3000	65.0000	52.6000	2.5848		0.1028		5.7196
				− 5.3237		− 0.0939	
1.3500	67.5000	52.7000	2.4426		0.0909		8.5558
				− 4.6578		− 0.0828	
1.4000	70.0000	52.8000	2.3129		0.0807		11.3682
				− 4.0822		− 0.0734	
1.4500	72.5000	52.9000	2.1943		0.0720		14.1507

An alternative approach, necessary in cases where the jump exists between two profiles, is to plot the conjugate depth curve for the jump on the profile to find the points of intersection, as shown in Figure 5.23.

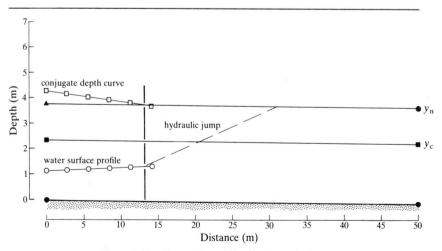

Figure 5.23   Composite profiles for Example 5.8.

The standard step method (for regular channels). Equation (5.30) may be rewritten in finite difference form as

$$\Delta E_S = \Delta x (S_0 - S_f)_{\text{mean}} \qquad (5.33)$$

where 'mean' refers to the mean values for the interval Δx.

This form of the equation may be used to determine the depth at given distance intervals. The solution method is an iterative procedure as follows:

(1) assume a value for depth (y)
(2) calculate the corresponding specific energy $(E_{S_{xG}})$
(3) calculate the corresponding friction slope (S_f)
(4) calculate ΔE_S over the interval Δx using (5.33)
(5) calculate $E_{S_{x+\Delta x}} = E_{S_x} + \Delta E$
(6) compare $E_{S_{x+\Delta x}}$ and $E_{S_{xG}}$
(7) if $E_{S_{x+\Delta x}} \neq E_{S_{xG}}$, then return to 1

Hence, to determine the depth at a given distance may require several iterations. However, this method has the advantage over the direct step method that depth is calculated from distance, which is the more usual problem. Example 5.9 illustrates the method by solving again the problem of Example 5.7.

Example 5.9   Determining a backwater profile by the standard step method

Solve Example 5.7 again by using a standard step length of 150 m.

Solution

The establishment of a control point and the normal and critical depths are as in Example 5.7. A tabular solution is shown in Table 5.5. Only the correct values of depth (y) are shown in the table (to save space). The solution is most easily carried out by computer, and an example program is included in Appendix B.

The standard step method (for natural channels). A common application of flow profiles is in determining the effects of channel controls in natural channels. For several reasons this application is more complex than the preceding cases. The discharge is generally more variable and difficult to quantify, and the assessment of Manning's n is less accurate. These aspects are dealt with elsewhere (see Section 5.6 and Ch. 10).

The other main difficulty lies in relating areas and perimeters to depth. This can only be accomplished by a detailed cross-sectional survey at known locations. In this way, tables of area and perimeter for a given stage may be prepared. Notice that depth must be replaced by stage, as depth is not

Table 5.5 Computer solution of Example 5.9.

INPUT Q,N,SO,B,Y1,DX,L 600,.04,.002,50,6.5, − 150,1950

Y(m)	A(m^2)	P(m)	EG(m)	S_f	$(S_0 - S_f)$ mean	ΔE(m)	EC(m)	X(m)
6.5000	325.0000	63.0000	6.6737	0.0006			6.6737	0
					0.0014	− 0.2032		
6.2848	314.2386	62.5695	6.4706	0.0007			6.4705	− 150
					0.0013	− 0.1928		
6.0795	303.9756	62.1590	6.2781	0.0008			6.2778	− 300
					0.0012	− 0.1814		
5.8852	294.2583	61.7703	6.0971	0.0008			6.0967	− 450
					0.0011	− 0.1692		
5.7027	285.1330	61.4053	5.9283	0.0009			5.9279	− 600
					0.0010	− 0.1561		
5.5328	276.6424	61.0657	5.7726	0.0010			5.7722	− 750
					0.0009	− 0.1424		
5.3765	268.8232	60.7529	5.6304	0.0011			5.6302	− 900
					0.0009	− 0.1283		
5.2351	261.7527	60.4701	5.5029	0.0012			5.5020	− 1050
					0.0008	− 0.1141		
5.1078	255.3910	60.2156	5.3891	0.0013			5.3887	− 1200
					0.0007	− 0.1002		
4.9958	249.7889	59.9916	5.2899	0.0014			5.2890	− 1350
					0.0006	− 0.0867		
4.8975	244.8755	59.7950	5.2035	0.0015			5.2032	− 1500
					0.0005	− 0.0740		
4.8134	240.6717	59.6269	5.1302	0.0015			5.1295	− 1650
					0.0004	− 0.0623		
4.7416	237.0775	59.4831	5.0680	0.0016			5.0679	− 1800
					0.0003	− 0.0518		
4.6818	234.0911	59.3636	5.0167	0.0017			5.0162	− 1950

a meaningful quantity for natural sections. Alternatively, the cross-sectional data may be stored on a computer, values of area and perimeter being computed as required for a given stage.

The solution technique is to use (5.29) in finite difference form as

$$\Delta H = \Delta x(-S_f)_{mean} \tag{5.34}$$

where 'mean' refers to the mean value over the interval Δx.

This may be solved iteratively to find the stage at a given distance in a similar manner to the standard step method for regular channels, if depth is replaced by stage and specific energy by total energy.

Several commercially available mathematical models have been developed for application to natural channels. One such model (FLUCOMP1&2),

developed by the Hydraulics Research Station (England), allows for calibration of Manning's n values from recorded discharge and water levels, incorporates the effects of weirs and bridge piers, and includes lateral inflows. Details of the model are given in Price (1977).

5.11 Unsteady flow

Types of unsteady flow

Unsteady flow is the normal state of affairs in nature, but for many engineering applications the flow may be considered to be steady. However, under some circumstances, it is necessary to consider unsteady flow. Such circumstances may include the following:

Translatory waves. The movement of flood waves down rivers.

Surges and bores. Produced by sudden changes in depth and/or discharge (e.g. tidal effects or control gates).

Oscillatory waves. Waves produced by vertical movement rather than horizontal movement (e.g. ocean waves).

Translatory waves are considered in a simplified way in Chapter 10 (Hydrology, flood routing) and they are considered in more detail later in this section as an example of gradually varied unsteady flow. Surge waves are an example of rapidly varied unsteady flow and a simple treatment is presented in the following section. Oscillatory waves are considered in Chapter 8.

Surge waves

A typical surge wave is shown in Figure 5.24a, in which a sudden increase in the downstream depth has produced a steep-fronted wave moving upstream at velocity V. The wave is, in fact, a moving hydraulic jump, and the solution for the speed of the wave is most easily obtained by transposing the problem to that of the hydraulic jump. This may be achieved by the use of the technique of the travelling observer, as shown in Figure 5.24b. To an observer travelling on the wave at velocity V, the wave is stationary but the upstream and downstream velocities are increased to $V_1 + V$ and $V_2 + V$, respectively (and the river bed is moving at velocity V). The travelling observer sees a (stationary) hydraulic jump. Hence, the hydraulic jump equation may be applied as follows:

$$\frac{y_2}{y_1} = \frac{1}{2}(\sqrt{1 + 8\mathrm{Fr}_1^2} - 1) \tag{5.21}$$

or

$$\text{Fr}_1^2 = \frac{1}{2}\frac{y_2}{y_1}\left(\frac{y_2}{y_1}+1\right) \quad \text{(see derivation of (5.21))}$$

i.e.

$$\frac{(V_1+V)^2}{gy_1} = \frac{1}{2}\frac{y_2}{y_1}\left(\frac{y_2}{y_1}+1\right) \tag{5.35}$$

This equation may be solved for known upstream conditions and the downstream depth. If y_2 is unknown, then the continuity equation may be used:

$$(V_1+V)y_1 = (V_2+V)y_2$$

In this case, Q_2 must be known for solution of V.

Considering (5.35) in more detail, three useful results may be obtained. First, as y_2/y_1 tends to unity, then (V_1+V) tends to $\sqrt{gy}$. As $(V_1+V)=c$ (the wave speed relative to the water) this confirms the result given in Section 5.7 that small disturbances are propagated at a speed $c=\sqrt{gy}$ (for rectangular channels).

Secondly, for $y_2/y_1 > 1$, then $(V_1+V) > \sqrt{gy_1}$. This implies that surge waves can travel upstream even for supercritical flow. This is confirmed by Example 5.10. However, after the passage of such a surge wave, the resulting flow is always subcritical.

Thirdly, for $(V_1+V) > \sqrt{gy_1}$, then the surge wave will overtake any upstream disturbances, and conversely any downstream disturbances will overtake the surge. Hence, this type of surge wave remains stable and steep-fronted.

There are, in fact, four possible types of surge waves; upstream and downstream, each of which may be positive or negative. Positive surges result in an increase in depth and are stable and steep fronted (as discussed above). Negative surges result in a decrease in depth. These are unstable and tend to die out as disturbances travel faster than the surge wave.

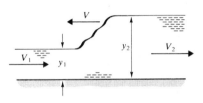

(a) Moving surge

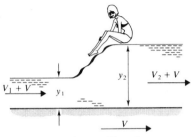

(b) View of a travelling observer

Figure 5.24   Surge waves.

Example 5.10 Speed of propagation of an upstream positive surge wave

Uniform flow in a steep rectangular channel is interrupted by the presence of a hump in the channel bed which produces critical flow for the initial discharge of 4 m³/s. If the flow is suddenly reduced to 3 m³/s some distance upstream, determine the new depth and speed of propagation of the surge wave. (*Channel data:* $b = 3$ m, $n = 0.015$, $S_0 = 0.01$, $\Delta z = 0.083$ m.)

Solution

Referring to Figure 5.12, this problem is an example of what happens when $\Delta z > E_{S1} - E_{S2}$ and $y_1 < y_c$. In this case, for the initial discharge flow over the hump is critical, but for the reduced discharge E_{S1} (for $y_1 = y_n$) is insufficient. The result is that choked flow is produced by y_1 increasing to a new value above critical depth, which produces a positive surge wave travelling upstream.

For $Q = 4$ m³/s:

$y_n = 0.421$ m   (using Manning's equation);
$y_c = 0.566$ m   (using (5.16));
$E_{Sc} = 0.849$ m   (using (5.18));
$E_{Sn} = 0.932$ m   (using (5.13));
$E_{Sn} - E_{Sc} = 0.083 = \Delta z$ (i.e. critical flow is produced).

For $Q = 3$m³/s:

$y_n = 0.348$ m;
$y_c = 0.467$ m;
$E_{Sc} = 0.701$ m;
$E_{Sn} = 0.769$ m;
$E_{Sn} - E_{Sc} = 0.068 < \Delta z$.

Hence, normal depth cannot be maintained. The new depth upstream of the hump (y_2) is found from

$$E_{S2} = E_{Sc} + \Delta z$$

i.e.

$$E_{S2} = 0.784 \text{ m}$$

giving

$$y_2 = 0.67 \text{ m (by iteration).}$$

The equation of the surge wave may now be applied:

$$\frac{(V_1 + V)^2}{gy_1} = \frac{1}{2}\left[\frac{y_2}{y_1}\left(\frac{y_2}{y_1} + 1\right)\right]$$

where

$$y_1 = y_n = 0.348 \text{ m} \qquad y_2 = 0.67 \text{ m}$$

$$V_1 = Q/A_1 = 3/(3 \times y_n) = 2.874 \text{ m/s}$$

Substituting into (5.35) yields

$$V = 0.23 \text{ m/s}$$

This example shows that a surge wave may travel upstream even for supercritical flow, but leaves behind subcritical flow. The surge wave will proceed upstream until either it submerges the upstream control or until a stationary hydraulic jump forms (by y_2 decreasing to the sequent depth).

Gradually varied unsteady flow: the equations of motion

This type of flow implies translatory wave motion of long wavelength and low amplitude. In this case the assumption of parallel streamlines and hydrostatic pressure distributions is reasonable. To define the motion, two equations must be derived relating to force/momentum/energy and continuity.

Referring to Figure 5.25, showing an elemental slice in a natural channel, the forces acting are the pressure forces and the resistance forces. These are now considered in turn.

Pressure forces (F_p). These are most easily accounted for by considering, directly, the difference in pressure between the upstream and downstream faces of the slice. Along any horizontal line the difference in pressure (δp) is given by

$$\delta p = - \rho g \delta h$$

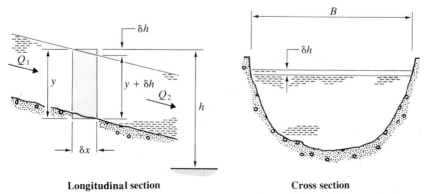

Longitudinal section   Cross section

Figure 5.25 Derivation of the dynamic equation for gradually varied unsteady flow.

(δh being defined as positive when increasing from the upstream to the downstream face).

As this pressure difference applies at all depths, then the net pressure force (F_p) acting on the cross section in the x-direction is given by

$$F_p = \sum \delta A \, \delta p$$

Hence

$$F_p = -\rho g A \, \delta h$$

This is only correct if y is much larger than both δh and δz, a condition which is met for gradually varied flow (steady or unsteady).

Resistance forces (F_s). Assuming a mean shear stress τ_0 (as previously)

$$F_s = \sum(-\tau_0 \, \delta x \, \delta P) = -\tau_0 \, \delta x P$$

For unsteady flow, the sum of the forces acting is equal to the mass ($\rho A \, \delta x$) × acceleration (dV/dt) of the slice. In this case, $V = f(x, t)$, and hence

$$\delta V = \frac{\partial V}{\partial x} \delta x + \frac{\partial V}{\partial t} \delta t$$

or

$$\frac{dV}{dt} = \frac{\partial V}{\partial x} \frac{dx}{dt} + \frac{\partial V}{\partial t} = \frac{\partial V}{\partial x} V + \frac{\partial V}{\partial t}$$

Hence

$$-\rho g A \, \delta h - \tau_0 \, \delta x \, P = \rho A \, \delta x \left(V \frac{\partial V}{\partial x} + \frac{\partial V}{\partial t} \right)$$

Substituting $\tau_0 P = \rho g A S_f$ ((5.5) with S_f replacing S_0) and simplifying,

$$S_f = -\frac{\partial h}{\partial x} + \frac{V}{g} \frac{\partial V}{\partial x} + \frac{1}{g} \frac{\partial V}{\partial t}$$

As $h = y + z$,

$$\frac{\partial h}{\partial x} = \frac{\partial}{\partial x}(y + z) = \frac{\partial y}{\partial x} - S_0$$

and by substitution

$$S_f = S_0 - \frac{\partial y}{\partial x} - \frac{V}{g}\frac{\partial V}{\partial x} - \frac{1}{g}\frac{\partial V}{\partial t} \qquad (5.36)$$

steady uniform flow

steady non-uniform flow

unsteady non-uniform flow

This is the dynamic equation of gradually varied unsteady flow. It may be rewritten in terms of Q, rather than V, as

$$\frac{\partial Q}{\partial t} + \frac{\partial}{\partial x}\left(\frac{Q^2}{A}\right) + gA\frac{\partial y}{\partial x} - gA(S_0 - S_f) = 0 \qquad (5.37)$$

This is often referred to as the momentum equation for unsteady flow.

The continuity equation is derived as follows. The change in discharge (δQ) with distance only is

$$\delta Q = \frac{\partial Q}{\partial x}\delta x$$

or, in volume terms,

$$\delta Q\,\delta t = \frac{\partial Q}{\partial x}\delta x\,\delta t$$

the corresponding change in storage volume ($\delta(\text{Vol})$) is

$$\delta(\text{Vol}) = B\frac{\partial h}{\partial t}\delta t\,\delta x$$

Applying the conservation of mass (continuity) principle,

$$\delta(\text{Vol}) + \delta Q\,\delta t = 0$$

or

$$B\frac{\partial h}{\partial t}\delta t\,\delta x + \frac{\partial Q}{\partial x}\delta x\,\delta t = 0$$

Simplifying,

$$B\frac{\partial h}{\partial t} + \frac{\partial Q}{\partial x} = 0 \qquad (5.38)$$

These two equations ((5.37) and (5.38)) were first derived by Saint-Venant and are sometimes referred to as the Saint-Venant equations.

Solution of the gradually varied unsteady flow equations. The solution of (5.37) and (5.38) is a complicated matter, even for rectangular channels. General solutions are only possible by using numerical methods. Detailed discussions of such methods are beyond the scope of the present text, and the interested reader is directed to the references at the end of the chapter. However, to end the discussion at this point would be unsatisfactory, so some remarks concerning the available methods follow, together with an example solution for flood waves in rectangular channels.

There are three principal types of numerical methods currently used in the study of open channel flow; namely, finite differences, characteristics and finite elements.

Finite difference models are currently the most common for this type of problem. The method of characteristics has advantages over. finite differences in solution of the boundary conditions, and can be used for other unsteady flow problems (e.g. dam burst flood waves) where finite difference models are inapplicable. Finite element models have the advantage of easily accommodating irregular boundaries, but they have not yet been used extensively for open boundary hydraulics.

Concentrating on finite difference models, these may be explicit (in which Q and y at the next time step are expressed in terms of Q and y at the current time step) or implicit (in which Q and y are related to subsequent and previous time steps).

Explicit models are simpler, but must conform to the Courant stability criteria ($c \, \Delta t / \Delta x < 1$, where c is the wave speed), which imposes a severe limitation on the time–distance grid. Implicit models are more complex, requiring the solution of a set of simultaneous equations through time and distance, but are unconditionally stable.

Example 5.11   Translatory wave in a rectangular channel

Determine the water surface profile through time, the maximum water depths and the outflow hydrograph for the following channel given the inflow hydrograph.

Channel: rectangular width (B) = 10 m   length = 10 km
         Manning's $n = 0.025$
         bed slope(S_0) = 0.002 m/m
         initial depth of flow = 1.0 m

Inflow hydrograph (in addition to steady flow)

Time (h)	0	0.5	1.0	1.5	2.0	2.5
Discharge (m³/s)	0	50	37.5	25	12.5	0

Solution

One method of solution is to use an explicit finite difference scheme for the solution of (5.37) and (5.38).

Following the method given by Koutitas (1983) and using a centred grid (Fig. 5.26), (5.38) may be expressed using a forward difference equation as

$$\frac{y_i^{n+\frac{1}{2}} - y_i^{n-\frac{1}{2}}}{\Delta t} + \frac{1}{B}\left(\frac{Q_{i+1}^n - Q_i^n}{\Delta x}\right) = 0$$

in which $y_i^{n+\frac{1}{2}}$ may be found explicitly. (*Note.* For a regular channel, y may be substituted for h in (5.38).) Equation (5.37) may be expressed, using a central difference equation and using the Lax–Wendroff method for the time derivative, as

$$Q_i^{n+1} = \frac{(QQ_2 + QQ_1)}{2} - \Delta t\left(\frac{QQ_2^2}{yy_2B} - \frac{QQ_1^2}{yy_1B}\right)\bigg/\Delta x - g\,\Delta t\,yyB\,\frac{(yy_2 - yy_1)}{\Delta x}$$

$$- gyyB(S_f - S_0)\,\Delta t$$

where

$$yy = (y_i^{n+\frac{1}{2}} + y_i^{n-\frac{1}{2}} + y_{i-1}^{n+\frac{1}{2}} + y_{i-1}^{n-\frac{1}{2}})/4$$

$$S_f = (Q_i^n)^2\,n^2(B + 2yy)^{1.333}/(Byy)^{3.333}$$

$$yy_2 = (y_i^{n+\frac{1}{2}} + y_i^{n-\frac{1}{2}})/2 \qquad yy_1 = (y_{i-1}^{n+\frac{1}{2}} + y_{i-1}^{n-\frac{1}{2}})/2$$

$$QQ_2 = (Q_{i+1}^n + Q_i^n)/2 \qquad QQ_1 = (Q_i^n + Q_{i-1}^n)/2$$

in which Q_i^{n+1} may be found explicitly. For the solution to proceed, the boundary conditions must be specified. At the upstream boundary $Q_1^n = $ inflow hydrograph + steady flow. At the downstream boundary, assuming the flow is unrestricted, then

$$V = \sqrt{gy}$$

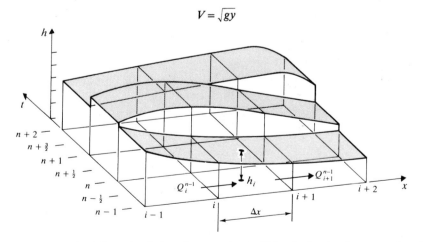

Figure 5.26 Three-dimensional view of a centred x,t grid for h, showing the progression of a surge wave.

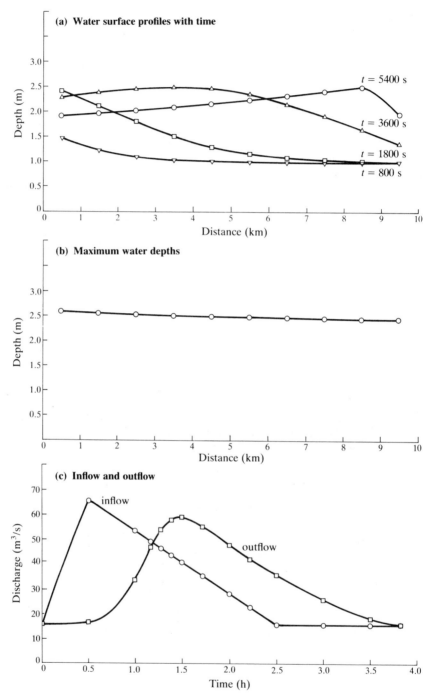

Figure 5.27 Solution to Example 5.11.

and

$$\delta Q = \delta A \sqrt{gy}$$

Also, the stability criteria must be satisfied, i.e. $c\Delta t / \Delta x < 1$. Taking $\Delta x = 1000$ m and assuming the maximum depth is approximately 2.56 m (the normal depth for $Q = 65$ m^3/s), then

$$\Delta t < 220$$

Taking a conservative value of $\Delta t = 20$ s, the solution may now be found, using the computer program in Appendix B. The results are shown in Figure 5.27.

Some interesting conclusions may be gleaned from Example 5.11. First, the attenuation of the flood peak is only 10% in 10 km (and this becomes less for steeper bed slopes). Secondly, the maximum depths correspond very closely to the normal depths at the peak discharges. Thirdly, the speed of propagation is very closely estimated using the wave speed formula with the normal depth at maximum discharge.

The particular channel analysed is representative of a small to medium sized river, and one may therefore conclude that the use of a steady gradually varied flow model will often be sufficiently accurate.

References and further reading

Brebbia, C. A. and J. J. Connor 1976. *Finite element techniques for fluid flow*. London: Butterworth.

Chow, Ven te 1959. *Open channel hydraulics*. Tokyo: McGraw-Hill.

Cunge, J. A., F. H. Holly Jr and A. Verwey 1980. *Practical aspects of computational river hydraulics*. London: Pitman.

Henderson, F. M. 1966. *Open channel flow*. New York. Macmillan.

Jansen, P. Ph. *et al.* (eds) 1979. *Principles of river engineering*. London: Pitman.

Koutitas, C. G. 1983. *Elements of computational hydraulics*. Plymouth: Pentech Press.

Price, R. K. 1977. *A mathematical model for river flows*. Hydraulics Res. Stn Rep. INT127, 2nd edn.

6

Pressure surge in pipelines

6.1 Introduction

All of the pipe flows considered in Chapter 4 were steady flows, i.e. discharge was assumed to remain constant with time. This corresponds to the situation when the control valves in the system are held at a fixed setting. From time to time it will be necessary to alter the valve setting. Immediately following such an alteration the flow will be accelerating (or decelerating), and will therefore be unsteady. Unsteady (or 'surge') conditions often continue for only a very short period. Nevertheless, the effects on the system may, under some circumstances, be dramatic (including pipe bursts).

For the sake of simplicity, consider first a case involving flow of an incompressible liquid. Civil engineers commonly deal with pipelines of considerable length and diameter. Such a pipeline will contain a large mass of flowing fluid, and the momentum of the fluid will therefore be correspondingly large, as in the case of a long pipeline from a reservoir (Fig. 6.1a). Note that the velocity of flow is here denoted by 'u':

$$\text{mass of liquid in pipeline} = \rho AL$$

$$\text{momentum of liquid} = M = \rho ALu \quad (u = Q/A)$$

If the flow is retarded by a control valve at the downstream end, then the fluid undergoes a rate of change of momentum of

$$\frac{\mathrm{d}M}{\mathrm{d}t} = \rho AL \frac{\mathrm{d}u}{\mathrm{d}t} \tag{6.1}$$

A force, F, is required to produce this change of momentum. This force is applied to the fluid by the control valve. This force exists only during the deceleration of the fluid, and it is therefore only a short-lived phenomenon.

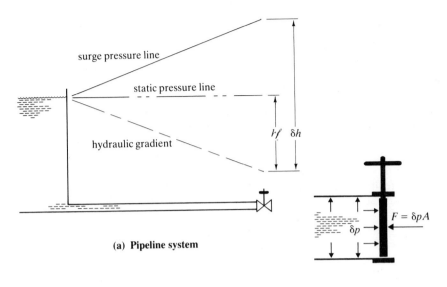

(a) **Pipeline system**

(b) **Surge pressure at valve**

Figure 6.1   Incompressible surge.

According to the laws of fluid pressure a force (or pressure) applied in one plane must be transmitted in all directions (Fig. 6.1b), therefore

$$\delta p A = F$$

where δp is the instantaneous rise in pressure corresponding to F (this is known as a 'surge pressure' or 'transient pressure'). Hence

$$F = \delta p A = \rho A L \frac{du}{dt}$$

and so

$$\delta p = \rho L \frac{du}{dt}$$

or

$$\frac{\delta p}{\rho g} = \delta h = \frac{L}{g} \frac{du}{dt} \tag{6.2}$$

As L is a constant for a given system, $\delta p = f(du/dt)$, i.e. δp is a function of the rate of closure of the valve. Upstream of the valve, δp (or δh) varies linearly with distance. An example will now be developed to illustrate the use of this approach.

Example 6.1   Incompressible unsteady flow

Water flows from a reservoir along a rigid horizontal pipeline. The pipe intake is 20 m below the water surface elevation in the reservoir. The pipe is 0.15 m in diameter and 1500 m long, and $\lambda = 0.02$. The pipe discharges to atmosphere through a valve at its downstream end. The rate of valve adjustment is such that it can be completely closed in 4 s and gives a uniform deceleration of the water in the pipe. Calculate the pressure just upstream of the valve and 500 m upstream of the valve if the valve aperture is adjusted from the fully open to the half-open position (in 2 s).

Solution

Before valve closure commences the velocity of flow is u_0, so

$$h_f = \frac{\lambda L u_0^2}{2gD} = \frac{0.02 \times 1500 \times u_0^2}{2 \times 9.81 \times 0.15} = 20 \text{ m}$$

Therefore

$$u_0 = 1.4 \text{ m/s}$$

During valve closure

$$\frac{du}{dt} = \frac{u_0}{t} = \frac{1.4}{4}$$

For half closure, $u = \frac{1}{2}u_0 = 0.7$ m/s, therefore

$$h_f = \frac{0.02 \times 1500 \times 0.7^2}{2 \times 9.81 \times 0.15} = 5 \text{ m}$$

The surge pressure (δh) is given by

$$\delta h = \frac{L}{g} \frac{\delta u}{\delta t} = \frac{1500}{9.81} \times \frac{1.4}{4} = 53.5 \text{ m}$$

Total head at valve = (static head + $\delta h - h_f$) is

$$20 + 53.5 - 5 = 68.5 \text{ m}$$

At a point 500 m upstream of the valve

$$\delta h = \frac{1000}{9.81} \times \frac{1.4}{4} = 35.7 \text{ m}$$

$$h_f = 5 \times \frac{1000}{1500} = 3.3 \text{ m}$$

Therefore, total head $= 20 + 35.7 - 3.3 = 52.4$ m.

This example presupposes that the liquid is incompressible, and that the pipe is rigid.

6.2 Effect of 'rapid' valve closure

Taking the pipe system used in the preceding example, consider now the effect of increasing the speed of valve closure:

Time of closure (s)	H(m)
4	53.5
3	71.3
2	107
1	214
0.5	428
0	∞

The above figures do not give a completely accurate picture. If pressure measurements were to be taken, it would be found that the increase in head was, in fact, self-limiting, and that the maximum head was 214 m, not infinity. Therefore, it is necessary to ask why the theory outlined above is so much in error for the fast valve closure times ($t < 1$ s) and yet is reasonably accurate at the slower closure times. The answer lies in the assumptions made about the liquid and the pipeline (i.e. incompressible liquid and rigid pipe material). When rapid changes occur in the flow, the system no longer behaves in this 'inelastic' manner. To elucidate this point, the case of a compressible liquid contained in a rigid pipe is now considered (the effect of pipe material elasticity is considered subsequently).

6.3 Unsteady compressible flow

General description

As the rate of valve closure (or opening) increases, so do the inertia forces. A point is reached at which the liquid is being subjected to pressures which are sufficient to cause it to compress. Once compressibility has been brought into play, a radical change takes place in the pressure surge process in the body of fluid. The rapid alteration of the valve setting does not cause a uniform deceleration along the pipeline. Instead, the alteration generates a shock wave in the fluid. A shock wave is a zone in which the fluid is rapidly compressed (i.e. p and ρ increase) (see Fig. 6.2). It travels through a fluid at the speed (or celerity) of sound (symbol c).

Through the mechanism of compressibility, the kinetic energy of the fluid before acceleration is transformed into elastic energy (thus upholding the principle of conservation of energy). Note that the fluid upstream of the wave is still flowing at its original velocity, pressure and density. In fact, one way of viewing the process is to regard the shock wave as a 'messenger' which is carrying the 'news' of the change in valve setting. This information is 'received' by the fluid at a given point only when the shock wave arrives at that point. The progress of the shock wave is now described, assuming a rigid pipe and complete valve closure, in a pipeline whose upstream end is connected to a reservoir.

(a) The valve closes, generating the shock wave (Fig. 6.3a).

(b) The shock wave travels back along the pipeline (Fig. 6.3b). Fluid between the wave and the valve is now compressed, with the pressure therefore raised to $p + \delta p$ (i.e. the head raised by δh).

(c) The wave arrives at the reservoir (Fig. 6.3c). The pipe is therefore now full of compressed and pressurised fluid.

(d) There is an inequilibrium at the pipe–reservoir interface since the reservoir head remains unchanged. The fluid in the pipe begins to discharge in reverse into the reservoir (Fig. 6.3d). A decompression wave is generated, which travels back towards the valve. The fluid between the reservoir and the decompression wave is therefore at the original p and ρ, but u is reversed. Once again, the fluid between the wave and the valve has not yet received the 'message' and is still at $(p + \delta p)$ and $(\rho + \delta \rho)$.

(e) The control valve (if closed) constitutes a dead end. When the decompression wave arrives, the reversed flow can proceed no further, so the fluid here cannot flow back to the reservoir. A negative pressure is generated at

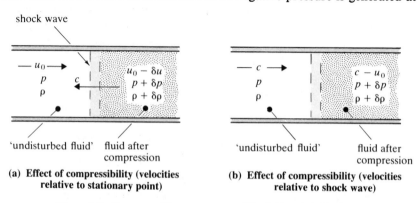

(a) Effect of compressibility (velocities relative to stationary point)

(b) Effect of compressibility (velocities relative to shock wave)

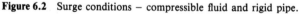

Figure 6.2 Surge conditions – compressible fluid and rigid pipe.

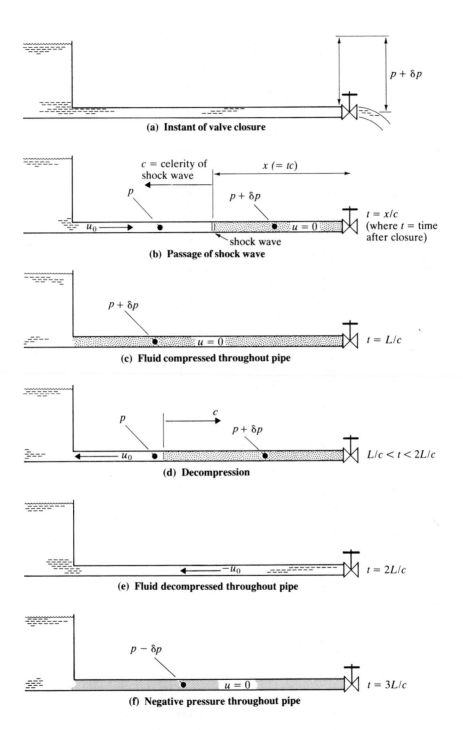

Figure 6.3 Propagation of shock waves in a pipeline.

the valve (Fig. 6.3e), which produces a 'negative' shock wave. This, in turn, is transmitted towards the reservoir. Fluid between the wave and the valve is at rest, but under a negative pressure (which is theoretically equal in magnitude to δp under (b), above). Fluid between the wave and the reservoir is still at p, ρ, and the flow is still reversed. (For simplicity, the possibility of cavitation has been ignored here.)

(f) When the negative wave arrives at the reservoir (Fig 6.3f), the pipeline pressure is lower than the reservoir pressure. Fluid therefore flows from the reservoir into the pipe. Theoretically, the velocity is now restored to its original magnitude and direction, since we have so far ignored all hydraulic losses.

The whole cycle (a) to (f) then repeats.

Strictly, the celerity of the shock wave is measured relative to the fluid, and therefore the shock wave travels past a stationary point at $(c - u)$. However, since $c \gg u$, $(c - u) \rightarrow c$. The time taken for the shock wave to traverse the length of the pipeline is therefore

$$t = L/c$$

At any point in the pipeline the timing of pressure and velocity changes may be conveniently measured in multiples of L/c.

Simple equations for 'instantaneous' alteration of valve setting in a rigid pipeline

This is the simplest case to treat. To develop the equations, consider the conditions on either side of a shock wave. In order to view these conditions, it is convenient to use co-ordinates which move with the shock wave. From this viewpoint, velocities are as shown in Figure 6.2b. Because of the compression, it is imperative that the equation of mass continuity (i.e. mass conservation) be used:

$$\text{mass entering control volume} = \text{mass leaving control volume}$$

$$\rho A c = (\rho + \delta\rho)(c - u_0)A$$

which may be rewritten as

$$\rho u_0 = \delta\rho(c - u_0) \tag{6.3}$$

Similarly, applying the momentum equation

$$\text{force} = \text{mass flow} \times \text{change in velocity}$$

$$(p + \delta p)A - pA = \rho A c[c - (c - u_0)]$$

Hence

$$\delta p = \rho c u_0 \tag{6.4}$$

The measure of the elasticity of a liquid is its bulk modulus (K):

$$K = -\frac{\delta p}{\delta V/V}$$

V being the volume of liquid. The negative sign arises because increasing pressure causes a reduction (negative increase) in volume.

By definition,

$$\rho = \frac{\text{mass}}{\text{volume}} = \frac{m}{V}$$

therefore

$$\frac{d\rho}{dV} = -\frac{m}{V^2} - \frac{\rho}{V}$$

so

$$\frac{\delta\rho}{\rho} = -\frac{\delta V}{V}$$

therefore

$$K = \frac{\delta p}{\delta\rho/\rho} \tag{6.5}$$

From (6.3),

$$\frac{\delta\rho}{\rho} = \frac{u_0}{(c - u_0)} \tag{6.6}$$

From (6.4),

$$u_0 = \frac{\delta p}{\rho c}$$

From (6.5),

$$\frac{\delta\rho}{\rho} = \frac{\delta p}{K}$$

Substituting for u_0 and $\delta\rho/\rho$ in (6.6),

$$\frac{\delta p}{K} = \frac{(\delta p/\rho c)}{[c - (\delta p/\rho c)]}$$

Rearranging,

$$\frac{\rho c^2 - \delta p}{K} = 1$$

If $\delta p \ll K$, then

$$\rho c^2/K = 1$$

therefore

$$K = \rho c^2 \qquad\qquad (6.7)$$

Equations (6.4) and (6.7) may then be solved simultaneously to obtain a value for δp. Note that no allowance has been made for straining of the pipe material, since the pipe is assumed to be rigid (i.e. $E \to \infty$). It should be further noted that the passage of events is extremely rapid, as will become apparent in the following example.

Example 6.2 Surge in a simple pipeline

A valve is placed at the downstream end of a 3 km long pipeline. Water is initially flowing along the pipe at a mean velocity of 2.5 m/s. What is the magnitude of the surge pressure generated by a sudden and complete valve closure? Sketch the variation in pressure at the valve and at the mid-point of the pipeline after valve closure. Take celerity of sound as 1500 m/s.

Solution

Increase in pressure is estimated from (6.4):

$$\delta p = \rho c u_0 = 1000 \times 1500 \times 2.5 = 3.75 \times 10^6 \text{ N/m}^2$$

To sketch the pressure variation, it is necessary to know the transmission time for the shock wave:

$$t = L/c = 3000/1500 = 2 \text{ s}$$

Following the cycle of events (a) to (f), above, at the valve the increase in pressure will be maintained while the shock wave travels to the reservoir and while the decompression wave returns to the valve, this takes $2L/c = 4$ s (Fig. 6.4a). Similarly, the pressure will be negative for another 4 s, etc.

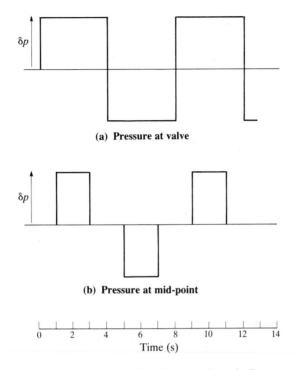

(a) **Pressure at valve**

(b) **Pressure at mid-point**

Figure 6.4 Fluctuation of pressure in a pipeline.

At the mid-point, it will take $t = L/2c$ ($= 1$ s) for the shock wave 'messenger' to arrive and 'deliver' the pressure increase. The initial increase is then maintained while the shock wave travels to the reservoir and the reflected decompression wave returns (this takes $2 \times (L/2c) = L/c$), whereupon the pressure reverts to its original value. The negative pressure from the valve arrives $L/2c (= 1$ s) after it has been generated at the valve, and so on (Fig. 6.4b).

Equations for 'instantaneous' valve closure in an elastic pipeline

We now proceed to a slightly more realistic setting for the pressure surge phenomenon, viz. a compressible fluid flowing in an elastic pipeline. The effect of the pressure increase behind the shock wave is:

(a) to compress the fluid;
(b) to strain the pipe walls.

Due to the strain effect, the pipe cross section behind the shock wave is greater than the unstrained cross section ahead of the wave (Fig. 6.5a). The mass continuity equation must therefore be written as

$$\rho A c = (\rho + \delta\rho)(A + \delta A)(c - u_0)$$

which may be rearranged to give

$$\rho u_0 (A + \delta A) - \rho c\, \delta A = \delta\rho (c - u_0)(A + \delta A) \qquad (6.8)$$

The momentum equation is

$$(p + \delta p)(A + \delta A) - pA = \rho A c u_0$$

Ignoring small quantities, this may be written as

$$\delta p = \rho c u_0$$

which is equation (6.4). The equation relating to the elasticity of the fluid (Equation (6.5)) remains unchanged:

$$K = \frac{\delta p}{\delta\rho/\rho}$$

From (6.8),

$$\frac{\delta\rho}{\rho} = \frac{u_0(A + \delta A) - c\,\delta A}{(c - u_0)(A + \delta A)} \qquad (6.9)$$

From (6.4),

$$u_0 = \frac{\delta p}{\rho c}$$

From (6.5),

$$\frac{\delta\rho}{\rho} = \frac{\delta p}{K}$$

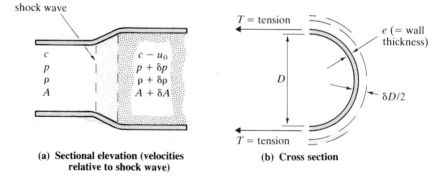

(a) Sectional elevation (velocities
relative to shock wave)

(b) Cross section

Figure 6.5 Surge conditions – compressible fluid and elastic pipe material.

Substituting the above two equations in (6.9),

$$\frac{\delta p}{K} = \frac{(\delta p/\rho c)(A + \delta A) - c\,\delta A}{[c - (\delta p/\rho c)](A + \delta A)} = \frac{\delta p(A + \delta A) - \rho c^2\,\delta A}{(\rho c^2 - \delta p)(A + \delta A)}$$

Therefore

$$\frac{\delta p}{K}\rho c^2(A + \delta A) + \rho c^2\,\delta A = \delta p(A + \delta A) + \frac{\delta p^2}{K}(A + \delta A)$$

and rearranging,

$$\rho c^2 = \frac{\delta p\left(1 + \dfrac{\delta A}{A}\right)\left(1 + \dfrac{\delta p}{K}\right)}{\dfrac{\delta p}{K}\left(1 + \dfrac{\delta A}{A}\right) + \dfrac{\delta A}{A}}$$

If $\delta p \ll K$ and $\delta A/A \ll 1$, then the above equation simplifies to

$$\rho c^2 = \frac{\delta p}{(\delta p/K) + (\delta A/A)} \tag{6.10}$$

Note that if the $\delta A/A$ term is ignored, this becomes identical to (6.7). However, as the elasticity of the pipe is to be taken into account, it is necessary to find some relationship between δA and A in terms of the pipe characteristics. In developing the relationship it is assumed that the pipe is axially rigid (due, say, to pipe supports) but free to move radially.

Referring to Figure 6.5b, an increase in pressure δp inside a pipe produces a bursting (or 'hoop') tension T. Thus, for a pipe of length δx and diameter D,

$$\delta p\, D\, \delta x = 2T = 2\sigma e\,\delta x$$

where σ is the hoop stress in the pipe wall. Since

$$\frac{\text{stress}}{\text{strain}}\left(= \frac{\sigma}{s}\right) = E$$

then

$$s = \sigma/E$$

Therefore

$$\frac{\delta p\, D\, \delta x}{E} = \frac{2\sigma e\,\delta x}{E}$$

so

$$\frac{\delta p \, D}{2Ee} = s \qquad (6.11)$$

But

$$s = \frac{\text{change in length}}{\text{original length}} = \frac{\pi \, \delta D}{\pi D} = \frac{\delta D}{D}$$

and

$$\frac{\delta A}{A} = \frac{\pi (D + \delta D)^2 - \pi D^2}{\pi D^2} = \frac{2 \, \delta D}{D} = 2s$$

Therefore

$$\frac{\delta A}{A} = \frac{\delta p \, D}{Ee}$$

$$\rho c^2 = \frac{\delta p}{(\delta p / K) + (\delta p \, D / Ee)} = \frac{1}{(1/K) + (D/Ee)} \qquad (6.12)$$

Hence, if K, ρ, E and e are known, c can be calculated.

The only difference between the rigid pipe case and the elastic pipe case is therefore the speed at which the shock wave travels down the pipe. In every other respect, the solutions follow the pattern of Example 6.2.

Some installations incorporate supports which allow some axial movement. When this is so, a hoop strain will produce a corresponding strain in the axial direction. This will slightly influence the magnitude of c. An appropriate allowance for this effect may be made by incorporating a term with the Poisson ratio in (6.11) (and hence in (6.12)).

6.4   Analysis of more complex problems

Description

For a realistic system under surge conditions the analysis must take account of:

(a) elasticity (of fluid and pipe material);
(b) effect of hydraulic losses;
(c) non-instantaneous valve movement.

The method given in the preceding section can be suitably extended, but it becomes rather cumbersome. It is therefore worth turning to a method which is readily adapted for computer use. Such a method is now outlined, and although the actual process of developing the equations can seem rather long, it should be borne in mind that we are really only interested in the final equations. These will introduce the 'method of characteristics' which may, in principle, be applied to any unsteady flow. To clarify matters as far as possible, the equations are developed step by step from first principles.

It is worth reviewing the events which take place as a pressure wave passes along a short length of pipe (Fig. 6.6a). The valve closure (or change in setting) is now not instantaneous, though it is fast enough to generate a shock wave. However, since the change in setting takes a finite time, it will also take a finite time for the pressure rise δp to be generated. Consequently, the shock wave now occupies a significant length of pipeline, δx. This length will clearly be directly related to the pressure rise time ($\delta x/c = \delta t$). Due to the increase in pressure, two changes will occur:

(a) the fluid will compress;
(b) the (elastic) pipe will experience a strain.

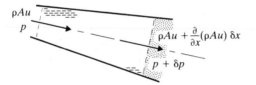

(a) Sectional elevation

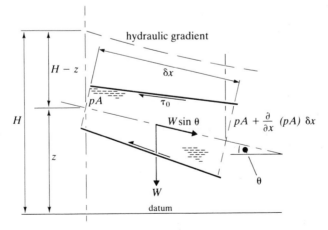

(b) Forces acting on fluid

Figure 6.6   Surge conditions – compressible fluid, elastic pipe and 'rapid' (but not instantaneous) valve closure.

Differential equations may therefore be developed based on the conservation of mass and momentum equations.

Conservation of mass.   For the control volume shown in Figure 6.6a,

(mass flow in) − (mass flow out) = (rate of change of mass in control volume)

$$\rho A u - \left[\rho A u + \frac{\partial}{\partial x} (\rho A u) \, \delta x \right] = \frac{\partial}{\partial t} (\rho A \, \delta x)$$

therefore

$$- \frac{\partial}{\partial x} (\rho A u) \, \delta x = \frac{\partial}{\partial t} (\rho A \, \delta x)$$

Expanding this equation by the 'product rule' gives

$$- \left(\rho A \frac{\partial u}{\partial x} \, \delta x + \rho u \frac{\partial A}{\partial x} \, \delta x + A u \frac{\partial \rho}{\partial x} \, \delta x \right) = \rho \frac{\partial A}{\partial t} \, \delta x + A \frac{\partial \rho}{\partial t} \, \delta x$$

Dividing throughout by $(\rho A \, \delta x)$ and rearranging,

$$\underbrace{\frac{u}{A} \frac{\partial A}{\partial x} + \frac{1}{A} \frac{\partial A}{\partial t}} + \underbrace{\frac{u}{\rho} \frac{\partial \rho}{\partial x} + \frac{1}{\rho} \frac{\partial \rho}{\partial t}} + \frac{\partial u}{\partial x} = 0$$

The first two pairs of terms are total differentials, so the equation may be written

$$\frac{1}{A} \frac{dA}{dt} + \frac{1}{\rho} \frac{d\rho}{dt} + \frac{\partial u}{\partial x} = 0 \qquad (6.13)$$

As it stands, this is not a convenient form, since there are too many variables. The first two terms can, however, each be presented in a different form.

The term $(1/A)(dA/dt)$ is clearly related to the elastic strain in the pipe walls. Referring to (6.11),

$$s = \frac{\delta p D}{2Ee}$$

and therefore

$$\frac{ds}{dt} = \frac{D}{2Ee} \frac{dp}{dt}$$

Therefore,

$$\text{rate of change of pipe circumference} = \pi D \frac{ds}{dt}$$

$$\text{change in diameter} = (1/\pi) \times \text{change in circumference} = \delta D$$

$$\text{change in pipe area} = \pi D \, \delta D / 2$$

Therefore,

$$\frac{dA}{dt} = \frac{\pi D^2}{2} \frac{ds}{dt} = \frac{\pi D^2}{2} \frac{D}{2Ee} \frac{dp}{dt}$$

so

$$\frac{1}{A} \frac{dA}{dt} = \frac{4}{\pi D^2} \frac{dA}{dt} = \frac{D}{Ee} \frac{dp}{dt} \tag{6.14}$$

The term $(1/\rho)(d\rho/dt)$ is related to the compressibility of the liquid. Referring to (6.5),

$$\frac{\delta \rho}{\rho} = \frac{\delta p}{K}$$

therefore

$$\frac{1}{\rho} \frac{d\rho}{dt} = \frac{dp}{dt} \frac{1}{K} \tag{6.15}$$

Substituting (6.14) and (6.15) into (6.13),

$$\frac{D}{Ee} \frac{dp}{dt} + \frac{1}{K} \frac{dp}{dt} + \frac{\partial u}{\partial x} = \left(\frac{D}{Ee} + \frac{1}{K} \right) \frac{dp}{dt} + \frac{\partial u}{\partial x} = 0$$

Dividing by $\rho [(D/Ee) + (1/K)]$ and using (6.12),

$$\frac{1}{\rho} \frac{dp}{dt} + \frac{1}{\rho [(D/Ee) + (1/K)]} \frac{\partial u}{\partial x} = \frac{1}{\rho} \frac{dp}{dt} + c^2 \frac{\partial u}{\partial x} = 0 \tag{6.16}$$

The momentum equation.   For the control volume shown in Figure 6.6b,

$$\text{Sum of forces} = \text{rate of change of momentum}$$

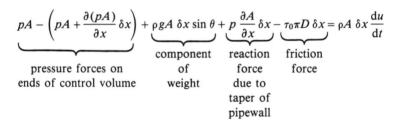

$$pA - \left(pA + \frac{\partial(pA)}{\partial x}\,\delta x\right) + \rho g A\,\delta x\,\sin\theta + p\,\frac{\partial A}{\partial x}\,\delta x - \tau_0 \pi D\,\delta x = \rho A\,\delta x\,\frac{du}{dt}$$

| pressure forces on ends of control volume | component of weight | reaction force due to taper of pipewall | friction force |

This can be slightly simplified by taking

$$\frac{\partial(pA)}{\partial x} = A\,\frac{\partial p}{\partial x} + p\,\frac{\partial A}{\partial x}$$

and by dividing throughout by $\rho A\,\delta x$ $(A = \pi D^2/4)$, which leads to

$$-\frac{1}{\rho}\frac{\partial p}{\partial x} + g\sin\theta - \frac{4\tau_0}{\rho D} = \frac{du}{dt} \qquad (6.17)$$

Equations (6.16) and (6.17) must be solved simultaneously. However, before proceeding to the solution method, the equations will be restated in terms of the head H, rather than pressure, since engineers conventionally use head.

For equation (6.16). From the definition of a total differential:

$$\frac{dp}{dt} = \frac{\partial p}{\partial x}\frac{dx}{dt} + \frac{\partial p}{\partial t} = \frac{\partial p}{\partial x}u + \frac{\partial p}{\partial t}$$

Also,

$$p = \rho g(H - z)$$

therefore

$$\frac{\partial p}{\partial x} \simeq \rho g\left(\frac{\partial H}{\partial x} - \frac{\partial z}{\partial x}\right) \text{ and } \frac{\partial p}{\partial t} \simeq \rho g\left(\frac{\partial H}{\partial t} - \frac{\partial z}{\partial t}\right)$$

But

$$-\frac{\partial z}{\partial x} = \sin\theta$$

and

$$\frac{\partial z}{\partial t} = 0$$

(as the pipe is presumably stationary!) so dividing through by g and substituting in (6.16),

$$u \frac{\partial H}{\partial x} + u \sin \theta + \frac{\partial H}{\partial t} + \frac{c^2}{g} \frac{\partial u}{\partial x} = 0 \tag{6.18}$$

For equation (6.17). The same substitution for $\partial p/\partial x$ may be made as above. From the Darcy equation,

$$\tau_0 = \rho \lambda u^2/8$$

However, under surge conditions the direction of u may be positive (i.e. in its original direction of flow) or negative. In order to obtain the correct sign for τ_0 it is usual to write

$$\tau_0 = \rho \lambda u \, | \, u \, |/8$$

where $| \, u \, |$ is the modulus (positive value) of u
Substituting in (6.17) for $\partial p/\partial x$ and τ_0,

$$g \frac{\partial H}{\partial x} + \frac{\lambda u \, | \, u \, |}{2D} + \frac{du}{dt} = 0 \tag{6.19}$$

6.5 The method of characteristics

It will be noted that (6.18) and (6.19) are both equal to zero. If (6.18) is labelled 'L_1', and (6.19) 'L_2', then the statement $L_1 + CL_2 = 0$ must also be valid (C is some arbitrary multiplier). Therefore

$$\left(u \frac{\partial H}{\partial x} + u \sin \theta + \frac{\partial H}{\partial t} + \frac{c^2}{g} \frac{\partial u}{\partial x} \right) + C \left(g \frac{\partial H}{\partial x} + \frac{\lambda u \, | \, u \, |}{2D} + \frac{du}{dt} \right) = 0$$

Since

$$\frac{du}{dt} = u \frac{\partial u}{\partial x} + \frac{\partial u}{\partial t}$$

the equation may be rearranged as

$$\left[\frac{\partial H}{\partial x}(u + Cg) + \frac{\partial H}{\partial t} \right] + C \left[\frac{\partial u}{\partial x} \left(u + \frac{c^2}{Cg} \right) + \frac{\partial u}{\partial t} \right] + u \sin \theta + \frac{C \lambda u \, | \, u \, |}{2D} = 0 \tag{6.20}$$

This is the basic 'characteristic equation'.

The two sets of terms bracketed [ ] may be further simplified. If $(u + Cg) = dx/dt$, then the first bracketed term becomes

$$\left\{ \frac{\partial H}{\partial x}\frac{dx}{dt} + \frac{\partial H}{\partial t} \right\} = \frac{dH}{dt} \qquad (6.21)$$

If $(u + (c^2/Cg)) = dx/dt$, then the second bracketed term becomes

$$\left\{ \frac{\partial u}{\partial x}\frac{dx}{dt} + \frac{\partial u}{\partial t} \right\} = \frac{du}{dt} \qquad (6.22)$$

Substituting for (6.21) and (6.22) in (6.20),

$$\frac{dH}{dt} + C\frac{du}{dt} + u\sin\theta + \frac{C\lambda u|u|}{2D} = 0$$

or,

$$dH + C\,du + u\sin\theta\,dt + \frac{C\lambda u|u|}{2D}\,dt = 0 \qquad (6.23)$$

For (6.23) to be true, then

$$\frac{dx}{dt} = (u + Cg) = \left(u + \frac{c^2}{Cg} \right) \qquad (6.24)$$

and, for this to be so,

$$C = \pm c/g \qquad (6.25)$$

Therefore, from (6.24), if $u \ll c$, $dx/dt \to c^2/Cg$.

What do these equations mean? Well, if a graph of events is depicted in the t–x plane, where t represents time (say from the start of a valve closure) and x represents distance (say along a pipeline), then a figure such as Figure 6.7 results. The differential dx/dt represents the slope of the lines, and may be positive or negative, as shown. The implication of this is that if conditions (head, velocity, etc.) are known at Station 1 at time t, then the conditions at Station 2 may be calculated for time $(t + \delta t)$ using (6.23) and (6.25). Alternatively, the conditions at Station 2 could also be calculated for time $(t + \delta t)$ if conditions were known at Station 3 for time t. Both of these calculations must give the same answer.

This is the essence of the 'method of characteristics'.

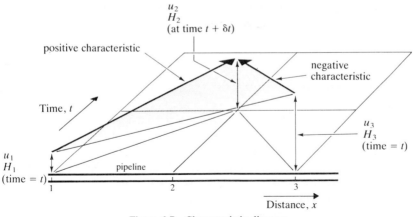

Figure 6.7   Characteristic diagram.

Finite difference form of the characteristic equation

The characteristic equation (6.23) may be stated in the finite difference form which is adopted for computer application:

$$\delta H + C\,\delta u + u \sin\theta\,\delta t + C\lambda u\,|u|\,\delta t/2D = 0 \tag{6.26}$$

For the positive characteristic line $\delta H = (H_2 - H_1)$ and $\delta u = (u_2 - u_1)$ so (6.26) becomes

$$(H_2 - H_1) + C(u_2 - u_1) + u_1 \sin\theta\,\delta t + \frac{C\lambda u_1\,|u_1|}{2D}\,\delta t = 0 \tag{6.27}$$

Similarly, for the negative characteristic with $\delta H = (H_2 - H_3)$, etc.,

$$(H_2 - H_3) - C(u_2 - u_3) + u_3 \sin\theta\,\delta t - \frac{C\lambda u_3\,|u_3|}{2D}\,\delta t = 0 \tag{6.28}$$

Adding (6.27) and (6.28) and rearranging,

$$H_2 = \frac{1}{2}\left(H_1 + H_3 - C(u_3 - u_1) - \sin\theta(u_3 + u_1)\,\delta t - \frac{C\lambda}{2D}(u_1\,|u_1| - u_3\,|u_3|)\,\delta t\right) \tag{6.29}$$

while subtracting (6.27) from (6.28) yields

$$u_2 = \frac{1}{2}\left(\frac{H_1 - H_3}{C} + u_1 + u_3 + \frac{\sin\theta(u_3 - u_1)\,\delta t}{C} - \frac{\lambda}{2D}(u_1\,|u_1| + u_3\,|u_3|)\,\delta t\right) \tag{6.30}$$

Note that H_2 and u_2 are conditions at $(t + \delta t)$ at Station 2, while H_1, H_3, u_1 and u_3 refer to conditions at time t at Stations 1 and 3 (in accordance with Figure 6.7).

6.6 Characteristic equations for system boundaries

It was stated at the beginning of the chapter that as long as the liquid in a pipeline is in equilibrium (stationary or in steady flow), then surge conditions cannot arise. It is the rapid alteration of conditions at a boundary (such as a valve or pump) which generates shock waves. Reflections of the shock waves also occur only at boundaries (another valve, a change of pipe geometry or material, a pipe–reservoir interface, etc.). The characteristic equations for some of the important boundaries are now derived.

Valve (Fig. 6.8a)

A valve is an orifice of variable area and so $Q = C_d A_V \sqrt{2gH}$ (where A_V is the valve orifice area and H is the pressure head at the valve). In the following analysis, the subscript 'o' refers to the fully open setting. Thus,

$$\frac{\text{discharge with valve partly open}}{\text{discharge with valve fully open}} = \frac{Q}{Q_o} = \frac{C_d A_V \sqrt{2gH}}{C_d A_{V,o} \sqrt{2gH_o}} = A_R \sqrt{\frac{H}{H_o}}$$

where $A_R = A_V / A_{V,o}$. Hence, the pipeline velocity ratio is

$$\frac{u}{u_o} = \frac{Q/A}{Q_o/A} = A_R \sqrt{\frac{H}{H_o}} \tag{6.31}$$

where u and u_o are the velocities in the pipe just upstream of the valve.

From Figure 6.8a, the valve is assumed to lie at the downstream end of the pipeline, and therefore the positive characteristic equation (6.27) is applicable. Equations (6.27) and (6.31) are solved simultaneously. From (6.31)

$$\left(\frac{u_{2\,(t+\delta t)}}{u_{2\,(o)}}\right)^2 = A_R^2 \frac{H_{2\,(t+\delta t)}}{H_{2\,(o)}}$$

From (6.31) therefore

$$H_{2\,(t+\delta t)} = \left(\frac{u_{2\,(t+\delta t)}}{u_{2\,(o)}}\right)^2 \frac{H_{2\,(o)}}{A_R^2}$$

It has been necessary to use a double subscript notation here since the velocity and head at time $(t + \delta t)$ are a function (a) of the conditions at the

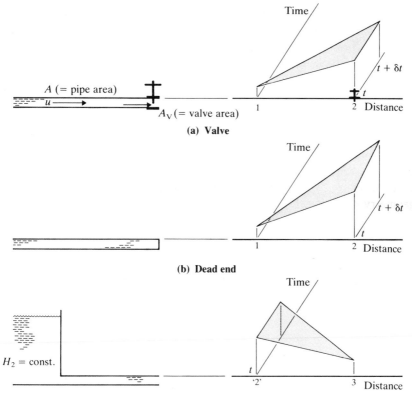

(a) Valve

(b) Dead end

$H_2 = \text{const.}$

(c) Reservoir

Figure 6.8 Typical system boundaries.

commencement of closure (time 0), (b) of conditions at the beginning of the time interval (time t) and (c) of station number (i.e. position). Substituting in (6.27),

$$\left[\left(\frac{u_{2\,(t+\delta t)}}{u_{2\,(o)}}\right)^2 \frac{H_{2\,(o)}}{A_R^2} - H_{1\,(t)}\right] + C(u_{2\,(t+\delta t)} - u_{1\,(t)})$$

$$+ u_{1\,(t)}(\sin\theta)\,\delta t + \frac{C\lambda u_{1\,(t)}\,|u_{1\,(t)}|}{2D}\,\delta t = 0$$

Multiplying through by $u_{2(o)}^2 A_R^2/H_{2(o)}$ and rearranging,

$$u_{2(t+\delta t)}^2 + \frac{Cu_{2(o)}^2 A_R^2}{H_{2(o)}}\,u_{2(t+\delta t)}$$

$$+ \frac{u_{2(o)}^2 A_R^2}{H_{2(o)}}\left(u_{1(t)}(\sin\theta)\,\delta t + \frac{C\lambda u_{1(t)}\,|u_{1(t)}|}{2D}\,\delta t - Cu_{1(t)} - H_{1(t)}\right) = 0$$

This is a quadratic equation in $u_{2(t+\delta t)}$ which may be solved to yield

$$u_{2(t+\delta t)} = -\frac{Cu_{2(0)}^2 A_R^2}{2H_{2(0)}} + \left[\left(\frac{Cu_{2(0)}^2 A_R^2}{2H_{2(0)}}\right)^2 \right.$$
$$\left. -\frac{u_{2(0)}^2 A_R^2}{H_{2(0)}}\left(u_{1(t)}(\sin\theta)\,\delta t + \frac{C\lambda u_{1(t)}|u_{1(t)}|}{2D}\,\delta t - Cu_{1(t)} - H_{1(t)}\right)\right]^{\frac{1}{2}} \quad (6.32)$$

In other words, conditions known at the beginning of a time interval are used to predict conditions at the end of the interval.

The double subscript notation is rather clumsy in appearance, so in the boundary characteristics which follow, H_2 represents $H_{2(t+\delta t)}$, H_1 represents $H_{1(t)}$, H_3 represents $H_{3(t)}$, and so on (refer to Fig. 6.7).

Dead end (Fig. 6.8b)

This could be a closed valve or a stopped-off pipe, therefore $u_2 = 0$. Like the valve, the dead end has a positive characteristic, so (6.27) is used:

$$H_2 - H_1 + C(0 - u_1) + u_1(\sin\theta)\,\delta t + \frac{C\lambda u_1|u_1|}{2D}\,\delta t = 0$$

Therefore

$$H_2 = H_1 + Cu_1 - u_1(\sin\theta)\,\delta t - \frac{C\lambda u_1|u_1|}{2D}\,\delta t \quad (6.33)$$

Remember that subscript '1' refers to conditions at Station 1 and time t, whereas subscript '2' refers to Station 2 at time $t + \delta t$.

Reservoir (Fig. 6.8c)

At the pipe–reservoir interface, the head H_2 must remain constant. The reservoir is here assumed to lie at the upstream end of the pipeline. A negative characteristic solution is therefore assumed with Stations 2 and 3 positioned so as to correspond to (6.28). In running the procedure on a computer, conditions at Station 3 would already be calculated, and H_2 would be a known (and constant) magnitude, so (6.28) may be rearranged as

$$u_2 = \frac{H_2 - H_3}{C} + u_3 + \frac{u_3(\sin\theta)\,\delta t}{C} - \frac{\lambda u_3|u_3|}{2D}\,\delta t \quad (6.34)$$

where u_2 is at time $t + \delta t$, and H_3 and u_3 are at time t.

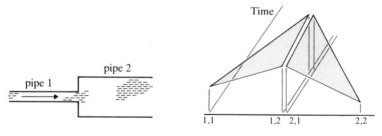

Figure 6.9 Change in pipe geometry or material.

Change in pipe geometry or material (Fig. 6.9)

Note that:
(a) It is necessary to use double subscript notation, e.g. head in pipe 1 at
    Station 2 is $H_{1,2}$, etc.;
(b) Pipe 1 is on + characteristic (Equation (6.27)),
    Pipe 2 is on − characteristic (Equation (6.28));
(c) By continuity,

$$A_1 u_{1,2} = A_2 u_{2,1} \tag{6.35}$$

(d) Ignoring losses,

$$p_{1,2} \simeq p_{2,1} \tag{6.36}$$

Rewriting (6.27),

$$(H_{1,2} - H_{1,1}) + C_1(u_{1,2} - u_{1,1}) + u_{1,1}(\sin \theta) \, \delta t + \frac{C_1 \lambda_1 u_{1,1} |u_{1,1}|}{2D_1} \delta t = 0$$

From (6.36), $H_{1,2} = H_{2,1}$, and therefore

$$u_{1,2} = \frac{(H_{1,1} - H_{2,1})}{C_1} + u_{1,1} - \frac{u_{1,1} \sin \theta}{C_1} \delta t - \frac{\lambda_1 u_{1,1} |u_{1,1}|}{2D_1} \delta t \tag{6.37}$$

By a similar procedure, (6.28) becomes

$$u_{2,1} = \frac{(H_{2,1} - H_{2,2})}{C_2} + u_{2,2} + \frac{u_{2,2} \sin \theta}{C_2} \delta t - \frac{\lambda_2 u_{2,2} |u_{2,2}|}{2D_2} \delta t \tag{6.38}$$

Substituting (6.37) and (6.38) into (6.35) and rearranging,

$$H_{2,1}\left(\frac{A_2}{C_2} + \frac{A_1}{C_1}\right) = A_1\left(\frac{H_{1,1}}{C_1} + u_{1,1} - \frac{u_{1,1} \sin \theta}{C_1} \delta t - \frac{\lambda_1 u_{1,1} |u_{1,1}|}{D_1} \delta t\right)$$

$$- A_2\left(-\frac{H_{2,2}}{C_2} + u_{2,2} + \frac{u_{2,2} \sin \theta}{C_2} \delta t - \frac{\lambda_2 u_{2,2} |u_{2,2}|}{D_2} \delta t\right) \tag{6.39}$$

Equation (6.39) presupposes changes in material and geometry.

If only the material changes, then $D_1 = D_2$ and $A_1 = A_2$.

If only the geometry changes, then $C_1 = C_2$.

Note that terms with subscripts '1,2' and '2,1' refer to time $t + \delta t$, whereas terms with subscripts '1,1' and '2,2' refer to time t.

As was stated at the outset of this section, the equations can appear rather awesome, so to illustrate the application of the equations a problem will now be partly solved. In order to elucidate the approach, the solution has been produced manually. However, this is emphatically not normal. By their very nature, the equations lend themselves to the computer-based approach. If the reader possesses a microcomputer, it is not too difficult to set up an appropriate program to solve this problem.

Example 6.3   Application of the method of characteristics

A reservoir supplies water through a horizontal pipeline, 1 km long and 500 mm internal diameter (Fig. 6.10). The discharge is controlled by a valve at the downstream end of the pipeline. If the valve closes in 4 s and gives a linear retardation, estimate the pressure rise at the valve and at the mid-point of the pipe over an 8 s period. The reservoir head is 100 m and the head at the valve before commencement of valve closure is 4.75 m. Friction factor $\lambda = 0.012$, and speed of sound $c = 1000$ m/s.

Solution

A time interval must first be adopted. The interval should normally be some convenient fraction of L/c. Here an interval of 0.5 s is used. This corresponds to the time for the shock wave to travel halfway along the pipe.

Initial conditions ($t = 0$):

$$h_f = 100 - 4.75 = \frac{0.012 \times 1000 \times u_0^2}{2 \times 9.81 \times 0.5}$$

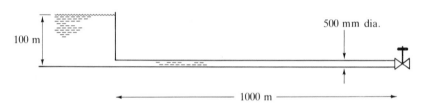

Figure 6.10   Reservoir and pipeline (Example 6.3).

so

$$u_0 = 8.824 \text{ m/s}$$

Therefore,

$$H_1 = 100 \text{ m} \qquad H_2 = 52.38 \text{ m} \qquad H_3 = 4.75 \text{ m}$$

$$C = \frac{c}{g} = \frac{1000}{9.81} \simeq 100$$

Pipe is horizontal, so $\theta = 0$.

First step. $t = 0.5$ s after commencement of valve closure:

$$\text{valve area reduction} = 0.5/4 = 0.125$$

Therefore

$$A_R = 0.875$$

Using (6.32), determine the velocity at the valve (for the computer solution the valve is, in effect, assumed to close in a series of discrete 'steps' or 'jerks', one step at each time interval).

$$u_3 = -\frac{100 \times 8.824^2 \times 0.875^2}{2 \times 4.75} + \left[\left(\frac{100 \times 8.824^2 \times 0.875^2}{2 \times 4.75} \right)^2 \right.$$
$$\left. -\frac{8.824^2 \times 0.875^2}{4.75} \left(\frac{100 \times 0.012 \times 8.824 \,|\, 8.824 \,|\, \times 0.5}{2 \times 0.5} - (100 \times 8.824) - 52.38 \right) \right]^{\frac{1}{2}}$$
$$= 8.81 \text{ m/s}$$

Notice that all of the data for evaluating the equation are derived from conditions at the preceding time ($t = 0$).

The pressure head is determined from (6.27), since the valve is assumed to have a positive characteristic. Rearranging (6.27) into a suitable form with appropriate subscripts,

$$H_3 = H_2 - C(u_3 - u_2) - u_2 \sin \theta \, \delta t - \frac{C\lambda u_2 \,|\, u_2 \,|}{2D} \delta t$$

Substituting for each term,

$$H_3 = 52.38 - 100(8.824 - 8.824) - 0 - \frac{100 \times 0.012 \times 8.824 \,|\, 8.824 \,|\, \times 0.5}{2 \times 0.5}$$

$$= 5.66 \text{ m}$$

Because of the assumption about the step closure pattern at the valve, the first shock wave has only just been generated. Conditions at Stations 1 and 2 therefore remain as at the preceding time. Summarising,

$$H_1 = 100 \text{ m} \qquad H_2 = 52.38 \text{ m} \qquad H_3 = 5.66 \text{ m}$$

$$u_1 = 8.824 \text{ m/s} \qquad u_2 = 8.824 \text{ m/s} \qquad u_3 = 8.81 \text{ m/s}$$

Second step. $t = 1.0$ s:

$$\text{valve area reduction} = 1/4 = 0.25$$

Therefore

$$A_R = 0.75$$

$$u_3 = -\frac{100 \times 8.81^2 \times 0.75^2}{2 \times 5.66} + \left[\left(\frac{100 \times 8.81^2 \times 0.75^2}{2 \times 5.66} \right)^2 - \frac{8.81^2 \times 0.75^2}{5.66} \right.$$

$$\left. \left(\frac{100 \times 0.012 \times 8.824 \mid 8.824 \mid \times 0.5}{2 \times 0.5} - (100 \times 8.824) - 52.38 \right) \right]^{\frac{1}{2}}$$

$$= 8.79 \text{ m/s}$$

$$H_3 = 52.38 - 100(8.79 - 8.824) - 0 - \frac{100 \times 0.012 \times 8.824 \mid 8.824 \mid \times 0.5}{2 \times 0.5} = 9.06 \text{ m}$$

The shock wave which was generated at the valve at $t = 0.5$ s has just reached the mid-point of the pipe. The velocity and head at the mid-point are evaluated using (6.30) and (6.29), respectively:

$$u_2 = \frac{1}{2} \left[\frac{100 - 5.66}{100} + 8.824 + 8.81 - \frac{0.012}{2 \times 0.5} (8.824 \mid 8.824 \mid + 8.81 \mid 8.81 \mid)0.5 \right]$$

$$= 8.82 \text{ m/s}$$

$$H_2 = \frac{1}{2} \left[100 + 5.66 - 100(8.81 - 8.824) - 0 \right.$$

$$\left. - \frac{100 \times 0.012(8.824 \mid 8.824 \mid - 8.81 \mid 8.81 \mid) \times 0.5}{2 \times 0.5} \right]$$

$$= 53.6 \text{ m}$$

Summarising:

$$H_1 = 100 \text{ m} \qquad H_2 = 53.6 \text{ m} \qquad H_3 = 9.06 \text{ m}$$

$$u_1 = 8.824 \text{ m/s} \qquad u_2 = 8.82 \text{ m/s} \qquad u_3 = 8.79 \text{ m/s}$$

Third step. $t = 1.5$ s

$$A_R = 0.625$$

From (6.23),

$$u_3 = 8.76 \text{ m/s (at the valve)}$$

From (6.27) (in its rearranged form),

$$H_3 = 12 \text{ m (at the valve)}.$$

At the pipe mid-point, $u_2 = 8.79$ m/s (from (6.30)) and $H_2 = 55.82$ m/s (from (6.29)).

The shock wave generated at $t = 0.5$ s has now reached the reservoir boundary, so the velocity must change correspondingly. From (6.34),

$$u_1 = \frac{H_1 - H_2}{C} + u_2 + 0 - \frac{\lambda u_2 |u_2|}{2D} \delta t$$

$$= \frac{100 - 53.6}{100} + 8.82 - \frac{0.012 \times 8.82 |8.82| \times 0.5}{2 \times 0.5}$$

$$= 8.817 \text{ m/s}$$

Summarising,

$$H_1 = 100 \text{ m} \qquad H_2 = 55.82 \text{ m} \qquad H_3 = 12 \text{ m}$$

$$u_1 = 8.817 \text{ m/s} \qquad u_2 = 8.79 \text{ m/s} \qquad u_3 = 8.76 \text{ m/s}$$

Notes on Example 6.3

(a) This example has been set up to highlight certain aspects of surge analysis, and not as an example of good design. It would not normally be considered sensible to run a system with such a high initial velocity, precisely because this would lead to surge problems.

(b) The solution has been completed by computer, and the results are presented in graphical form in Figure 6.11. The violent fluctuation in pressure should be noted. This would almost certainly cause a rupture in the pipeline.

(c) In order to avoid excessive complexity, the equations used here have been developed on the basis of the assumption that the liquid always forms a continuum. This is not necessarily true. Once the negative (gauge) pressure approaches the vapour pressure of the liquid, a vapour 'cavity' (or 'cavities') will begin to form. Formation usually begins at the controlling

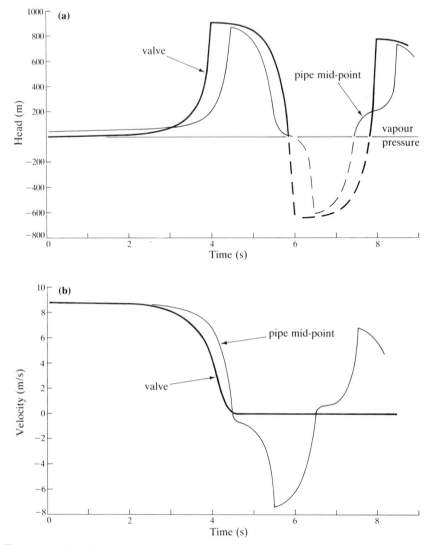

Figure 6.11   Graphical representation of solution to Example 6.3. (a) Variation of head with time. (b) Variation of velocity with time.

boundary (the valve in this case) and a measurable 'gap' thus develops between the boundary and the column of liquid in the pipeline. When the next positive pressure wave returns, the liquid is driven back across the cavity and strikes the boundary. The resulting impact can generate tremendous surge pressures (sometimes greater than the original surge). Thus, the negative pressures are limited to the liquid vapour pressure, but the peak positive pressures are not limited (except by the failure of the pipe).

(d) The presence of minute quantities of dissolved gas in the liquid will considerably influence the propagation of shock waves. This has not been allowed for in the equations, or in the solution to the example.

(e) For the sake of simplicity, the characteristic equations have been developed for a 'three-station' system. This also applies to the example. Real systems may require a greater number of stations, depending on the complexity of the system (number of branches, number of boundaries, etc.). A computer program must therefore be written in sufficiently general terms. It is usual to write the heads as H (I), velocities as u (I), etc., where 'I' is a counter. The distance between stations must be selected with care. In many cases it is best to use a 'time centred' system, i.e. positions of stations are selected such that the time ($t = x/c$) for the passage of a shock wave between any two stations is constant for the whole system.

6.7 Concluding remarks

(1) The intensity of surge pressures depends on the rate of change at a (variable) controlling boundary. If the rate of change is sufficiently fast, then the liquid will behave in the elastic mode. Therefore, it is important to have some general rules which indicate whether shock waves are likely to be generated. To illustrate this point, consider the effect of the time T required for complete closure of a valve at the downstream end of a pipeline of length L:

(a) If $T < 2L/c$, then the maximum surge pressure is incurred, since a returning decompression wave cannot reach the valve before complete closure. Closure is therefore effectively instantaneous;
(b) at the other extreme, if $T > 20L/c$, then there will be little or no shock wave activity, and the liquid will behave in the incompressible mode – this is 'slow' closure;
(c) for $2L/c < T < 20L/c$, conditions are intermediate between (a) and (b), above – this is known as a 'rapid' valve closure.

(2) In this chapter it has been assumed that the surge conditions have been generated by the closing of a valve. However, any variable boundary can produce a surge. An example of a boundary which is variable (but might not be thought of as such) is a pump under power failure conditions. Downstream of the pump, a cavitation bubble may occur, the effects of which have already been discussed.

(3) In designing a major pipeline system, it is necessary to evaluate the surge pressures under all foreseeable changes of boundary conditions. If excessively high values are predicted, then it is necessary to incorporate a system for reducing the surge to an acceptable level. This aspect of unsteady flow is discussed in Section 12.6.

(4) To assist those who wish to write a computer program based on the method of characteristics, a flow diagram is presented in Figure 6.12. Further information on this topic is available in Wylie (1983) and in Streeter and Wylie (1978).

(5) The treatment of unsteady flows as presented here is only introductory. For a more advanced approach, reference may be made to Fox (1977) and to Streeter and Wylie (1978). Parmakian (1963) gives some alternative methods, together with a number of worked examples. Kranenburg (1974) may be consulted regarding the effects of cavitation and gas release during surge.

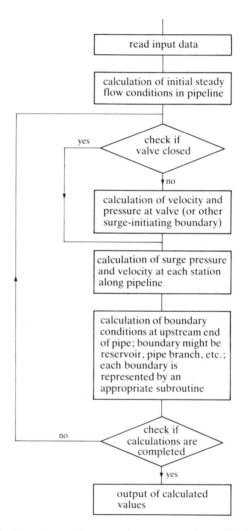

Figure 6.12   Flow diagram for computer program using method of characteristics.

References and further reading

Fox, J. A. 1977. *Hydraulic analysis of unsteady flow in pipe networks*. London: Macmillan.

Kranenburg, C. 1974. Gas release during transient cavitation in pipes. *Am. Soc. Civ. Engrs, J. Hydraulics Divn.* **100** (HY 10) 1383–98.

Parmakian, J. 1963. *Waterhammer analysis*. New York: Dover.

Streeter, V. L. and E. B. Wylie, 1978. *Hydraulic transients*. New York: McGraw-Hill.

Wylie, E. B. 1983. The microcomputer and pipeline transients. *Am. Soc. Civ. Engrs, J. Hydraulic Engng* **109** (12) 1723–39.

7

Hydraulic machines

7.1 Classification of machines

Civil engineers are not usually involved in the design of hydraulic machines. However, they are frequently called upon to design systems of which these machines form an integral part. Examples of such systems are pipelines for water supply, for surface and foul water drainage, for hydroelectric schemes, and so on. It is therefore important that the civil engineer should have a sound basic understanding of hydraulic machines and their applications.

It is convenient to divide these machines into two categories:

(a) pumps, i.e. machines which transform a power input (e.g. from an electric motor) into an hydraulic power output;
(b) turbines, i.e. machines which transform an hydraulic power input into a mechanical power output which is mainly utilised for generation of electrical power.

Machines can be further subdivided into 'positive displacement' units and 'continuous flow' units. Typical positive displacement units are piston and diaphragm pumps, which are often used for pumping out groundworks. This chapter concentrates on the continuous flow units, which are by far the most widely used for permanent installations.

7.2 Continuous flow pumps

Continuous flow or 'rotodynamic' pumps comprise two elements:

(a) a rotating element (impeller) which transfers energy to the fluid;
(b) some form of casing which encloses the impeller and to which the pipeline is connected.

In many cases the impeller is driven by an electric motor.

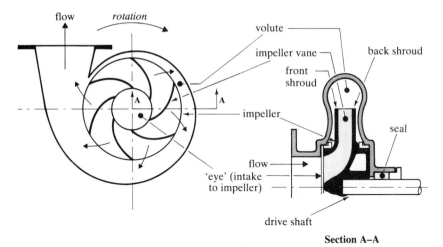

Figure 7.1 Centrifugal pump.

The radial flow (centrifugal) pump

This is one of the most widely utilised hydraulic machines (Fig. 7.1). Like most continuous flow machines, it has the advantage of being able to pass either liquids or liquids with suspended solids (there are, of course, limits to the size and nature of the solids). These machines can thus be applied in such environments as sewage plants, drainage schemes and irrigation schemes. The casing must initially be full of liquid (i.e. the pump must be 'primed') in order to function. When the impeller rotates at its design speed, it imparts a radial component of motion to the volume of fluid trapped in the passages between the vanes. That fluid therefore passes outwards into the outer part of the casing (the volute) and then out into the high pressure (delivery) pipeline. As fluid continually evacuates the impeller, replenishment fluid is continually drawn from the low pressure (suction) pipeline into the casing.

Energy transfer in radial flow pumps. An insight into the energy transfer process may be obtained by producing vector diagrams representing the flow as it enters and as it leaves the impeller. The flow through a radial flow pump impeller provides a relatively straightforward example. It is usually assumed that the incoming liquid enters the casing axially and then turns outwards, so that the flow is in the radial plane as it approaches the impeller. The curvature of the impeller vanes is designed to produce a change in the direction of the flow as the fluid passes outwards through the impeller. Vectors may be constructed which represent:

(a) the absolute velocity of flow at a point, i.e. the velocity relative to a stationary point;

(b) the relative velocity of flow, i.e. the velocity as 'seen' from a point on the impeller.

The principal velocity vectors are

(a) the tangential velocity of the impeller itself at radius r

$$V_I = 2\pi rn$$

where n is the rotational speed of impeller (rev/s);
(b) the absolute velocity V_A of the fluid at radius r;
(c) the relative velocity, V_R (i.e. the velocity of the fluid relative to the impeller);
(d) the tangential component of V_A is V_W, often known as the 'whirl' velocity;
(e) the radial component of V_A is V_F.

In order to produce the vector diagrams, it is simplest to assume that, as the flow passes through the impeller, the relative velocity traces a path which is everywhere congruent to the vane. Vector diagrams (Fig. 7.2) can then be drawn.

Based on these diagrams, equations describing the energy transfer are now developed. At any radius r in the impeller, the flow will possess a tangential component of velocity V_W. It will therefore have a corresponding tangential momentum flux, $\rho Q V_W$. The moment of this flux about the impeller axis is $\rho Q V_W r$. Just as force equals change of momentum flux, so torque equals change of moment of momentum flux. In this case, the change takes place between radii r_1 and r_2, so

$$\text{torque } (T) = \rho Q V_{W2} r_2 - \rho Q V_{W1} r_1$$

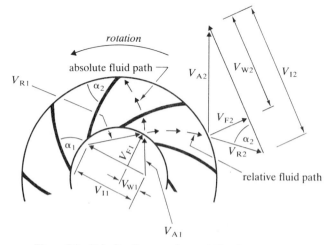

Figure 7.2   Velocity diagrams for centrifugal pump.

$$\text{power} = 2\pi n T = 2\pi n (\rho Q V_{w2} r_2 - \rho Q V_{w1} r_1)$$

But $2\pi rn = V_I$, therefore

$$\text{power} = \rho Q (V_{w2} V_{I2} - V_{w1} V_{I1}) \tag{7.1}$$

Therefore ideal energy head, H_{ideal}, imparted to the fluid is

$$H_{\text{ideal}} = \frac{\text{power}}{\rho g Q} = \frac{1}{g}(V_{w2} V_{I2} - V_{w1} V_{I1}) \tag{7.2}$$

If V_{A1} is radial in direction, then $V_{w1} = 0$.

Energy losses in radial flow pumps. The overall rate of energy transfer actually attained is always less than the ideal values predicted by (7.2). There are a number of contributory factors, of which the major ones are:

(a) During its passage through the impeller and volute, the fluid will suffer both frictional and local energy losses.

(b) The liquid in the volute is at a much higher pressure than the liquid at the entry so, unless the internal sealing arrangements are extremely good, some of the pressurised fluid leaks back to the impeller eye and then recirculates through the impeller. This clearly wastes a certain amount of energy.

(c) In the vector analysis, it had to be assumed that the flow through the impeller was axisymmetric. However, in order to produce a tangential acceleration, the force (and therefore pressure) must be higher on the front of each vane than on the back (Fig. 7.3). At any given radius in the impeller, the pressure is therefore non-uniform. This results in a non-uniform relative velocity distribution, with the highest velocities at the back of the vanes and the lowest velocities at the front of the vanes. The effect of this is to distort the pattern of the vectors at the exit from the impeller and to reduce V_{w2}.

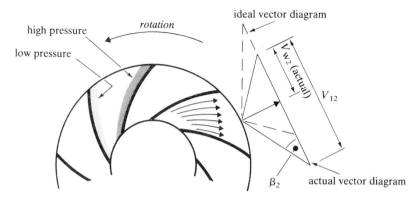

Figure 7.3 Effect of pressure distribution on velocity diagram for centrifugal pump.

(d) If the pressure in the suction line is too low, cavitation problems may arise. The low pressure zone near the backs of the vanes is particularly prone to cavitation, and if the problem is severe, then the pump performance will fall off dramatically.

Example 7.1   Centrifugal pump

A centrifugal pump is required to deliver 350 l/s against a 15 m head. The impeller is of 400 mm outer diameter and 200 mm inner diameter, and rotates at 1400 rev/min. The intake and delivery pipelines are of the same diameter. The water flows in a radial direction as it approaches the impeller, and the radial velocity component remains constant at 5 m/s as the flow passes through the impeller. Assume that the hydraulic losses account for a 10% head loss. Calculate the required vane angles at the outlet of the impeller. If the mechanical efficiency is 88%, estimate the power requirement for an electric motor.

Solution

(a) The ideal head = $15/0.9 = 16.67$ m, since the 15 m head was a net head (i.e. head after deduction of losses), and the hydraulic efficiency = $(100 - 10)\% = 90\%$.

(b) From (7.2)

$$H_{ideal} = (1/g)(V_{w2}V_{12} - V_{w1}V_{11})$$

Since the incoming flow is in a radial direction, $V_{11} = 0$. Therefore,

$$H_{ideal} = 16.67 \text{ m} = (1/g)(V_{w2}V_{12})$$

Now,

$$V_{12} = 2\pi r_2 n = 2\pi \times \frac{0.4}{2} \times \frac{1400}{60} = 29.32 \text{ m/s}$$

Hence

$$V_{w2} = 5.58 \text{ m/s}$$

(c) From Figure 7.2, the outlet vane angle may be derived by taking

$$\tan \alpha_2 = \frac{V_{12} - V_{w2}}{V_{F2}} = \frac{29.32 - 5.58}{5} = 4.748$$

therefore,

$$\alpha_2 = 78.1°$$

(d) Power absorbed by impeller is

$$\rho g Q H_{\text{ideal}} = 1000 \times 9.81 \times 0.35 \times 16.67$$

$$= 57\,236 \text{ W}$$

Since mechanical efficiency is 88%, the power absorbed by the pump will be

$$\frac{57\,236}{0.88} = 65\,041 \text{ W}$$

say, 65 kW.

Axial flow machines

The axial flow pump comprises an impeller in the form of a propeller sited on the pipeline axis (Fig. 7.4). The pipeline itself therefore forms a 'casing'. A cross section through one of the propeller blades has a shape which resembles a section through an aerofoil (aircraft wing). The section is arranged at some angle to the plane of the pipe cross section. This angle and the blade section shape combine to impart the required tangential acceleration to the fluid as it passes through the rotating impeller. Vector diagrams can be constructed to represent the flow at impeller entry and exit, using exactly the same principles as for the radial flow machine. However, for almost all practical cases, the exit vector diagram varies with radius, so the procedure for estimating rate of transfer of energy is more complicated. In any case, modern design techniques are based on an adaptation of aerofoil theory known as 'cascade theory', rather than on vector diagrams. A study of cascade theory is beyond the scope of this chapter, so the discussion will be curtailed here.

For axial flow machines with fixed blades, high efficiency is attained only over a fairly narrow range of discharge. For applications where a wider operational range is required, pumps with variable blade angle are available. Such machines are clearly more expensive, since a mechanical control system is required, together with more sophisticated instrumentation.

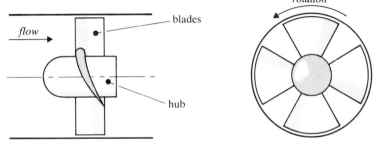

Figure 7.4 Axial flow pump.

'Mixed flow' machines

'Mixed flow' machines are, in effect, a compromise between axial and radial flow machines. The impeller induces a three-dimensional flow pattern such that the path of a fluid particle would lie approximately along the surface of a diverging cone. The performance characteristics of these units lie between those of radial and axial flow machines. Some fairly large installations have been designed to incorporate mixed flow pumps.

7.3 Performance data for continuous flow pumps

The equations developed in Section 7.2 cannot readily be extended to enable realistic estimates of pump performance to be made. Actual performance data are invariably derived from a comprehensive laboratory testing programme. Such tests provide values of pump head, H_p, and power input, P, for the range of discharge, Q, at the design speed of the impeller. The energy imparted to the fluid is $\rho g Q H_p$, so that a pump efficiency, η, may be derived:

$$\eta = \rho g Q H_p / P \tag{7.3}$$

The performance data may be presented in tabular or graphical form. Figures 7.5 and 7.6 show typical performance curves for a centrifugal pump and an axial flow pump, respectively. In this connection, some general observations may be made:

(a) The performance of a radial pump (or a fixed blade axial pump) depends substantially on the exit vane angle α_2. For many radial pumps, $\alpha_2 < 90°$, and such designs have a peak power requirement corresponding to a particular value of Q. The electrical (or other) drive motor must be rated accordingly.

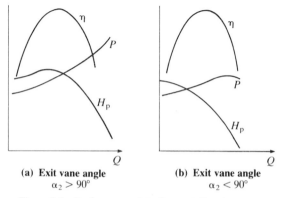

(a) Exit vane angle
     $\alpha_2 > 90°$

(b) Exit vane angle
     $\alpha_2 < 90°$

Figure 7.5 Performance data for centrifugal pump.

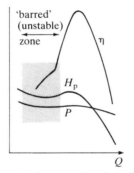

Figure 7.6 Performance data for axial pump.

(b) Where machines are designed with $\alpha_2 > 90°$, then the power requirement increases with Q. This can pose difficult motor selection problems.

(c) Pressure at the intake to a centrifugal pump should not be allowed to fall below about 3 m absolute head of water, or cavitation problems will arise. Axial machines have a much more limited suction capacity, since the flow around the blade section is very sensitive to low pressure. Too low an intake pressure therefore provokes flow separation and consequent loss of energy.

(d) The relationship between H_p and Q for some axial designs is such that there may be parts of the range which are 'barred'. The reason for this is that the pump would be prone to unsteady flow in the barred range.

(e) In order to obtain the pump performance required by the system, it is sometimes necessary to use a multiple array of pumps. The pumps may be arranged either in series or in parallel. This is discussed further in Section 12.5.

(f) In order to function satisfactorily, the pump must be matched to the characteristics of the pipeline system in which it is installed. Again, this is covered in Section 12.5.

7.4 Pump selection

An indication of the type of pump which is appropriate to a particular duty may be obtained by reference to a performance parameter known as the 'specific speed' N_S. For pumps,

$$N_S = NQ^{1/2}/H_p^{3/4}$$

The derivation and significance of N_S is discussed in Section 11.6. The 'specific speed' is not really representative of any meaningful or measurable speed in a machine, so some practitioners refer to it as 'type number', since it is used in selection of pump type. The value of N_S for a particular machine is calculated for the conditions obtaining at its point of optimum efficiency, since ideally this should coincide with the installed operating point of the pump.

Unfortunately, a wide variety of units are used in calculating values of N_S:

N may be in rad/s, rev/s or rev/min;
Q may be in m^3/s, m^3/min, l/s or l/min, UK gallons/min, US gallons/min, etc.;
H_p may be in m or ft.

The engineer needs to check this point carefully with the pump manufacturer.

Taking N in rev/min, Q in m^3/s and H in m; a very rough guide to the range of duties covered by the different machines is as follows:

$10 < N_S < 70$ centrifugal (high head, low to moderate discharge);
$70 < N_S < 165$ mixed flow (moderate head, moderate discharge);
$110 < N_S$ axial flow (low head, high discharge).

In making the selection, the points raised in the preceding section should also be borne in mind.

7.5 Hydro-power turbines

There are many hydroelectric power installations throughout the world. The various modern water turbine designs are capable of covering a very wide range of operating conditions. Thus, hydroelectric schemes may now be found in mountain, river or estuarine environments. Rudimentary water power machines have been in use for centuries, but it was not until the later part of the 19th century that efficient designs emerged. Subsequently, the range of machines has widened, but for large scale power generation, three basic types are dominant. These are the radial flow (Francis) turbine, the axial flow turbine and the Pelton turbine.

Of the three machines mentioned above, the Francis turbine (Fig. 7.7) may be regarded as a centrifugal pump operating 'in reverse'. The water enters the outer part of the casing and flows radially inwards through the rotating element (this time called a 'runner'), transferring energy to the runner. The water then drains out of the casing through the central outlet passage. There are two major differences between the turbine and the pump:

(a) The rotating element is surrounded by guide vanes, so that the direction of the absolute velocity vector of the water can be adjusted. This adjustable angle provides for optimum flow conditions over a wide

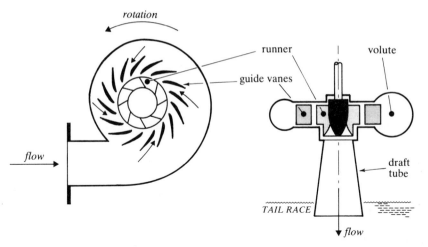

Figure 7.7   Radial flow (Francis) turbine.

range of discharge and power output (N.B. electrical requirements vary over a 24 h period).

(b) The water leaving the turbine passes through a 'draft tube', which is tapered so as to minimise loss of kinetic energy. This ensures that the maximum energy is available at the turbine.

The axial flow turbine (Fig. 7.8) may similarly be regarded as the reversal of the axial flow pump. Once again, guide vanes are installed at the turbine inlet for the same reasons as outlined above. Virtually all modern units incorporate mechanical means of varying the propeller blade angle of the runner, so that high runner efficiency may be maintained over the specified range of discharge and power output. Variable angle designs are called 'Kaplan' turbines. In recent years, a number of 'pumped storage' schemes have utilised a special type of Kaplan unit. This unit operates as a turbine

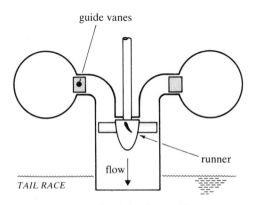

Figure 7.8   Axial flow turbine.

during periods when the demand for electricity is high. However, demand is low during the night. The blade angle is then reversed and the unit is motor-driven so as to act as a pump. Water is then pumped from low level back up to the high level reservoir in readiness for the next period of high demand.

Both the Kaplan and Francis units run full of water. There is, therefore, a continuous flow between the reservoir and the tailwater. The gross head, H, is the difference between the reservoir and the tail water levels. Ideally, all of this energy would be converted by the turbine into mechanical energy. In practice, hydraulic losses are incurred in the pipeline system, so the head available at the turbine is H_t $(= H - \text{losses})$. Losses also occur in the machine itself, and there are, additionally, mechanical losses (bearing friction, etc.). Nevertheless, turbines are very efficient, commonly attaining figures in excess of 90%. Kaplan units are generally applied where there is a large discharge at low pressure, whereas the Francis unit is appropriate for moderate discharge at moderate head.

The Pelton Wheel (Fig. 7.9) is applied where a high pressure water supply is available. By contrast with the other turbines, the casing does not run full of water. Essentially, the runner comprises a disc, with a series of flow deflectors or 'buckets' around its periphery. The water enters the casing through one or more nozzles as a high velocity jet. As the runner rotates, the jet impinges upon each bucket in turn. The water is deflected by the buckets, and the change in its momentum transfers a force to the bucket, and hence a torque to the drive shaft. The jet incorporates a streamlined

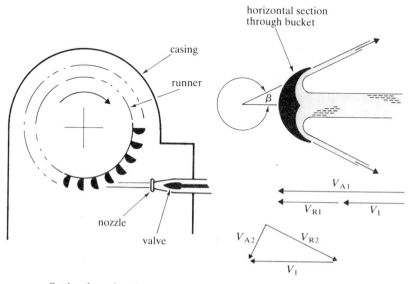

Section through casing   Velocity diagram

Figure 7.9   The 'Pelton Wheel' turbine.

control valve which is used to regulate the discharge, and hence the power output. Peak efficiency coincides with a particular runner speed.

The gross head for a Pelton Wheel is due only to the height of the reservoir above the nozzle(s). The head, H_t, is again defined as gross head minus losses.

Readers who require a slightly more detailed coverage of pump and turbine analysis (together with worked examples) are referred to Dugdale and Morfett (1983). For a more advanced treatment of pumps, see Stepanoff (1957), and for a broad practical treatment of both pumps and turbines see Davis and Sorensen (1969).

7.6 Turbine selection

Turbines, like pumps, may be selected on the basis of the specific speed appropriate to their duty. For turbines, the definition of specific speed is

$$N_S = NP^{1/2}/H^{5/4}$$

(see Section 11.6) for which, again, there are a variety of units.

Taking N in rev/min, power (P) in kW and H in m, the range of duties appropriate to each of the different types of turbine is:

$8 < N_S < 80$ Pelton Wheel (high head, low discharge);
$55 < N_S < 320$ radial flow (Francis) (moderate head, moderate discharge);
$250 < N_S < 750$ axial flow (low head, high discharge).

7.7 Cavitation in hydraulic machines

Before closing the discussion, it is worth taking a slightly more detailed glance at the subject of cavitation. It is possible for cavitation to occur in both pumps and turbines if the liquid pressure falls sufficiently far below atmospheric pressure at any point in the machine. The vacuum head, H_C, at which cavitation will commence is defined by the equation

$$H_C = (p_{atm} - p_{vap})/\rho g$$

where p_{vap} is absolute vapour pressure and p_{atm} is the absolute pressure of the atmosphere. For the case of a pump sited above a sump (see Fig. 12.6), the suction head at the pump equals the sum of the height of pump above sump, the friction loss in suction pipe and the kinetic head in suction pipe or, expressed algebraically,

$$H_{ps} = H_s + h_{fs} + V_s^2/2g$$

The 'net positive suction head' (NPSH) is defined as

$$\text{NPSH} = H_C - H_{ps}$$

A cavitation parameter σ_{Th} which was suggested by Thoma (a German engineer) relates NPSH to pump head

$$\sigma_{Th} = \text{NPSH}/H_p$$

For a turbine, the definition of NPSH is usually related to the height of the turbine (z) above tailrace, so that $\text{NPSH} = (H_C - z)$ and

$$\sigma_{Th} = \text{NPSH}/H_t$$

where H_t is the head available at the turbine ($=$ gross head minus losses).

A considerable amount of data has been amassed on cavitation, and critical values of σ_{Th} ($= \sigma_{crit}$) can be estimated for pumps or turbines. For freedom from cavitation, the operational σ_{Th} should exceed σ_{crit}. Some typical equations for σ_{crit} are as follows:

for a single intake radial flow pump,

$$\sigma_{crit} = \left(\frac{NQ^{1/2}}{H_p \times 191} \right)^{4/3} = \left(\frac{N_S}{191} \right)^{4/3}$$

for a radial flow turbine,

$$\sigma_{crit} = 0.006 + 0.55 \ (N_S/381)^{1.8}$$

For an axial flow turbine,

$$\sigma_{crit} = 0.10 + 0.3(N_S/381)^{2.5}$$

These values give only an approximate guide, and for more detailed information refer to Davis and Sorensen (1969).

References and further reading

Davis, C. V. and K. E. Sorensen 1969. *Handbook of applied hydraulics*. New York: McGraw-Hill.

Dugdale, R. H. and J. C. Morfett 1983. *Mechanics of fluids*. London: George Godwin.

Stepanoff, A. J. 1957. *Centrifugal and axial flow pumps*, 2nd edn. New York: Wiley.

8

Wave theory

8.1 Wave motion

Various special types of waves have already been discussed in Chapter 5, e.g. surge and flood waves. These are examples of solitary waves, and are not examined further. Periodic waves are those in which the fluid particles undergo motions which are repeated continuously. Progressive waves are those in which the wave fronts move with time.

Ocean waves are examples of periodic progressive waves, and as they approach the shores they are of particular interest to civil engineers involved in the design of marine structures or coastal protection works. Consequently, this chapter is concerned with the theories of periodic progressive waves and the interaction of waves with shorelines and coastal structures.

Ocean waves are mainly generated by the action of wind on water. The waves are formed initially by a complex process of resonance and shearing action, in which waves of differing wave height, length and period are produced. Once formed, ocean waves can travel for vast distances, spreading in area and reducing in height, but maintaining wavelength and period. This process is called dispersion. For example, waves produced in the gales of the 'roaring forties' have been monitored all the way north across the Pacific Ocean to the shores of Alaska (a distance of 10 000 km).

Casual observation of waves at sea may lead one to the conclusion that translation of water is taking place. This is incorrect, in that there is little net translation of water. Individual water particles describe elliptical orbits in a vertical plane, and only the wave fronts (e.g. wave crests) move forward.

As waves approach a shoreline their height and wavelength are modified by the processes of refraction and shoaling before breaking on the shore. Therefore, it is necessary to differentiate between deep, transitional and shallow water waves. Deep water waves become transitional water waves when their wavelength (L) is approximately half the water depth (d), and shallow waves when $d/L < 0.04$ (as will be shown in the next section). As deep water waves produced by gale force winds may have wavelengths in excess of 100 m, transitional water depth effects may occur many kilometres out to sea (for example, in the English Channel).

The following sections describe some of the more useful aspects of wave theory. Many results are quoted without derivation, as the derivations are often long and complex. The interested reader should consult the references at the end of the chapter for further details.

8.2   Airy waves

The mathematical description of periodic progressive waves is complex, and is still the subject of research. Fortunately, the earliest (and simplest) description, attributed to Airy, is sufficiently accurate for many engineering purposes. Other examples of wave theories are Stokian (second-, third-, and fifth-order) and Cnoidal, both of which give better approximations to the waveform of steep waves. Surprisingly, the Airy waveform appears to give more accurate predictions of orbital velocities.

The Airy wave was derived using the concepts of two-dimensional ideal fluid flow. This is a reasonable starting point for ocean waves which are not greatly influenced by either viscosity or surface tension.

Figure 8.1 shows a sinusoidal wave of wavelength L, height H and period T. The variation of surface elevation with time, from the still water level, is denoted by η and given by

$$\eta = \frac{H}{2} \cos 2\pi \left(\frac{x}{L} - \frac{t}{T} \right) \tag{8.1}$$

The wave celerity, c, is given by

$$c = L/T \tag{8.2}$$

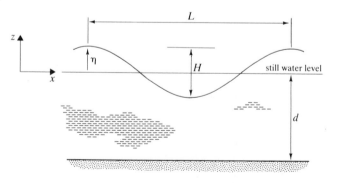

Figure 8.1   Definition sketch for a sinusoidal wave.

Substitution of (8.1) into the Laplace equation for two-dimensional irrotational flow yields (after considerable manipulation) the result

$$c = (gT/2\pi)\tanh(2\pi d/L) \qquad (8.3)$$

provided that $H \ll L$ and $H \ll d$.

The corresponding equations for the horizontal and vertical displacements of a particle at a mean depth $-z$ below the still water level are

$$x = \frac{H}{2} \frac{\cosh 2\pi(z+d)/L}{\sinh 2\pi d/L} \sin 2\pi\left(\frac{x}{L} - \frac{t}{T}\right) \qquad (8.4)$$

and

$$z = \frac{H}{2} \frac{\sinh 2\pi(z+d)/L}{\sinh 2\pi d/L} \cos 2\pi\left(\frac{x}{L} - \frac{t}{T}\right) \qquad (8.5)$$

Equations (8.4) and (8.5) describe an ellipse in which the vertical and horizontal excursions decrease with depth.

Deep water approximations

For $d/L > 0.5$, $\tanh(2\pi d/L) \simeq 1$. Hence, (8.3) reduces to

$$c_0 = gT/2\pi \qquad (8.6)$$

where the subscript 0 refers to deep water. Alternatively, using (8.2),

$$c_0 = (gL_0/2\pi)^{\frac{1}{2}}$$

Thus, the deep water wave celerity and wavelength are determined solely by the wave period.

It should be noted that the Airy wave equations only apply to waves of very small height and do not give any information as to the wave height. For deep water waves this is a function of wind speed, and fetch length and width. Further discussion of this important aspect is delayed until Section 8.7.

The particle displacement equations ((8.4) and (8.5)) describe approximately circular patterns of motion in deep water. At a depth $(-z)$ of $L/2$, the diameter is only 4% of the surface value. Hence deep water waves are unaffected by depth and have little influence on the sea bed (see Fig. 8.2).

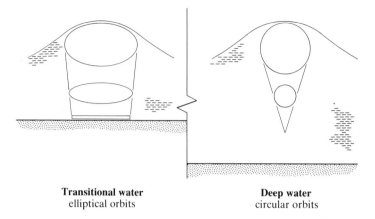

Transitional water   **Deep water**
elliptical orbits   circular orbits

Figure 8.2 Particle displacements for deep and transitional waves.

Shallow water approximations

For $d/L < 0.04$, $\tanh(2\pi d/L) \simeq 2\pi d/L$. Hence (8.3) reduces to

$$c = gTd/L$$

and substituting this into (8.2) gives $c = \sqrt{gd}$.

This may be compared with the result quoted in Chapter 5 ($c = \sqrt{gy}$). Thus, the shallow water wave celerity is determined by depth, and not by wave period.

Transitional water depth

This is the zone between deep water and shallow water, i.e. $0.5 > d/L > 0.04$. In this zone $\tanh(2\pi d/L) < 1$, hence

$$c = \frac{gT}{2\pi} \tanh\left(\frac{2\pi d}{L}\right) = c_0 \tanh\left(\frac{2\pi d}{L}\right) < c_0$$

This has important consequences, exhibited in the phenomena of refraction and shoaling, which are discussed separately.

In addition, the particle displacement equations show that at the sea bed vertical components are suppressed, so only horizontal displacements now take place (see Fig. 8.2). This has important implications regarding sediment transport and the phenomenon of littoral drift, which are discussed in Chapter 15.

A complete summary of the Airy wave characteristics for deep, transitional and shallow water is given by Muir Wood and Fleming (1981).

8.3   Wave energy and wave power

The energy contained within a wave is the sum of the potential, kinetic and surface tension energies of all the particles within a wavelength. For Airy waves, the potential (E_P) and kinetic (E_K) energies are equal and $E_P = E_K = \rho g H^2 L/16$. Hence, the energy (E) per unit area of ocean is

$$E = \rho g H^2/8 \tag{8.7}$$

(ignoring surface tension energy which is negligible for ocean waves). This is a considerable amount of energy. For example, a (Beaufort) Force 8 gale blowing for 24 h will produce a wave height in excess of 5 m, giving a wave energy exceeding $30\ kJ/m^2$.

One might expect that wave power (or the rate of transmission of wave energy) would be equal to wave energy times the wave celerity. This is incorrect, and the derivation of the equation for wave power leads to an interesting result which is of considerable importance. Wave energy is transmitted by individual particles which possess potential, kinetic and pressure energy. Summing these energies and multiplying by the particle velocity in the x-direction for all particles in the wave gives the rate of transmission of wave energy or wave power (P), and leads to the result (for an Airy wave).

$$P = \frac{\rho g H^2}{8}\frac{c}{2}\left(1 + \frac{4\pi d/L}{\sinh 4\pi d/L}\right) \tag{8.8}$$

or

$$P = EC_G$$

where C_G is the group wave celerity, given by

$$C_G = \frac{c}{2}\left(1 + \frac{4\pi d/L}{\sinh 4\pi d/L}\right) \tag{8.9}$$

In deep water $(d/L > 0.5)$ the group wave velocity $C_G = c/2$, and in shallow water $C_G = c$.

Hence, in deep water wave energy is transmitted forward at only half the wave celerity. This is a difficult concept to grasp, and therefore it is useful to examine it in more detail.

Consider a wave generator in a model bay supplying a constant energy input of 100 units. After one wave period, half of this energy (50 units) will have been transmitted one wavelength, as shown in Table 8.1. After two wave periods, half of the energy at one wavelength from the generator (25

Table 8.1   Wave generation: to show group wave speed.

Wave period	\multicolumn{11}{c}{Wave energy at various wavelengths from generator}

Wave period	0	1	2	3	4	5	6	7	8	9	10
1	100.0	50.0									
2	100.0	75.0	25.0								
3	100.0	87.5	50.0	12.5							
4	100.0	93.7	68.8	31.2	6.3						
5	100.0	96.9	81.2	50.0	18.7	3.1					
6	100.0	94.4	89.1	65.6	34.4	10.9	1.6				
7	100.0	99.2	93.7	77.3	50.0	22.7	6.3	0.8			
8	100.0	99.6	96.5	85.5	63.7	36.3	14.5	3.5	0.4		
9	100.0	99.8	98.0	91.0	74.6	50.0	25.4	9.0	2.0	0.2	
10	100.0	99.9	98.9	94.5	82.8	62.3	37.7	17.2	5.5	1.1	0.1

units) will have been transmitted a further wavelength (i.e. two wavelengths in total). Also, the energy at one wavelength will have gained another half of the energy from the generator (50 units) and lost half of its previous energy in transmissions (25 units). Hence, the energy level at one wavelength after two wave periods will be $50 + 50 - 25 = 75$ units. The process may be repeated indefinitely. Table 8.1 shows the result after ten wave periods. This demonstrates that although energy has been radiated to a distance of ten wavelengths, only 62% of the energy at the generator has reached a distance of five wavelengths.

The appearance of the waveform to an observer, therefore, is one in which the leading wave front moves forward but continuously disappears. If the wave generator were stopped after ten wave periods, the wave group (of ten waves) would continue to move forward but, in addition, would transmit wave energy to the trailing edge in the same way as to the leading edge. Thus, the wave group would appear to move forward at half the wave celerity, with individual waves appearing at the rear of the group and moving through the group to disappear again at the leading edge.

If this analysis is extended to 100 wave periods, then the wave height formed at the generator would have progressed nearly 50 wavelengths. The group wave celerity is thus half the wave celerity.

Returning to our example of a Force 8 gale, a typical wave celerity is 14 m/s (for a wave period of 9 s), the group wave celerity is thus 7 m/s, giving a wave power of 210 kW/m^2.

8.4   Refraction and shoaling

As waves approach a shoreline, they enter the transitional depth region in which the wave motions are affected by the sea bed. These effects include reduction of the wave celerity and wavelength, and thus alteration of the direction of the wave crests (refraction) and wave height (shoaling).

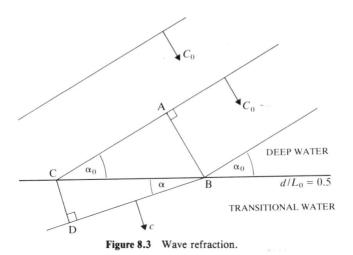

Figure 8.3   Wave refraction.

Refraction

Wave celerity and wavelength are related through two equations ((8.2) and (8.3)) to wave period (which is the only parameter which remains constant). This can be appreciated by postulating a change in wave period (from T_1 to T_2) over an area of sea. The number of waves entering the area in a fixed time t would be t/T_1, and the number leaving would be t/T_2. Unless T_1 equals T_2 the number of waves within the region could increase or decrease indefinitely. Thus,

$$c/C_0 = \tanh (2\pi d/L) \text{ (from (8.3))}$$

and

$$c/C_0 = L/L_0 \text{ (from (8.2))}.$$

To find the wave celerity and wavelength at any depth d, these two equations must be solved simultaneously. The solution is always such that $c < C_0$ and $L < L_0$ for $d < d_0$ (where the subscript 0 refers to deep water conditions).

Consider a deep water wave approaching the transitional depth limit ($d/L = 0.5$), as shown in Figure 8.3. A wave travelling from A to B (in deep water) traverses a distance L_0 in one wave period T. However, the wave travelling from C to D traverses a smaller distance, L, in the same time, as it is in the transitional depth region. Hence, the new wave front is now BD, which has rotated with respect to AC. Letting the angle α represent the angle of the wave front to the depth contour, then

$$\sin \alpha = L/BC \text{ and } \sin \alpha_0 = L_0/BC$$

Combining,

$$\frac{\sin \alpha}{\sin \alpha_0} = \frac{L}{L_0}$$

Hence

$$\frac{\sin \alpha}{\sin \alpha_0} = \frac{L}{L_0} = \frac{c}{C_0} = \tanh\left(\frac{2\pi d}{L}\right) \qquad (8.10)$$

As $c < C_0$, then $\alpha < \alpha_0$, which implies that as a wave approaches a shoreline from an oblique angle the wave fronts tend to align themselves with the underwater contours. Figure 8.4 shows the variation of c/C_0 with d/L_0, and α/α_0 with d/L_0 (the latter specifically for the case of parallel contours). It should be noted that L_0 is used in preference to L as the former is a fixed quantity.

In the case of non-parallel contours, individual wave rays (i.e. the orthogonals to the wave fronts) must be traced. Figure 8.4 can still be used to find α at each contour if α_0 is taken as the angle (say α_1) at one contour and α is taken as the new angle (say α_2) to the next contour. The wave ray is usually taken to change direction midway between contours.

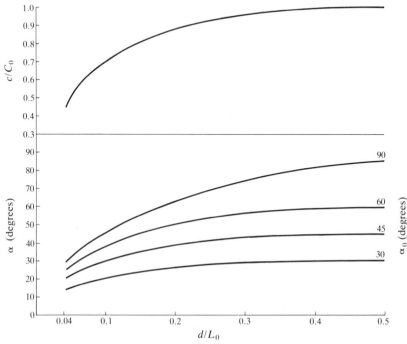

Figure 8.4 Variation of wave celerity and angle with depth.

This procedure may be carried out by hand using tables or figures (see Silvester 1974) or by computer (see Muir Wood and Fleming 1981).

Wave breaking

As the refracted waves enter the shallow water region, they break before reaching the shoreline. The foregoing analysis is not strictly applicable to this region, because the wave fronts steepen and are no longer described by the Airy waveform. However, as a general guideline, waves will break when

$$d_B = 1.28 \, H_B \qquad (8.11)$$

where the subscript B refers to breaking.

It is common practice to apply this refraction analysis up to the breaker line. This is justified on the grounds that the inherent inaccuracies are small compared with the initial predictions for deep water waves, and are within acceptable engineering tolerances.

To find the breaker line, it is necessary to estimate the wave height as the wave progresses inshore. This can be estimated from the refraction diagram in conjunction with shoaling calculations, as is now described.

Shoaling

Consider first a wave front travelling parallel to the sea bed contours. Making the assumption that wave energy is transmitted shorewards without loss due to bed friction or turbulence, then

$$\frac{P}{P_0} = 1 = \frac{EC_G}{E_0 C_{G0}} \qquad \text{(from (8.8)).}$$

Substituting

$$E = \rho g H^2 / 8 \qquad \text{(Equation (8.7))}$$

then,

$$\frac{P}{P_0} = 1 = \left(\frac{H}{H_0}\right)^2 \frac{C_G}{C_{G0}}$$

or

$$\left(\frac{H}{H_0}\right) = \left(\frac{C_{G0}}{C_G}\right)^{1/2} = K_S$$

where K_S is the shoaling coefficient.

The shoaling coefficient can be evaluated from the equation for the group wave celerity (Equation (8.9))

$$K_S = \left(\frac{C_{G0}}{C_G}\right)^{\frac{1}{2}} = \left(\frac{C_0/2}{\frac{c}{2}[1 + (4\pi d/L/\sinh 4\pi d/L)]}\right)^{\frac{1}{2}} \tag{8.12}$$

The variation of K_S with d/L_0 is shown in Figure 8.5.

Consider next a wave front travelling obliquely to the sea bed contours as shown in Figure 8.6. In this case, as the wave rays bend, they may converge or diverge as they travel shoreward. At the contour $d/L_0 = 0.5$,

$$BC = \frac{b_0}{\cos \alpha_0} = \frac{b}{\cos \alpha}$$

or

$$\frac{b}{b_0} = \frac{\cos \alpha}{\cos \alpha_0}$$

Again, assuming that the power transmitted between any two wave rays is constant, then

$$\frac{P}{P_0} = 1 = \frac{Eb}{E_0 b_0} \frac{C_G}{C_{G0}}$$

Substituting for E and b,

$$\left(\frac{H}{H_0}\right)^2 \frac{\cos \alpha}{\cos \alpha_0} \frac{C_G}{C_{G0}} = 1$$

or

$$\frac{H}{H_0} = \left(\frac{\cos \alpha_0}{\cos \alpha}\right)^{\frac{1}{2}} \left(\frac{C_{G0}}{C_G}\right)^{\frac{1}{2}}$$

Hence

$$H/H_0 = K_R K_S \tag{8.13}$$

where

$$K_R = \left(\frac{\cos \alpha_0}{\cos \alpha}\right)^{\frac{1}{2}}$$

and is called the refraction coefficient.

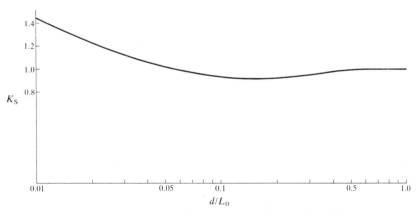

Figure 8.5 Variation of the shoaling coefficient with depth.

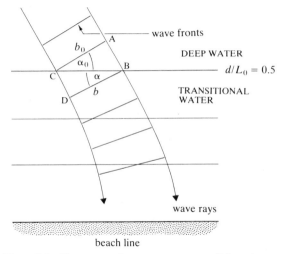

Figure 8.6 Divergence of wave rays over parallel contours.

For the case of parallel contours, K_R can be found using Figure 8.4. In the more general case, K_R can be found from the refraction diagram directly by measuring b and b_0.

Example 8.1   Wave refraction and shoaling

A deep water wave has a period of 8.5 s, a height of 5 m and is travelling at $45°$ to the shoreline. Assuming that the sea bed contours are parallel, find the height, depth, celerity and angle of the wave when it breaks.

Table 8.2 Tabular solution for breaking waves.

d/L_0	d(m)	c/C_0	c(m/s)	K_S	α(degrees)	K_R	H/H_0	H(m)	d_B(m)
0.1	11.3	0.7	9.3	0.93	30	0.9	0.84	4.2	5.4
0.05	5.6	0.52	6.9	1.02	22	0.87	0.89	4.45	5.7

Solution

(a) Find the deep water wavelength and celerity. From (8.6),

$$C_0 = gT/2\pi = 13.27 \text{ m/s}$$

From (8.2),

$$L_0 = C_0 T = 112.8 \text{ m}$$

(b) At the breaking point, the following conditions (from (8.11) and (8.13)) must be satisfied:

$$d_B = 1.28 H_B$$

and

$$H_B/H_0 = K_R K_S$$

For various trial values of d/L_0, H_B/H_0 can be found using Figures 8.4 and 8.5. The correct solution is when (8.11) and (8.13) are satisfied simultaneously. This is most easily seen by preparing a table, as shown in Table 8.2. For $d/L_0 = 0.05$, $d = 5.6$ m and $H = 4.45$, requiring a depth of breaking of 5.7 m. This is sufficiently accurate for an acceptable solution, so

$$H_B = 4.45 \text{ m} \qquad c = 6.9 \text{ m/s} \qquad d_B = 5.7 \text{ m} \qquad \alpha_B = 22°$$

8.5 Standing waves

Waves incident on solid vertical boundaries (e.g. harbour walls and sea walls) are reflected such that the reflected wave has the same phase and substantially the same amplitude as the incident wave. The resulting wave pattern set up is called a standing wave.

For an incident wave front at right angles to a boundary, the reflected wave is of opposite direction and the resulting waveform is shown in Figure 8.7. The equation of the standing wave (subscript s) may be found by adding the two waveforms of the incident (subscript i) and reflected (subscript r) waves. Thus,

$$\eta_i = \frac{H_i}{2} \cos 2\pi \left(\frac{x}{L} - \frac{t}{T} \right)$$

$$\eta_r = \frac{H_r}{2} \cos 2\pi \left(\frac{x}{L} + \frac{t}{T} \right)$$

$$\eta_s = \eta_i + \eta_r$$

Taking

$$H_r = H_i = H_s/2,$$

then

$$\eta_s = H_s \cos(2\pi x/L)\cos(2\pi t/T) \tag{8.14}$$

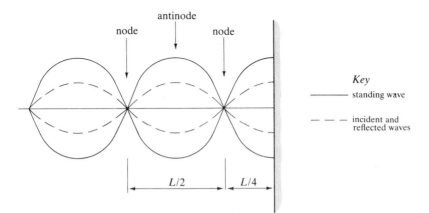

Figure 8.7 Standing waves.

Photograph 3 Wave impact and reflection during a severe storm.

At the nodal points there is no vertical movement with time. By contrast, at the antinodes, crests and troughs appear alternately. If the reflected wave has a similar amplitude to the incident wave, then the advancing and receding crests collide in a spectacular manner, forming a plume known as a clapotis. Standing waves can cause considerable damage to maritime structures, and bring about substantial erosion because the wave height in the vicinity of the structure has been doubled, compared with the deep water wave height.

Clapotis gaufre

When the incident wave is at an angle α to the normal from a vertical boundary, then the reflected wave will be in a direction α on the opposite side of the normal. This is illustrated in Figure 8.8. The resulting wave motion (the clapotis gaufre) is complex, but essentially consists of a diamond pattern of island crests which move parallel to the boundary. It is sometimes referred to as a short crested system. The crests form at the intersection of the incident and reflected wave fronts. The resulting particle displacements are also complex, but include the generation of a pattern of moving vortices. A detailed description of these motions may be found in Silvester (1974). The consequences of this in terms of sediment transport may be severe. Very substantial erosion and littoral drift may take place. Considering that oblique wave attack to sea walls is the norm rather than the exception, the existence of the clapotis gaufre has a profound influence on the long term stability and effectiveness of coastal protection works (Photograph 3). This does not seem to have been fully understood in traditional designs of sea walls, with the result that collapsed sea walls and eroded coastlines have occurred.

8.6   Wave diffraction

This is the process whereby waves bend round obstructions by radiation of the wave energy. Figure 8.9 shows an oblique wave train incident on the tip of a breakwater. There are three distinct regions:

(a) the shadow region in which diffraction takes place;
(b) the short crested region in which incident and reflected waves form a clapotis gaufre;
(c) an undisturbed region between (a) and (b).

In region (a), the waves diffract with the wave fronts forming circular arcs centred on the point of the breakwater. When the waves diffract, the wave heights diminish as the energy of the incident wave spreads over the region. Values for diffraction coefficients (ratio of diffracted wave height to incident wave height at various radii) have been determined for some specific cases (see Muir Wood and Fleming 1981).

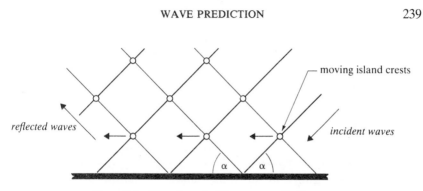

Figure 8.8   Plan view of oblique wave reflection.

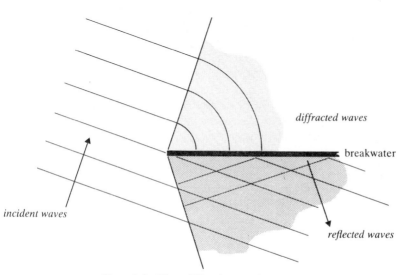

Figure 8.9   Wave diffraction at a breakwater.

The real situation is, of course, more complex than that presented in Figure 8.9. The reflected waves in region (c) will diffract into region (b) and extend the short crested system into region (b).

8.7   Wave prediction

Significant parameters

Wind-generated waves are complex, incorporating many superimposed components of wave periods and heights. If the sea state is recorded in a storm zone, then the resulting wave trace appears to consist of random periodic fluctuations. To find order in this apparent chaos, considerable research and measurement has been, and is being, undertaken.

For analytical purposes, the significant parameters for wind are speed (U), duration (D) and fetch length (F), and for waves they are period (T) and height (H).

Storm waves

For a given wind speed, the waves produced will depend on the duration and fetch. The longer the fetch or duration, the bigger the waves produced. However, as the wind contains only a given amount of energy, the wave heights will approach some limiting value (when the rate of transfer of energy to the waves equals the energy dissipation by wave breaking and friction) for particular values of fetch and duration. This sea state is referred to as the fully arisen sea (FAS). Measurements of the FAS for various wind speeds indicate that the wave energy spectrum is similar for all FAS's, independent of location or wind speed.

Perhaps the form of a wave energy spectrum which is most easily appreciated is that of the energy distribution curve, in which H^2 per unit time is plotted against T. This ensures that the total area under the curve is proportional to the total wave energy per unit area of ocean. Figures 8.10 and 8.11 show energy distribution curves for FASs with varying wind speed, and for developing seas with varying ratios of fetches and durations, respectively.

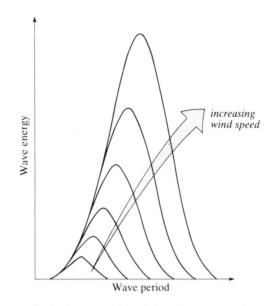

Figure 8.10 Energy distribution curves for a fully arisen sea at various wind speeds.

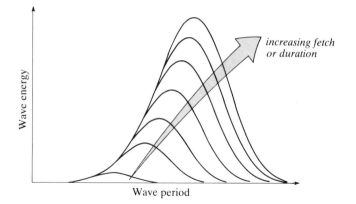

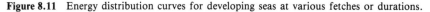

Figure 8.11 Energy distribution curves for developing seas at various fetches or durations.

Swell waves

Outside the storm zone the wave energy is radiated over a large area of ocean without significant loss. However, due to dispersion, the wave energy is reduced at any particular location downwind of the storm zone, and hence the wave heights decrease. The waves with longer period also travel more quickly than those with shorter period, so the wave spectrum tends to split into its component parts. Such waves are called swell waves, their distinguishing features being that they are longer crested and have longer periods than storm waves. The distinction is an important one when predicting waves incident on maritime structures and the effects of waves on coastlines.

Methods of predicting waves from wind records

Wind is recorded for meteorological purposes at many sites and is, therefore, far more frequently recorded than are waves. Given that storm waves are dependent solely on wind speed, duration and fetch, the most common method used to predict wave climate is based on the use of wind records. This technique is generally referred to as hindcasting. As no purely theoretical means of predicting waves from wind has yet been devised, empirical relationships have been derived. The simplest of these provide a prediction in the form of a single wave height and period for a given wind speed, rather than a wave spectrum. The characteristic wave height most commonly used is that of the significant wave height (H_S). This corresponds to the average of the highest one-third of recorded waves for any given wind condition.

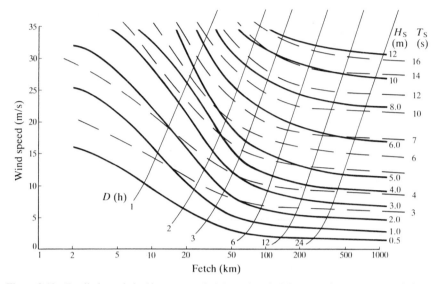

Figure 8.12 Prediction of significant wave height and period for oceanic waters around the UK.

Tables and charts relating H_S to wind speed, fetch and duration have been devised by various people after considerable and painstaking research. Notable examples of these include the Darbyshire and Draper (1963) charts for oceanic and coastal waters around the UK, and the SMB method given in the US Shore Protection Manual (Sverdrup *et al.* 1975). Figure 8.12 represents the Darbyshire and Draper Charts. To use these charts, the required wind speed is selected and the corresponding wave height (and period) is found at the first intersection of this wind speed with either the known fetch or duration. Thus the chart also determines whether the sea state is fetch or duration limited.

Example 8.2   Use of the Darbyshire and Draper charts

Using Figure 8.12, (a) Find the significant wave height (H_S) and period (T_S), and state whether the fully arisen sea has been reached under the following conditions:

wind speed (U) = 20 m/s (i.e. Beaufort Force 8);
fetch (F) = 100 km
storm duration (D) = 6 h.

(b) Find the new significant wave height if the fetch is now increased to 500 km (i.e. corresponding to a storm from a different direction).

Solution

(a) The intersection of the two lines representing $U = 20$ m/s and $F = 100$ km on Figure 8.12 give:

$$H_S \simeq 5.75 \text{ m} \qquad T_S \simeq 7 \text{ s} \qquad D \simeq 4 \text{ h}$$

Hence, as the storm has a duration of 6 h, the fully arisen sea condition has been reached and the waves are fetch limited.

(b) In this case, intersection of the line representing $U = 20$ m/s with that representing $D = 6$ h occurs before that with the line representing $F = 500$ km. Hence, the sea state is not fully arisen and the waves are duration limited, giving $H_S \approx 6$ m.

In cases where the full wave spectrum is required (for example, physical model testing of breakwaters) then the Pierson–Moskowitz (PM) spectrum is often used. For a fully arisen sea, this may be stated in terms of wave period as

$$SE(T) = \frac{0.0081 g^2}{(2\pi)^4} \, T^3 \exp\left[-0.74 \left(\frac{gT}{2\pi U_{19.5}} \right)^4 \right] \tag{8.15}$$

where SE (T) is the spectral wave energy function (in m^2/s) and $U_{19.5}$ is the wind speed (in m/s) at 19.5 m from the sea surface. The PM spectrum was derived from measurements of ocean waves. It does not describe conditions in fetch-limited seas.

More-recent work on wave energy spectra has been based on observations in the North Sea (UK) resulting in the JONSWAP spectrum. Details of this and more advanced numerical methods of modelling wave characteristics can be obtained from the Marine Information and Advisory Service, Institute of Oceanographic Sciences, Wormley, UK.

8.8 Analysis of wave records

Wave records are available for certain locations. These are normally gathered by either ship borne wave recorders (for fixed locations) or wave rider buoys (which may be placed at specific sites of interest). These records generally consist of a wave trace for a short period (typically 10 min) recorded at fixed intervals (normally 3 h). In this way, the typical sea state may be inferred without the necessity for continuous monitoring. An example wave trace is shown in Figure 8.13.

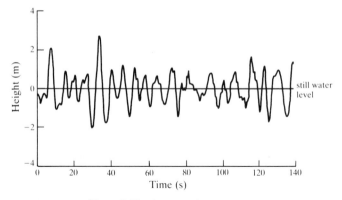

Figure 8.13 An example wave trace.

Using such records the wave energy spectrum at a particular site may be derived directly, and suitable design waves for maritime structures can be synthesised.

For a given wave record (e.g. a 10 min record representing a 3 h period) the following parameters may be directly derived:

(a) H_{rms} (root-mean-square wave height);
(b) H_{max} (maximum difference between adjacent crest and trough);
(c) H_S (mean height of the highest one-third of the waves);
(d) $H_{1/10}$ (mean height of the highest one-tenth of the waves);
(e) T_C (mean period between wave crests);
(f) T_Z (mean period between zero upward crossings).

Based on previous experience, wave record analysis is greatly simplified if some assumptions regarding the probability distributions of wave heights and periods are made. For example, the distribution of wave heights is often assumed to follow the Rayleigh distribution, thus

$$P(h \geqslant H) = \exp\left[(-2H/H_S)^2\right] \qquad (8.16)$$

where $P(h \geqslant H)$ is the probability that h will equal or exceed a given value H.

For a wave record of N waves, taking

$$P(h \geqslant H) = i/N$$

where i is the rank number and

$$i = \begin{cases} 1, & \text{for the largest wave} \\ N, & \text{for the smallest wave} \end{cases}$$

then rearranging (8.16) and substituting for P gives

$$H = H_S [\tfrac{1}{2} \ln(N/i)]^{1/2}$$

For the case of $i = 1$, corresponding to $H = H_{max}$,

$$H_{max} = H_S (\tfrac{1}{2} \ln N)^{1/2}$$

Other useful results which have been derived include:

$$H_S \simeq 2^{1/2} H_{rms} \simeq 1.414 H_{rms}$$

$$H_{1/10} \simeq 1.27 H_S \simeq 1.8 H_{rms}$$

$$H_{1/100} \simeq 1.67 H_S \simeq 2.36 H_{rms}$$

$$H_{max} \simeq 1.6 H_S \text{ (for a typical 10 min wave trace)}$$

Thus, if the value of H_{rms} is calculated from the record, the values of H_S, etc., may easily be estimated.

However, the Rayleigh distribution is only a good fit to sine waves of single period but varying amplitude. This is more characteristic of swell waves than storm waves. The PM spectrum for storm waves (in its frequency form) is a close fit to a narrow-band Gaussian distribution (i.e. a normal distribution with small standard deviation). To determine what type of distribution is applicable, the parameter ε, known as the spectral width, may be calculated:

$$\varepsilon^2 = 1 - \left(\frac{T_C}{T_Z}\right)^2 \tag{8.17}$$

For the Rayleigh distribution (i.e. a single period) $T_C = T_Z$, hence $\varepsilon = 0$. For a Gaussian distribution (containing many frequencies) the period between adjacent crests is much smaller than the period between zero upward crossings, hence $\varepsilon \to 1$.

Actual measurements of swell and storm waves give the following results:

	Swell waves	Storm waves
ε	$\simeq 0.3$	$\simeq 0.6$ to 0.8
H_S	$\simeq 1.42 H_{rms}$	$\simeq 1.48 H_{rms}$
$H_{1/10}$	$\simeq 1.8 H_{rms}$	$\simeq 2.0 H_{rms}$

8.9   Extension of wave records

If wave records are collected for an extended period (e.g. for a season or several years), then the computed values of relevant parameters for each wave recording period (e.g. H_S, T_S or ε) may be presented in the form of histograms or percentage exceedence graphs. These graphs can provide very useful information to contractors undertaking marine construction works. For some purposes it may be useful to extend these records to estimate the one-in-T-year event (e.g. the maximum wave height expected once in 100 years). For further details of these and other techniques, the reader should consult the references and further reading list.

References and further reading

Darbyshire, M and L. Draper 1963. Forecasting wind-generated sea waves. *Engineering* **195** (April).

Muir Wood, A. M. and C. A. Fleming 1981. *Coastal hydraulics*, 2nd edn. London: Macmillan.

Silvester, R. 1974. *Coastal engineering, 1*. Oxford: Elsevier.

Sverdrup, H. U., W. H. Munk and C. L. Bretschneider 1975. SMB method for predicting waves in deep water. *Shore Protection Manual*. Washington: US Army Coastal Engineering Research Centre.

9

Sediment transport

9.1 Introduction

Sediment transport governs or influences many situations that are of importance to mankind. Silt deposition reduces the capacity of reservoirs, interferes with harbour operation and closes or modifies the path of watercourses. Erosion or scour may undermine structures. In rivers and on coastlines, sediment movements form a part of the long term pattern of geological processes. Clearly, sediment transport occurs only if there is an interface between a moving fluid and an erodible boundary. The activity at this interface is extremely complex. Once sediment is being transported, the flow is no longer a simple fluid flow, since two materials are involved.

Sediment transport may be conceived of as occurring in one of two modes:

(a) by rolling or sliding along the floor (bed) of the river or sea – sediment thus transported constitutes the bedload;
(b) by suspension in the moving fluid (this is usually applicable to finer particles) which is the suspended load.

9.2 The threshold of movement

Description of threshold of movement

If a perfectly round object (a cylinder or sphere) is placed on a smooth horizontal surface, it will readily roll on application of a small horizontal force. In the case of an erodible boundary, of course, the particles are not perfectly round, and they lie on a surface which is inherently rough and may not be flat or horizontal. Thus, the application of a force will only cause motion when it is sufficient to overcome the natural resistance to motion of the particle. The particles will probably be non-uniform in size. At the interface, a moving fluid will apply a shear force (Fig. 9.1a), which implies that a proportionate force will be applied to the exposed surface of a particle.

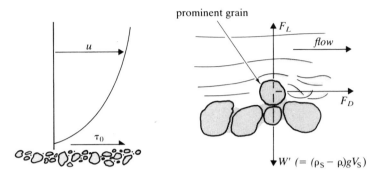

(a) Shear force on granular bed   (b) Force on 'prominent' grain

Figure 9.1   Fluid forces causing sediment movement.

Observations by many experimenters have confirmed that if the shear force is gradually increased from zero, a point is reached at which particle movements can be observed at a number of small areas over the bed. A further small increase in τ_0 (and therefore u) is usually sufficient to generate a widespread sediment motion (of the bedload type). This describes the 'threshold of motion'. After further increments in τ_0, another point is reached at which the finer particles begin to be swept up into the fluid. This defines the inception of a suspended load.

Parameters of sediment transport

Some idea of the problems which face the engineer may be gained by considering the case of a channel flow over a sandy bed which is initially level. Once the shear stress is sufficient to cause transport, 'ripples' will form in the bed (Photograph 4). These ripples may grow into larger 'dunes'. In flows having quite moderate Froude Numbers, the dunes will migrate downstream. This is due to sand being driven from the dune crests and then being deposited just downstream on the lee side. Once the flow is sufficient to bring about a suspended load, major changes occur at the bed as the dunes will be 'washed out'. A relationship probably exists between the major parameters of the transport process (Froude Number, sediment properties, fluid properties, shear stress, bed roughness or dune size, and rate of sediment transport). A multitude of attempts have been made to develop a rational theory, so far with limited success. Most of the equations in current use have been developed on the basis of a combination of dimensional analysis, experimentation and simplified theoretical models.

Photograph 4 Sediment ripple formation.

The entrainment function

A close inspection of an erodible granular boundary would reveal that some of the surface particles were more 'prominent' or 'exposed' (and therefore more prone to move) than others (Fig. 9.1b). The external force on this particle is due to the separated flow pattern. The other force acting on the particle is related to its submerged self-weight, W' (where $W' = \pi D^3 g\,(\rho_S - \rho)/6$ for a spherical particle) and to the angle of repose, ϕ. The number of prominent grains in a given surface area is related to the areal grain packing ($=$ area of grains/total area $= A_p$). As the area of a particle is proportional to the square of the typical particle size (D^2), the number of exposed grains is a function of A_p/D^2. The shear stress at the interface, τ_0, is presumably the sum of the forces on the individual particles, with the contribution due to prominent grains dominating, so the total force on each prominent grain in unit area may be expressed as

$$F_D \propto \tau_0 D^2/A_p$$

At the threshold of movement $\tau_0 = \tau_{CR}$, so

$$\tau_{CR}\frac{D^2}{A_p} \propto (\rho_S - \rho)g\,\frac{\pi D^3}{6}\tan\phi$$

This can be rearranged to give a dimensionless relationship

$$\frac{\tau_{CR}}{(\rho s - \rho)gD} \propto \frac{\pi A_p}{6} \tan \phi \tag{9.1}$$

In 1936, A. Shields published the results of some pioneering research. He reasoned that the particle entrainment must be some function of the Reynolds Number, and that the form of the Reynolds Number used should relate to conditions at the grain, rather than to the general fluid flow. The formulation used was

$$Re_* = \rho u_* D / \mu$$

where u_* is the friction velocity (see Section 3.4). The left-hand side of (9.1) is the ratio of shear force to gravity force, and is known as the entrainment function, F_S. Shields plotted F_S against Re_*, and demonstrated that the nature and effects of the transport process were a function of position on the F_S–Re_* diagram (Fig. 9.2). The line corresponding to the threshold of movement was added later (by Rouse). This showed that the threshold F_S was constant (at approx. 0.056) beyond a certain 'critical' value of Re_*. Initially this critical value was taken as 400, though recent publications indicate values in the range 70–150. Taking $F_S = 0.056$ and substituting in (9.1):

$$\tau_{CR}/(\rho s - \rho)gD = 0.056 \tag{9.1a}$$

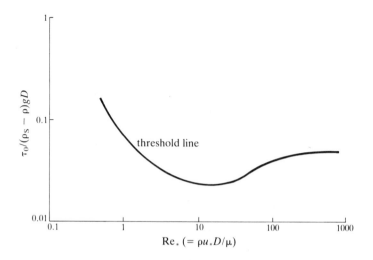

Figure 9.2 Shields' diagram.

Also, from the Chézy formula,

$$\tau_0 = \rho g R S_0$$

where R is the hydraulic radius. Therefore, at the threshold condition,

$$\frac{1000 \times 9.81 \times R \times S_0}{(2650 - 1000) \times 9.81 \times D} = 0.056$$

$$RS_0/D = 0.0924 \tag{9.1b}$$

This may be used to estimate (a) the minimum stable particle size for a given channel or (b) the critical shear stress for a given particle size. This treatment is, naturally, only as good as the Chézy formula and the other approximations made in this development. Nevertheless, many channels have been successfully designed with it. The Chézy formula assumes a generally turbulent flow, however the effects of turbulence at the boundary itself are more difficult to analyse than is suggested by the above treatment.

9.3 A general description of the mechanics of sediment transport

Remarks on the approach used in the rest of this chapter

It should already be evident that this area of study is fraught with difficulty. In consequence, the pattern of study for the remainder of the chapter departs somewhat from that used elsewhere in the text. To begin with, rigorous proofs of some equations are not possible. Where proofs are available, they are often long, and would occupy considerable space. Furthermore, in many cases, the nature of the algebra would add little to a physical understanding of the phenomena. In general, therefore, proofs are only given for the simpler equations. The more complex equations will simply be stated, together with brief details of their provenance, the underlying principles and limitations (insofar as they are currently known), and examples of their application. It is possible to give only a small selection from the array of equations developed by various researchers. No special merits are claimed for the selection adopted here, though it is hoped that, at the very least, a path will have been cleared through the 'jungle' sufficient to enable the reader to use some of the more advanced texts.

Conditions at the interface between a flowing fluid and a granular boundary

For most practical cases, channel flows are turbulent. This means that eddy transport processes are embedded in the general body of the flow. A close look in the region of the granular boundary at the bed would reveal the

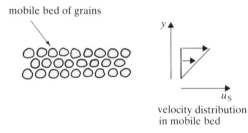

mobile bed of grains

u_S

velocity distribution
in mobile bed

Figure 9.3   Linear velocity distribution for idealised bed motion.

existence of a sub-layer comprising 'pools' of stationary or slowly moving
fluid in the interstices. This sublayer zone is not stable, since eddies (with
high momentum) from the turbulent zone periodically penetrate the sub-
layer and eject the (low momentum) fluid from the 'pools'. The momentum
difference between the fluid from the two zones itself generates a shearing
action, which in turn generates more eddies, and so on. Grains are thus
subjected by the fluid to a fluctuating impulsive force. Once the force is sufficient to dislodge the more prominent grains (i.e. when $\tau_0 \geqslant \tau_{CR}$) they will
roll over the neighbouring grain(s). As sediment movement becomes more
widespread, the pattern of forces becomes more complex as moving particles collide with each other and with stationary particles. As τ_0 increases
further, granular movement penetrates more deeply into the bed. Bed
movement may most simply be represented as a series of layers in relative
sliding motion (Fig. 9.3), with a linear velocity distribution (note the
analogy with laminar flow). This model forms the basis of the 'tractive
force' equations of Du Boys, Shields, etc. (see Section 9.4). In reality, the
pattern of movement is much more irregular.

The mechanics of particle suspension

If sediment grains are drawn upward from the channel bed and into suspension, it must follow that some vertical (upward) force is being applied to
the grains. The force must be sufficient to overcome the immersed self-
weight of the particles. Consider a particle suspended in a vertical flask (Fig.
9.4). If the fluid is stationary, then the particle will fall due to its self-weight
(assuming $\rho_S > \rho$), accelerating up to a limiting (or 'terminal') velocity v_{FS}
at which the self-weight will be equal in magnitude to the drag force, F_D,
acting on the particle. Since drag force $= C_D \times (\frac{1}{2}\rho A U_\infty^2)$ (see Section 3.6),

$$(\rho_S - \rho)gV_S = C_D \times \tfrac{1}{2}\rho A_S v_{FS}^2 = F_D \qquad (9.2)$$

where V_S and A_S are the volume and cross-sectional area of the particle.

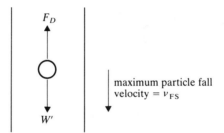

Figure 9.4   Forces acting on a falling particle.

If a discharge is now admitted at the base of the flask, the fluid is given a vertical upward velocity v. As $v \to v_{FS}$, the particle will cease to fall and will appear to be stationary. If $v > v_{FS}$ then the particle can be made to travel upward.

From this argument, it must follow that the suspension of sediment in a channel flow implies the existence of an upward velocity component. In fact, this should not come as a complete surprise (as a review of Section 3.4 will reveal). Fluctuating vertical (and horizontal) components of velocity are an integral part of a turbulent flow. Flow separation over the top of a particle provides an initial lift force (Fig. 9.1b) which tends to draw it upwards. Providing that eddy activity is sufficiently intense, then the mixing action in the flow above the bed will sweep particles along and up into the body of the flow (Fig. 9.5a). Naturally, the finer particles will be most readily suspended (like dust on a windy day).

The Prandtl model of turbulence (Section 3.4) can be used to illustrate the mixing ('lifting') action of turbulence, and to predict the variations of the sediment concentration C (volume of sediment/volume of fluid

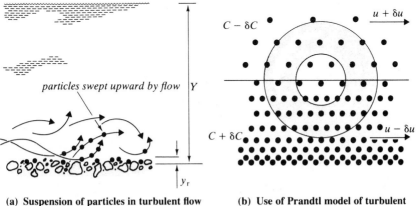

(a) **Suspension of particles in turbulent flow**

(b) **Use of Prandtl model of turbulent eddy to estimate concentration**

Figure 9.5   Suspension of sediment.

+ sediment) with distance y from the bed. If the concentration varies from $(C + \delta C)$ to $(C - \delta C)$ over the height of one eddy (Fig. 9.5b), and if $u' = v'$, then in 1 s,

volume of particles transferred from lower to upper layer $= u' \, \delta A \, (C + \delta C)$

volume of particles transferred from upper to lower layer $= u' \, \delta A \, (C - \delta C)$

therefore total upward rate of transport $= 2u' \, \delta A \, 2 \, \delta C$

If $u' = \delta y \, du/dy$ and $\delta C = \delta y \, dC/dy$, then

$$\text{total upward rate of transport} = 2 \, \delta A \left(\delta y \, \frac{du}{dy} \right) \left(\delta y \, \frac{dC}{dy} \right)$$

and, by comparison with (3.4), this may be written

$$\text{total upward rate of transport} = \frac{\varepsilon}{\rho} \frac{dC}{dy} \tag{9.3}$$

In a steady two-dimensional channel flow, the shear stress may be assumed to vary roughly as

$$\frac{\tau_y}{\tau_0} = 1 - \frac{y}{Y}$$

where Y is the depth of flow. From (3.6a),

$$\frac{du}{dy} = \frac{u_*}{Ky}$$

From (3.4),

$$\tau_y = \varepsilon \frac{du}{dy}$$

These three equations may be combined to give

$$\frac{\varepsilon}{\rho} = Kyu_* \left(1 - \frac{y}{Y} \right) \tag{9.4}$$

If the natural velocity of fall of a particle is v_{FS}, then the rate (volume/s) at which the sediment would sink in still fluid must be $- Cv_{\text{FS}}$ per unit volume. So if the sediment transport process is in equilibrium, upward transport must equal downward transport:

$$-Cv_{FS} = \frac{\varepsilon}{\rho}\frac{dC}{dy}$$

i.e.

$$Cv_{FS} + \frac{\varepsilon}{\rho}\frac{dC}{dy} = 0 \qquad (9.5)$$

Substituting (9.4) into (9.5) and integrating yields

$$\frac{C}{C_r} = \left(\frac{y_r(Y-y)}{y(Y-y_r)}\right)^{(v_{FS}/Ku_*)} \qquad (9.6)$$

where C_r is a reference concentration at height y_r.

As with the Prandtl model of shear stress, no further progress can be made unless experimental results are invoked. The equation can be made to fit such results quite well. However, the fit is only apparent, since values of the exponent (v_{FS}/Ku_*) must be suitably selected to obtain a good fit. One of the problems undoubtedly lies with the implied incorporation of ε in the equation. Turbulent eddies are generated at the (rough) boundary, but when there is sediment transport the eddies have to pass through the moving bedload. This has the effect of modifying the eddy patterns, together with the effective local density and the effective viscosity of the fluid.

9.4   Sediment transport equations

In practice, virtually all sediment transport occurs either as bedload or as a combination of bedload and suspended load (suspended load rarely occurs in isolation, except for certain cases involving very fine silts). The combined load is known as a total load.

Bed load formulae

Tractive force ('Du Boys type') equations. In 1879, Du Boys proposed an equation which related sediment transport to shear stress:

$$q_S = f(\tau_0)$$

where q_S is the volume of sediment transported per second per unit channel width. Subsequent researchers proposed that the function should be in the form of a power series:

$$q_S = K_1 + K_2\tau_0 + K_3\tau_0^2 + \ldots$$

If higher order terms are neglected, then the constants can be evaluated, since

$$q_S = 0 \text{ if } \tau_0 < \tau_{CR}, \qquad \text{therefore } K_1 = 0$$

$$q_S = 0 \text{ if } \tau_0 = \tau_{CR}, \qquad \text{therefore } K_2 = -K_3\tau_{CR}$$

Therefore

$$q_S = K_3\tau_0(\tau_0 - \tau_{CR}) \qquad (9.7)$$

The problem of evaluating K_3 now arises. One very simple equation, which is based only on limited data, is

$$K_3 = 0.54/g(\rho_S - \rho) \qquad \text{(metric units)} \qquad (9.8)$$

A more rational formulation was derived by Shields (1936) for a level bed:

$$q_S = \frac{10qS_0\rho^2}{\rho_S} \frac{(\tau_0 - \tau_{CR})}{(\rho_S - \rho)^2 gD} \text{ m}^3/\text{s} \qquad (9.9)$$

Here, again, the equation offers a solution for K_3.

Probabilistic equations. It has already been pointed out that grain movement is brought about by the impulsive force of turbulent eddies. Eddy action does not occur uniformly with time or space. It might therefore be thought that the incidence of an eddy capable of transporting a particular grain is some statistical function of time. H. A. Einstein (1942) proposed just such a probabilistic model of bedload for the case of a level bed of grains.

The basic ideas underlying Einstein's equation are the following:

(a) For an individual grain, migration will take place in a series of jumps (Fig. 9.6) of length $L = K_L D$. During a time T a series of n such jumps will occur, so that the particle will travel a total distance nL.

(b) The probability, p, that a grain will be eroded during the typical time scale, T, must be some function of the immersed self-weight of the particle and the fluid lift force acting on the particle. The immersed self-weight is $(\rho_S - \rho)g(K_V D^3)$, and the lift force is $C_L\rho(K_A D^2)u^2/2$, where grain area $A_S = K_A D^2$ and grain volume $V_S = K_V D^3$. Therefore,

$$p = f\left\{\frac{(\rho_S - \rho)g(K_V D^3)}{C_L\rho(K_A D^2)u^2/2}\right\} \qquad (9.10)$$

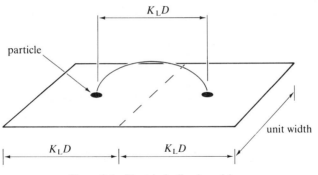

Figure 9.6 Einstein bedload model.

u is a 'typical' velocity at the sub-layer. Researchers have proposed that

$$u \simeq 11.6u_* \simeq 11.6\sqrt{gR'S_0}$$

where R' is that proportion of the hydraulic radius appropriate to sediment transport. Equation (9.10) is usually expressed as

$$p = f\{\mathbf{B}_* \cdot \Psi\} \tag{9.11}$$

where

$$\mathbf{B}_* = \frac{K_V}{C_L K_A 135/2} \qquad \Psi = \frac{(\rho_S - \rho)D}{\rho R' S_0} \tag{9.12}$$

(c) The number of grains of a given size in area A ($= K_L D \times 1$) is $K_L D / K_A D^2$, therefore the number of grains dislodged during time T will be $p K_L D / K_A D^2$. The volume of grains crossing a given boundary must therefore be

$$\frac{p K_L D}{K_A D^2} K_V D^3 = \frac{p K_L K_V D^2}{K_A} \tag{9.13}$$

The volume must also be given by $q_S T$. If the time T is some function of particle size and fall velocity, say $T = K_T D / v_{FS}$, then

$$q_S T = q_S \frac{K_T D}{v_{FS}} \tag{9.14}$$

Equating (9.13) and (9.14),

$$\frac{q_S K_T D}{v_{FS}} = \frac{p K_L K_V D^2}{K_A}$$

Therefore

$$p = \frac{q_S K_T K_A}{v_{FS} K_L K_V D} \qquad (9.15)$$

Equating (9.11) and (9.15), and evaluating v_{FS} from (9.2), leads, with some rearrangement, to

$$\Phi = q_S \sqrt{\frac{\rho}{(\rho_S - \rho)gD^3}} = f\{\mathbf{B}_*, \Psi\} \qquad (9.16)$$

where Φ is a dimensionless bedload function and $\mathbf{B}_*$ and Ψ have been defined above.

(d) Following Einstein, a number of researchers investigated the relationship between Φ and Ψ. A typical result is due to Brown (in Rouse 1950)

$$\Phi = 40(1/\Psi)^3 \qquad (9.17)$$

which is valid for $\Phi < 0.04$. As Φ (and therefore q_S) $\rightarrow 0$, $1/\Psi \rightarrow 0.056$, which corresponds to the Shields threshold condition.

Example 9.1   River bedload

A large European river has the following hydraulic characteristics: $Q = 450 \text{ m}^3/\text{s}$, width 50 m, depth 6 m, bed slope (S_0), 3×10^{-4}. The sediment has a typical diameter $D_{50} = 0.01$ m. Estimate the bedload transport, using the Shields and Einstein–Brown formulae.

Solution

From the data given,

$$R = \frac{A}{P} = \frac{50 \times 6}{50 + (2 \times 6)} = 4.84 \text{ m}$$

From (9.1a),

$$\tau_{CR} = 0.056(\rho_S - \rho)gD_{50} = 0.056(2650 - 1000) \times 9.81 \times 0.01$$

$$= 9.064 \text{ N/m}^2$$

Also,

$$\tau_0 = \rho g R S_0 = 1000 \times 9.81 \times 4.84 \times 3 \times 10^{-4}$$

$$= 14.245 \text{ N/m}^2$$

Using the Shields formula (equation (9.9)) with $q = 450/50 = 9$ m^3/s m

$$q_S = \frac{10 \times 9 \times 3 \times 10^{-4} \times 1000^2}{2650} \quad \frac{(14.245 - 9.064)}{(2650 - 1000)^2 \times 9.81 \times 0.01}$$

$$= 1.976 \times 10^{-4} \text{ m}^3/\text{s m}$$

Therefore

$$Q_S (= 50 \times q_S) = 0.01 \text{ m}^3/\text{s}$$

Using the Einstein–Brown formula, from (9.12),

$$\Psi = \frac{(\rho_S - \rho)D}{\rho R' S_0} = \frac{(2650 - 1000) \times 0.01}{1000 \times 4.84 \times 3 \times 10^{-4}} = 11.364$$

Notice that, in the absence of other information, it has been assumed that $R' = R$.

From (9.17),

$$\Phi = 40(1/11.364)^3 = 0.0273$$

and, using (9.16),

$$\Phi = q_S \sqrt{\frac{\rho}{(\rho_S - \rho)} \frac{1}{gD^3}} = q_S \sqrt{\frac{1000}{(2650 - 1000)} \frac{1}{9.81 \times 0.01^3}}$$

$$= q_S \times 248.6$$

Therefore

$$248.6 q_S = 0.0273$$

i.e.

$$q_S = 1.096 \times 10^{-4} \text{ m}^3/\text{s}$$

or

$$Q_S = 0.0055 \text{ m}^3/\text{s}$$

The substantial difference in the two estimates is not surprising. Of the two, the Einstein–Brown method is probably the more reliable. The answers obtained are the maximum probable values, the actual load may be less. It should also be emphasised that most bed load formulae apply primarily to coarse sand, and perhaps to some gravels.

Total load formulae. It is possible to calculate the total load from the sum of the bedload and suspended load. Separate equations are available for bedload (e.g. those in the preceding section) and for suspended load. However, experimental data are still rather sparse, and it is very difficult to separate bed and suspended load from these data. For this reason, some researchers have tackled directly the problem of total load. Two examples of total load formulae are outlined below.

Energy (stream power) formula (Bagnold 1966)

The equation is based on the immersed weight of sediment per unit bed area:

$$\text{total weight transport per unit channel width } (W_{\mathrm{T}}'U) = W_{\mathrm{b}}'U_{\mathrm{b}} + W_{\mathrm{s}}'U_{\mathrm{s}}$$

$$(9.18)$$

where b refers to bedload and s to suspended load, and the U terms refer to 'typical' sediment transport velocities.

The bedload component. From a consideration of the forces acting on a particle (Fig. 9.1b), movement will occur when $F_D >$ resistance. The resistance is assumed to be a function of W' (the immersed weight of the particles) and ϕ:

$$F_D = W' \tan \phi$$

So if W_{b}' is the immersed weight of particles per unit bed area being transported as bedload, then the bedload work rate is

$$W_{\mathrm{b}}'U_{\mathrm{b}} \tan \phi$$

The power, P, required to maintain the bed movement is provided by the flow. For uniform flow in a channel of rectangular cross section:

$$P = \rho g V b Y S_0 \quad \text{or} \quad P/b = \rho g V Y S_0 \qquad (9.19)$$

where $V = Q/A$. Only a fraction of the stream power is absorbed in bed movement. Bagnold used an 'efficiency' e_{b} to estimate the bedload power:

$$e_{\mathrm{b}}P/b = e_{\mathrm{b}}\rho g V Y S_0 = W_{\mathrm{b}}'U_{\mathrm{b}} \tan \phi \qquad (9.20a)$$

The suspended load component. If a suspended load exists, then the fluid must supply an effective upward velocity, which must be equal and opposite to v_{FS}. The suspended load work rate is given by

$$W_s' v_{FS} = W_s' U_s \frac{v_{FS}}{U_s} = W_s' U_s \tan \phi_s$$

As some power has been absorbed in bed movement, the power remaining is $(1 - e_b)P/b$. Power absorbed in suspended load transport is

$$(e_s P/b)(1 - e_b) = W_s' v_{FS} = W_s' U_s \tan \phi_s \qquad (9.20b)$$

where e_s is a suspended load 'efficiency'.

Total load. This is obtained by adding the two components

$$W_T' U = W_b' U_b + W_s' U_s = \frac{P}{b} \left\{ \frac{e_b}{\tan \phi} + \frac{e_s(1 - e_b)}{\tan \phi_s} \right\} \qquad (9.21)$$

($W_T' U = (\rho_S - \rho) g q_S$, where q_S is the total sediment transport per unit width.

This leaves the problem of evaluating e_b, e_s, $\tan \phi$, etc. Bagnold assumed a universal constant value for e_s ($= 0.015$) and hence estimated that $e_s(1 - e_b) \simeq 0.01$, both values being for fully turbulent flow. Some inevitable uncertainties were ignored in obtaining these values. The two remaining values have been reduced to analytical form.

tan ϕ.

$$\text{If } G^2 = \frac{\rho_S D^2 u_*^2}{\rho \times 14 \nu^2}$$

where G is analogous to the Reynolds Number, Bagnold's data lead to the following:

for $G^2 < 150$,

$$\tan \phi = 0.75 \qquad (9.22a)$$

for $150 < G^2 < 6000$,

$$\tan \phi = -0.236 \log G^2 + 1.25 \qquad (9.22b)$$

for $G^2 > 6000$,

$$\tan \phi = 0.374 \qquad (9.22c)$$

e_b. If $\rho_S = 2650$ kg/m^3 and $0.3 < V < 3.0$ m/s, then
for $0.015 < D$ mm < 0.06

$$e_b = -0.012 \log 3.28V + 0.15 \tag{9.23a}$$

for $0.06 < D$ mm < 0.2,

$$e_b = -0.013 \log 3.28V + 0.145 \tag{9.23b}$$

for $0.2 < D$ mm < 0.7,

$$e_b = -0.016 \log 3.28V + 0.139 \tag{9.23c}$$

for D mm > 0.7,

$$e_b = -0.028 \log 3.28V + 0.135 \tag{9.23d}$$

The Bagnold total load equation must thus be regarded as primarily a 'sand in water' transport equation for water depths $Y > 150$ mm and particle sizes limited to the range given above.

Ackers and White (A & W) formula (White 1972)

The A & W formula is one of the more recent developments in this field. Initially the underlying theoretical work was developed by considering the transport of coarse material (bedload) and fine material (suspended load) separately. Ackers and White then sought to establish 'transitional' relationships to account for the intermediate grain sizes. The functions which emerged are based upon three dimensionless quantities, G_{gr}, F_{gr} and D_{gr}: G_{gr} is the sediment transport parameter, which is based on the stream power concept. For bedload, the effective stream power is related to the velocity of flow and to the net shear force acting on the grains. Suspended load is assumed to be a function of total stream power, P. The particle mobility number, F_{gr}, is a function of shear stress/immersed weight of grains. The critical value of F_{gr} (i.e. the magnitude representing inception of motion) is denoted by A. Finally, a dimensionless particle size number, D_{gr}, expresses the relationship between immersed weight of grains and viscous forces.

The equations are then as follows:

$$G_{gr} = \frac{q_S D_m}{qD}\left[\frac{u_*}{V}\right]^n = C\left[\frac{F_{gr}}{A} - 1\right]^m \tag{9.24a}$$

$$F_{gr} = \frac{u_*^n}{\sqrt{gD[(\rho s/\rho) - 1]}} \left(\frac{V}{\sqrt{32} \log (10D_m/D)}\right)^{1-n} \qquad (9.24b)$$

$$D_{gr} = D\left(\frac{g[(\rho s/\rho) - 1]}{\nu^2}\right)^{1/3} \qquad (9.25)$$

(note that $V = Q/A$). The index n does have a physical significance, since its magnitude is related to D_{gr}. For fine grains $n = 1$, for coarse grains $n = 0$, and for transitional sizes $n = f(\log D_{gr})$.

The values for n, m, A, and C are as follows:

for $D_{gr} > 60$

$$n = 0, \; m = 1.5, \; A = 0.17, \; C = 0.025 \qquad (9.26a)$$

for $1 < D_{gr} < 60$

$$n = 1 - 0.56 \log D_{gr} \qquad (9.26b)$$

$$m = 1.34 + 9.66/D_{gr} \qquad (9.26c)$$

$$A = 0.14 + 0.23/\sqrt{D_{gr}} \qquad (9.26d)$$

$$\log C = 2.86 \log D_{gr} - (\log D_{gr})^2 - 3.53 \qquad (9.26e)$$

The equations have been calibrated by reference to a wide range of data, and good results are claimed − 'good results' in this context meaning that for 50% or more of the results,

$$\tfrac{1}{2} < \left(\frac{\text{estimated } q_S}{\text{measured } q_S}\right) < 2$$

Example 9.2   Siltation of reservoir

A reservoir having a capacity of 20×10^6 m³ is to be sited in a river valley. The river has the following characteristics: width 10 m; bed slope 1 in 3000; discharge 87 m³/s (assumed to be constant); depth 5 m. The river boundary is alluvial ($D_{50} = 0.3$ mm, $\rho s = 2650$ kg/m³). Estimate the time which would elapse before the reservoir capacity is reduced to half its original capacity. Assume a rectangular channel section.

Solution

Using Bagnold's method

Estimation of tan ϕ_s. The natural rate of fall v_{FS} may be obtained from the relationships

$$C_D = F/\tfrac{1}{2}\rho A v_{FS}^2$$

and

$$F = (\rho_S - \rho)gK_V D^3$$

(i.e. assuming that the only forces acting on the body are the immersed self-weight and the drag). For high Reynolds Numbers (and turbulent wake) $C_D = 0.44$ for a sphere.

If the sediment is assumed to be spherical, then the area $A = K_A D^2 = (\pi/4)D_{50}^2$, and the volume $= K_V D^3 = (\pi/6)D_{50}^3$. Therefore

$$0.44 = \frac{(2650 - 1000) \times 9.81 \times \pi \times (0.3 \times 10^{-3})^3 \times 4}{\tfrac{1}{2} \times 1000 \times \pi \times (0.3 \times 10^{-3})^2 \times v_{FS}^2 \times 6}$$

Hence

$$v_{FS} = 0.121 \text{ m/s}$$

Assuming that $U_s \rightarrow$ velocity of flow, then $U_s \simeq 87/(10 \times 5) = 1.74$ m/s $= V$. Therefore

$$\tan \phi_s = 0.121/1.74$$

Estimation of e_b and tan ϕ. For $D_{50} = 0.3$ mm, use (9.23c):

$$e_b = -0.016 \times \log(3.28 \times 1.74) + 0.139 = 0.127$$

$$R = 2.5 \text{ m}$$

$$u_*^2 = gRS_0 = 9.81 \times 2.5 \times 1/3000$$

Therefore

$$G^2 = \frac{2650 \times (0.3 \times 10^{-3})^2 \times 9.81 \times 2.5 \times 1}{14 \times 1000 \times (1.14 \times 10^{-6})^2 \times 3000} = 107.2$$

Therefore, from (9.22a), tan $\phi = 0.75$.

Estimation of stream power. From (9.19),

$P/b = \rho g VYS_0 = 1000 \times 9.81 \times 1.74 \times 5 \times 1/3000 = 28.45$ W/m. Thus, from (9.21),

$$W_{\mathrm{T}}^{\dagger}U = 28.45\left\{\frac{0.127}{0.75} + \left(0.01 \times \frac{1.74}{0.121}\right)\right\} = 8.91 \text{ N/ms}$$

Therefore, over 10 m width flux = $10 \times 8.91 = 89.1$ N/s. This represents a volume

$$Q_{\mathrm{S}} = \frac{89.1}{(2650 - 1000) \times 9.81 \times (1 - p_{\mathrm{S}})} \text{ m}^3/\text{s}$$

p_{S} is the porosity or voidage of the grains when packed closely together, and is usually approximately 0.3. Therefore, the rate at which the reservoir will fill is given by

$$\frac{89.1}{(2650 - 1000) \times 9.81 \times (1 - 0.3)} = 7.864 \times 10^{-3} \text{ m}^3/\text{s}$$

Therefore, the annual sediment volume = 247 320 m^3. So the sediment will have reduced the capacity to half its original value in

$$\frac{20 \times 10^6}{2 \times 247\,320} \simeq 40 \text{ years}$$

Using the A & W equations

$$u_* = \sqrt{\tau_0/\rho} = \sqrt{gRS_0} = \sqrt{9.81 \times 2.5 \times 1/3000}$$

$$= 0.0904 \text{ m/s}$$

From (9.25),

$$D_{\mathrm{gr}} = D\left(\frac{g[(\rho_{\mathrm{S}}/\rho) - 1]}{\nu^2}\right)^{\frac{1}{3}} = 0.3 \times 10^{-3}\left(\frac{9.81\,[(2650/1000) - 1]}{(1.1 \times 10^{-6})^2}\right)^{\frac{1}{3}}$$

$$= 7.12$$

Since D_{gr} is in the transitional range, (9.26b)–(9.26e) are used to calculate n, m, A and C:

from (9.26b)

$$n = 1 - 0.56 \log D_{\mathrm{gr}} = 1 - 0.56 \, (\log 7.12) = 0.5226$$

from (9.26c)

$$m = 1.34 + 9.66/D_{\mathrm{gr}} = 1.34 + 9.66/7.12 = 2.697$$

from (9.26d)

$$A = 0.14 + 0.23/\sqrt{D_{\mathrm{gr}}} = 0.14 + 0.23/\sqrt{7.12} = 0.2262$$

from (9.26e)

$$\log C = 2.86 \log D_{gr} - (\log D_{gr})^2 - 3.53$$

$$= 2.86(\log 7.12) - (\log 7.12)^2 - 3.53 = -1.8186$$

Therefore

$$C = 0.01518$$

Using (9.24b),

$$F_{gr} = \frac{u_*^n}{\sqrt{gD[(\rho s/\rho) - 1]}} \left(\frac{V}{\sqrt{32} \log(10D_m/D)} \right)^{1-n}$$

$$= \frac{0.0904^{0.5226}}{\sqrt{9.81 \times 0.3 \times 10^{-3}[(2650/1000) - 1]}} \left(\frac{1.74}{\sqrt{32} \log[10 \times 5/0.3 \times 10^{-3}]} \right)^{1-0.5226}$$

$$= 1.0574$$

Therefore, from (9.24a),

$$G_{gr} = \frac{q_s D_m}{qD} \left(\frac{u_*}{V} \right)^n = C \left(\frac{F_{gr}}{A} - 1 \right)^m$$

$$= \frac{q_s \times 5}{8.7 \times 0.3 \times 10^{-3}} \left(\frac{0.0904}{1.74} \right)^{0.5226} = 0.01518 \left(\frac{1.0574}{0.2262} - 1 \right)^{2.697}$$

Hence

$$q_s = 1.243 \times 10^{-3} \text{ m}^3/\text{s}$$

Therefore if the porosity, $p_s, = 0.3$, the reservoir will fill at a rate: $1.243 \times 10^{-3} \times 10/(1 - 0.3) = 17.757 \times 10^{-3} \text{ m}^3/\text{s}$

Therefore annual volume of sediment deposited is $559\,985$ m^3. Hence the reservoir capacity will be reduced by half in

$$\frac{20 \times 10^6}{2 \times 559\,985} \simeq 18 \text{ years}$$

A computer program for this method is given in Appendix B.

9.5 Concluding notes on transport formulae

For the sake of simplicity, this exposition of sediment transport has assumed that estimates may be based on a single 'typical' particle size (say D_{50}). In real situations, this may not give a true impression of the transport process, since the boundary will consist of a range of particle sizes. It is possible to extend the analysis to estimate the appropriate transport rate for each of the particle size fractions present. The extension is not as straightforward as it sounds, since:

(a) the exposed particles of a given size will constitute only a fraction of the total bed area;
(b) although each grain size has a theoretical threshold condition, some grains will be wholly or partly sheltered by surrounding grains, and will therefore move less readily than others.

Einstein and his co-workers produced an analysis which attempted to account for these conditions. A lucid account of their work is contained in Graf (1971).

All of the equations for sediment transport presuppose that the number and size of the particles eroded from a given area are in equilibrium with the incoming particle deposits supplied from upstream. This is not always the case. For example, where the finer fractions are eroded and not replaced, the nature of the channel surface changes. The remaining, coarser, particles are less readily eroded, so the channel becomes more stable and the sediment load in the water is reduced. This process is known as 'armouring'.

Where a mixture of sediments is present, transport will often be governed by one dominant fraction. This may not be the fraction present in greatest quantity. For example, the presence of just 10% clay may increase cohesion and reduce erosion quite substantially.

To the question 'which sediment transport equation is best?', there is as yet no clear answer. Further measurements, especially in the field, are still greatly needed. In the meantime, Einstein's work is generally held in high regard, though some of his later formulae are rather involved. Other researchers have also sought to refine the statistical approach further and to calibrate their formulae against measurements (see, for example, Cheong and Shen 1983). From the student's viewpoint, the Bagnold formulae have the advantage that they are clearly linked to the physics of the transport process. They seem to work well within their stated limits, but have been superseded by later work. The A & W equations also appear to work well, and they were calibrated against a wide range of laboratory flume data and field measurements. Research continues on an international scale, and further improvements will undoubtedly be published regularly. An excellent summary of formulae, with detailed comparisons of their performance, is given by White *et al.* (1973). For further reading, see Graf (1971), Raudkivi (1976) and Yalin (1977).

References and further reading

Bagnold, R. A. 1966. *An approach to the sediment transport problem from general physics*. Professional Paper, 422–I. Washington, DC: US Geological Survey.

Cheong, H. F. and H. W. Shen 1983. Statistical properties of sediment movement. *Am. Soc. Civ. Engrs, J. Hydraulic Engng* **109**(12), 1577–88.

Einstein, H. A. 1942. Formulas for the transportation of bedload. *Trans. Am. Soc. Civ. Engrs* **107**, 561–77.

Graf, W. H. 1971. *Hydraulics of sediment transport*. New York: McGraw-Hill.

Raudkivi, H. 1976. *Loose boundary hydraulics*, 2nd edn. Oxford: Pergamon.

Rijn, L. van 1984. Sediment transport (in 3 parts) *Am. Soc. Civ. Engrs, J. Hydraulic Engng*: Part I, **110**(10), 1431–56; Part II, **110**(11), 1613–41; Part III, **110**(12), 1733–54.

Rouse, H. (ed.) 1950. *Engineering hydraulics*. New York: Wiley.

Shields, A. 1936. Anwendung der Ahnlichkeitsmechanik und der Turbulenzforschung auf die Geschiebebewegung, Heft 26. Berlin: Preuss. Vers. für Wasserbau und Schiffbau.

White, W. R. 1972. *Sediment transport in channels, a general function*. Rep. Int. 102. Wallingford: Hydraulics Research.

White, W. R., H. Milli and A. D. Crabbe 1973. *Sediment transport, an appraisal of existing methods*. Rep. Int. 119. Wallingford: Hydraulics Research.

Yalin, M. S. 1977. *Mechanics of sediment transport*, 2nd edn. Oxford: Pergamon.

10

Flood hydrology

10.1 Classifications

Hydrology has been defined as the study of the occurrence, circulation and distribution of water over the world's surface. As such, it covers a vast area of endeavour and is not the exclusive preserve of civil engineers. Engineering hydrology is concerned with the quantitative relationship between rainfall and 'runoff' (i.e. passage of water on the surface of the Earth) and, in particular, with the magnitude and time variations of runoff. This is because all water resource schemes require such estimates to be made before design of the relevant structures may proceed. Examples include reservoir design, flood alleviation schemes and land drainage. Each of these examples involves different aspects of engineering hydrology, and all involve subsequent hydraulic analysis before safe and economical structures can be constructed.

Engineering hydrology is conveniently subdivided into two main areas of interest, namely, surface water hydrology and groundwater hydrology. The first of these is further subdivided into rural hydrology and urban hydrology, since the runoff response of these catchment types to rainfall is very different.

The most common use of engineering hydrology is the prediction of 'design' events. This may be considered analogous to the estimation of 'design' loads on structures. Design events do not mimic nature, but are merely a convenient way of designing safe and economical structures for water resources schemes. As civil engineers are principally concerned with the extremes of nature, design events may be either floods or droughts. The design of hydraulic structures will normally require the estimation of a suitable design flood (e.g. for spillway sizing) and sometimes a design drought (e.g. for reservoir capacity).

The purpose of this chapter is to introduce the reader to some of the concepts of engineering hydrology. The treatment is limited to the estimation of design floods for rural and urban catchments. This limitation is necessary because a complete introduction to hydrology would occupy a textbook in its own right. However, the chapter is considered useful, since hydrological design is the precursor to many hydraulic designs. Consequently, civil engineers should have an overall understanding of both subjects and their interactions.

10.2   Methods of flood prediction for rural catchments

Historically, civil engineers were faced with the problem of flood prediction long before the current methods of analysis were available. Two techniques in common use in the 19th and early 20th centuries were those of using the largest recorded 'historical' flood and the use of empirical formulae relating rainfall to runoff.

The former was generally the more accurate, but as runoff records were sparse, the latter was often used in practice. This was possible because rainfall records have been collected for much longer periods than have runoff records.

The occurrence of a series of catastrophic floods in the 1960s in the UK prompted the Institution of Civil Engineers to instigate a comprehensive research programme into methods of flood prediction. This was carried out at the Institute of Hydrology, and culminated in the publication of the *Flood studies report* (NERC 1975). The report represented a milestone in British hydrology, assimilating previous knowledge with new techniques which were comprehensively tested against an enormous data set of hydrological information. The result was the formulation of a new set of design methods which could be applied with greater confidence to a wide range of conditions.

There are fundamentally two types of flood prediction technique recommended in the *Flood studies report*. These are statistical methods (e.g. frequency analysis) and unit hydrograph methods. In addition, there are two types of catchment, those which are gauged (i.e. have recorded rainfall and runoff records) and those which are ungauged. One of the major aspects of the *Flood studies report* was the derivation of techniques which allow flood prediction for ungauged catchments. This involved finding quantitative relationships between catchment characteristics and flood magnitudes for large numbers of gauged catchments, and the application of these results to ungauged catchments by the use of multiple regression techniques.

In the following sections, the basic ideas of catchment characteristics, frequency analysis and unit hydrographs are introduced, and the application of these methods to gauged and ungauged catchments is discussed.

10.3 Catchment characteristics

A good starting point for a quantitative assessment of runoff is to consider the physical processes occurring in the hydrological cycle and within the catchment, as shown in Figures 10.1a and b.

Circulation of water takes place from the ocean to the atmosphere by evaporation, and this water is deposited on a catchment mainly as rainfall. From there, it may follow several routes, but eventually the water is returned to the sea via the rivers.

Within the catchment, several circulation routes are possible. Rainfall is initially intercepted by vegetation and may be re-evaporated. Secondly, infiltration into the soil or overland flow to a stream channel or river may occur. Water entering the soil layer may remain in storage (in the un-saturated zone) or may percolate to the groundwater table (the saturated zone). All subsurface water may move laterally and eventually enter a stream channel. The whole system may be viewed as a series of linked storage processes with inflows and outflows, as shown in Figure 10.1c. Such a representation is referred to as a conceptual model. If equations defining the storages and flows can be found, a mathematical catchment model can be constructed (see Fleming 1975).

Using this qualitative picture, a set of characteristics may be proposed which determine the response of the catchment to rainfall. These might include the following:

(a) catchment area;
(b) soil type(s) and depth(s);
(c) vegetation cover;
(d) stream slopes and surface slopes;
(e) rock type(s) and area(s);
(f) drainage network (natural and man-made);
(g) lakes and reservoirs;
(h) impermeable areas (e.g. roads, buildings, etc.).

In addition, different catchments will experience different climates, and hence the response of the catchment to rainfall will depend also on the prevailing climate. This may be represented by:

(a) rainfall (depth, duration and intensity);
(b) evaporation potential (derived from temperature, humidity, windspeed and solar radiation measurements or from evaporation pan record.

However, from an engineering viewpoint, qualitative measures of catchment characteristics are inadequate in themselves, and quantitative measures are necessary to predict flood magnitudes. This was one of the

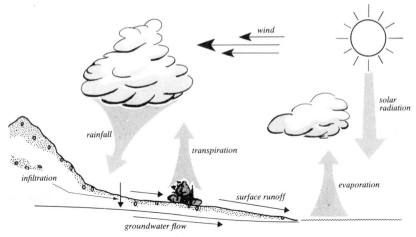

(a) The hydrological cycle

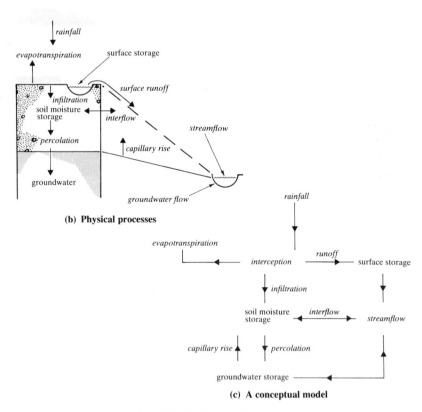

(b) Physical processes

(c) A conceptual model

Figure 10.1 Hydrological processes.

tasks performed by the Flood Studies team which led, for example, to the following equation for rural catchments:

$$\bar{Q} = \text{const.} \times \text{AREA}^{0.94} \ \text{STMFRQ}^{0.27} \ \text{S1085}^{0.16}$$

$$\text{SOIL}^{1.23} \ \text{RSMD}^{1.03} \ (1 + \text{LAKE})^{-0.85} \qquad (10.1)$$

where $\bar{Q}$ = the mean annual flood (m^3/s);
    const. = a number depending on location;
    AREA = the catchment area (km^2);
STMFRQ = the stream frequency (no. of stream junctions/AREA);
    S1085 = the slope of the main stream (m/km);
    SOIL = a number depending on soil type;
    RSMD = 1 day rainfall of 5 year return period minus the mean soil moisture deficit (mm);
    LAKE = the area of lakes or reservoirs (as a fraction of AREA).

Full details are given in the *Flood studies report* and in the *Guide to the flood studies report* issued subsequently (Sutcliffe 1978). This equation contains all the catchment characteristics which were found to be statistically significant, and may be applied to ungauged catchments. However, it should only be used for UK catchments, and be limited to those whose characteristics are within the bounds of the original data set from which it was derived. The equation will only give an approximate value for $\bar{Q}$, and this reflects the difficulty of predicting natural events with any certainty.

10.4 Frequency analysis

For gauged catchments with long records (e.g. greater than 25 years) the techniques of frequency analysis may be applied directly to determine the magnitude of any flood event (Q) with a specified return period (T_r). The concept of return period is an important one because it enables the determination of risk (economic or otherwise) associated with a given flood magnitude. It may be formally defined as the number of years, on average, between a flood event of magnitude (X) which is greater than or equal to a specified value (Q). The qualifier 'on average' is often misunderstood. For example, although a 100 year flood event will occur, on average, once every 100 years, it may occur at any time (i.e. today or in several years' time). Also, within any particular 100 year period, floods of greater magnitude may occur.

The probability of occurrence $P(X \geqslant Q)$ is inversely related to return period (T_r), i.e.

$$P(X \geqslant Q) = 1/T_r \qquad (10.2)$$

This relationship is the starting point of frequency analysis.

The annual maxima series

This is the simplest form of frequency analysis, in which the largest flood event from each year of record is abstracted. The resulting series, in statistical terms, is considered to be an independent series, and constitutes a random sample from an unknown population. The series may be plotted as a histogram, as shown in Figure 10.2a. Taking, as an example, a 31 year record, the annual maxima are divided into equal class intervals (0–10, 10–20, etc.). The probability that the discharge will exceed, say, 60 m^3/s is equal to the number of events greater than 60 m^3/s divided by the total number of events:

$$P(X \geqslant 60) = (4 + 3 + 2)/31 = 0.29$$

and the corresponding return period is

$$T_r = 1/P(X \geqslant 60) = 3.4 \text{ years}$$

If the histogram is now replaced by a smooth curve, as shown in Figure 10.2b, then

$$P(X \geqslant Q) = \int_Q^\infty f(x)\, dx$$

The function $f(x)$ is known as a probability density function (pdf) and, by definition,

$$\int_0^\infty f(x)\, dx = 1$$

(i.e. $P(X \geqslant 0) = 1$). Hence, the scale equivalent to the histogram ordinate is given by $n/N\Delta Q$ as

$$\sum_0^Q \left(\frac{n}{N\Delta Q} \Delta Q \right) = 1$$

The point of this analysis is that it makes it possible to estimate the probability that the discharge will exceed any given value greater than the maximum value in the data set (90 m^3/s in this case). Replacing the histogram with the pdf allows such estimates to be made.

The technique which is used in practice looks rather different from the histogram, so the method is now extended. If, instead of drawing a pdf, the cumulative probability of non-exceedence $(F(x))$ is drawn, i.e.

$$F(X) = P(X \leqslant Q) = \int_0^Q f(x)\, dx$$

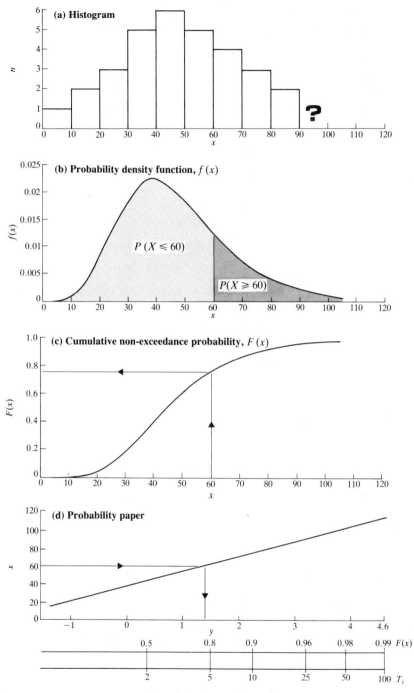

Figure 10.2 Frequency analysis.

then Figure 10.2c is the result. This curve may be transformed to a straight line by 'stretching' the $F(x)$ scale as shown in Figure 10.2d, with the discharge axis now drawn as ordinate and $F(x)$ represented by the reduced variate y, which has a linear scale. Figure 10.2d is known as probability paper. The probability scale is non-linear and depends on the shape of the original pdf. However, if the data are now plotted on this type of graph paper, it will fall on a straight line, and hence extrapolation beyond the data set is made easy. (Note: this is only true if the pdf is an exact fit to the data!)

Plotting positions

To plot the data on suitable probability paper requires the estimation of the probability of exceedence (or non-exceedence) of each event. As an initial estimate, one could say that

$$P(X \geqslant Q) = i/N$$

where i is the rank number and N is the total number of events, and where the data are ranked from the highest value, with $i = 1$, to the lowest value, with $i = N$.

However, this formula (known as the Weibull formula) is unsatisfactory because the data set is a sample drawn from an unknown population. The largest value in the sample may have a considerably higher return period than that suggested by this simple formula which exhibits bias at the extremes of the distribution. A more general formula is

$$P(X \geqslant Q) = \frac{i - a}{N + b}$$

where a and b are constants for particular pdfs.

For the case of extreme value distributions (discussed in the following section) a more unbiased plotting position (due to Gringorten) is given by

$$P(X \geqslant Q) = \frac{i - 0.44}{N + 0.12} \tag{10.3}$$

Hence

$$F(X) = P(X \leqslant Q) = 1 - \left(\frac{i - 0.44}{N + 0.12}\right)$$

For example, again using the data set with $N = 31$ for the largest discharge ($90 \ \text{m}^3/\text{s}$), $i = 1$ and

$$P(X \geqslant 90) = \frac{1 - 0.44}{31 + 0.12} = 0.018$$

$$F(X) = P(X \leqslant 90) = 1 - 0.018 = 0.982$$

and

$$T_r = 1/P(X \geqslant 90) = 55.6 \text{ years}$$

In other words, the expected return period of the largest discharge in a 31 year period is 55.6 years.

The extreme value type 1 distribution

Various pdfs have been tested to see if they fit hydrological data sets. A general result is that such data tends to be asymmetrical and that no single pdf is universally applicable. One family of pdfs which have been recommended in the *Flood studies report* is that of the 'general extreme value distributions' (GEVs). A particular example of this is the so-called extreme value type 1 (EV1).

For an EV1,

$$f(x) = \frac{1}{\alpha} \exp\left[-\left(\frac{x-u}{\alpha}\right) - e^{-[(x-u)/\alpha]} \right] \tag{10.4}$$

and

$$F(x) = \exp(- e^{-[(x-u)/\alpha]}) \tag{10.5}$$

To transpose the $F(x)$ to a linear scale, the reduced variate (y) is used, where

$$y = -\ln\{-\ln[F(x)]\} \tag{10.6}$$

and

$$x = u + \alpha y \tag{10.7}$$

(i.e. a straight line). The two parameters u and α are known as location and scale parameters, respectively. For an EV1 they define the shape of the pdf and may be estimated from the mean $(\bar{x})$ and standard deviation (s) of the data set, using the method of moments to give

$$u = \bar{x} - 0.45s \tag{10.8}$$

$$\alpha = 0.78s \tag{10.9}$$

Hence, for any given data set an EV1 distribution may be 'fitted' by estimating the location and scale parameters from (10.8) and (10.9). The

fitted EV1 may then be drawn as a straight line on EV1 probability paper
by substituting u and α in (10.7) for any two values of y to find the corresponding values of x. The procedure is illustrated in Example 10.1 and
a computer program is included in Appendix B.

General extreme value distributions

The EV1 distribution is the simplest of the family of GEVs. There are two
more members of this family – the EV2 and EV3. These two require the
estimation of a third parameter (k), known as the shape parameter.
Unfortunately, it is not possible to estimate this parameter with any
certainty, except for very long records. Consequently, the *Flood studies
report* recommends that k should be selected according to region (see Fig.
10.3b), from results derived from regional pooling of data (*Flood studies
report* Vol. I, table 2.38). This may then be combined with the location and
scale parameters (u and α) derived from the record.

The reduced variate form of a GEV is given by

$$x = u + \alpha W$$

where $W = (1 - e^{-ky})/k$. For $k = 0$, this reduces to the EV1 case. For $k < 0$,
an EV2 results and for $k > 0$, an EV3. All three cases may be plotted on
EV1 probability paper as shown in Figure 10.4.

Application of frequency analysis to ungauged catchments

Obviously, frequency analysis may not be directly applied to ungauged
catchments. A very powerful technique designed to overcome this problem,
developed in the *Flood studies report,* is that of the so-called region curves.
This allows estimation of the magnitude of the flood peak of any return
period for ungauged catchments. A region curve is a dimensionless plot of
the ratio of flood peak (Q_{Tr}) of return period T_r to mean annual flood $(\bar{Q})$
against return period (T_r). By combining the records of gauged catchments
in a particular region, a single region curve may be plotted. The region
curves and geographical locations for the UK are shown in Figures 10.3a
and b.

For ungauged catchments $\bar{Q}$ may be estimated from (10.1) using catchment characteristics and $Q_{Tr}/\bar{Q}$ from the region curve. Additionally, for
short records frequency analysis is unreliable, and hence, in this case $\bar{Q}$ may
be estimated from the record and Q_{Tr} can be found using the region curves.

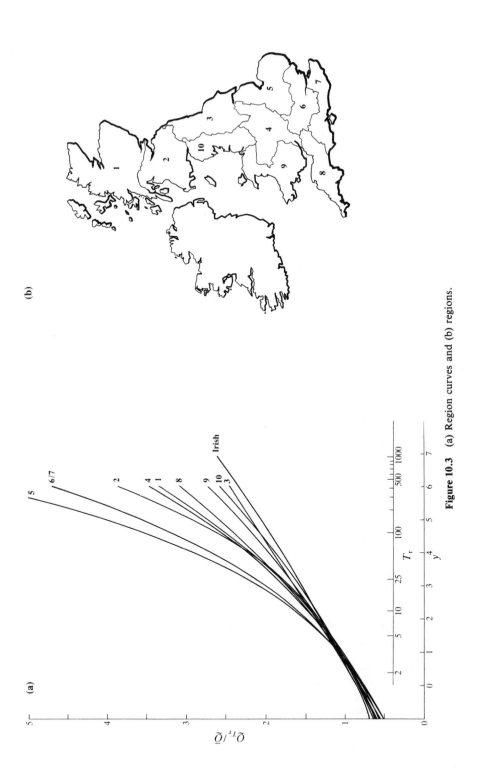

Figure 10.3 (a) Region curves and (b) regions.

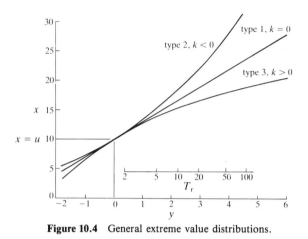

Figure 10.4   General extreme value distributions.

10.5   Unit hydrograph theory

Components of a natural hydrograph

Figure 10.5a shows a flood hydrograph and the causative rainfall. The hydrograph is composed of two parts, the surface runoff, which is formed directly from the rainfall, and the base flow. The latter is supplied from groundwater sources which do not generally respond quickly to rainfall. The rainfall may also be considered to be composed of two parts. The net or effective rainfall is that part which forms the surface runoff, while the rainfall losses constitute the remaining rainfall (this is either evaporated or enters soil moisture and groundwater storages). Simple techniques for separating runoff and rainfall have been developed, and are described elsewhere (see Shaw 1983, Wilson 1983).

Unit hydrograph principles

The net rainfall and corresponding surface runoff are shown in Figure 10.5b. The purpose of unit hydrograph theory is to be able to predict the relationships between the two for any storm event. A unit hydrograph is thus a simple model of the response of a catchment to rainfall.

This concept was first introduced by Sherman in 1932, and rests on three assumptions as shown in Figures 10.6a–c. These are:

(a) any uniform net rainfall having a given duration will produce runoff of specific duration, regardless of intensity;
(b) the ratios of runoff equal the ratios of net rainfall intensities, provided that the rainfalls are of equal duration;
(c) the hydrograph representing a combination of several runoff events is the sum of the individual contributory events, i.e. the principle of superposition may be applied.

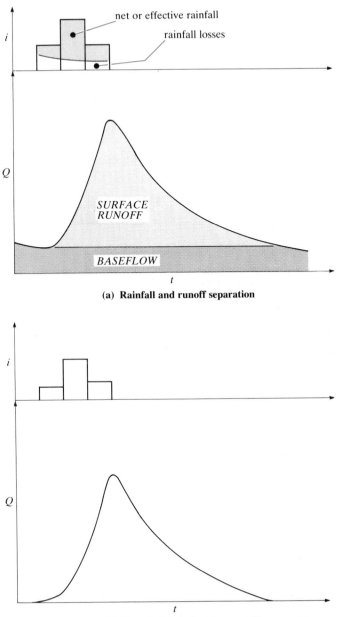

net or effective rainfall

rainfall losses

i

Q

SURFACE
RUNOFF

BASEFLOW

t

(a) Rainfall and runoff separation

i

Q

t

(b) Net rainfall and surface runoff

Figure 10.5 Hydrograph analysis.

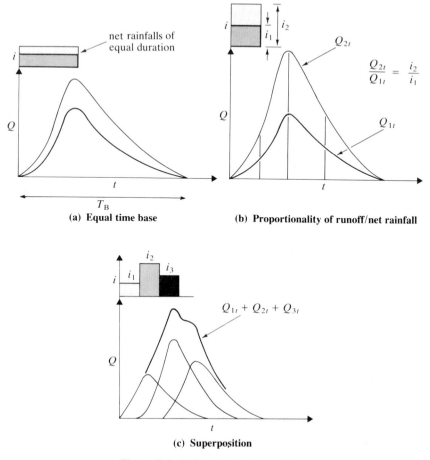

Figure 10.6 Unit hydrograph principles.

These assumptions all imply that the response of a catchment to rainfall is linear, which is not true. However, unit hydrographs have been found to work reasonably well for a wide range of conditions. Unit hydrograph theory has, therefore, been used extensively for design flood prediction.

Unit hydrograph definition and convolution

The foregoing principles are embodied in the definition of the unit hydrograph and its application.

The P mm, D hour unit hydrograph is the hydrograph of surface runoff produced by P mm of net rainfall in D hours, provided the net rainfall falls uniformly over the catchment in both space and time. Both P and D may have any values, but commonly P is taken as 10 mm and D as 1 h.

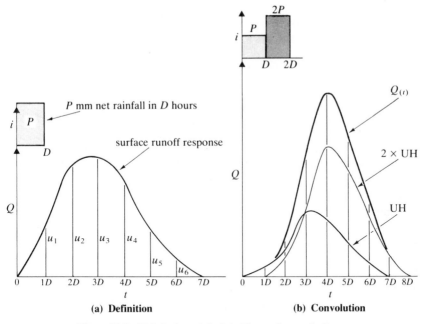

Figure 10.7 Unit hydrograph definition and convolution.

Table 10.1 Tabular and matrix methods of convolution.

Time	Rainfall	u_1	u_2	u_3	u_4	u_5	u_6	Surface runoff
				Unit hydrograph ordinates				
1	P_1	p_1u_1						q_1
2	P_2	p_2u_1	p_1u_2					q_2
3			p_2u_2	p_1u_3				q_3
4				p_2u_3	p_1u_4			q_4
5					p_2u_4	p_1u_5		q_5
6						p_2u_5	p_1u_6	q_6
7							p_2u_6	q_7

Equivalent matrix form

$$
\begin{vmatrix}
p_1 & 0 & 0 & 0 & 0 & 0 \\
p_2 & p_1 & 0 & 0 & 0 & 0 \\
0 & p_2 & p_1 & 0 & 0 & 0 \\
0 & 0 & p_2 & p_1 & 0 & 0 \\
0 & 0 & 0 & p_2 & p_1 & 0 \\
0 & 0 & 0 & 0 & p_2 & p_1 \\
0 & 0 & 0 & 0 & 0 & p_2
\end{vmatrix}
\cdot
\begin{vmatrix}
u_1 \\ u_2 \\ u_3 \\ u_4 \\ u_5 \\ u_6
\end{vmatrix}
=
\begin{vmatrix}
q_1 \\ q_2 \\ q_3 \\ q_4 \\ q_5 \\ q_6 \\ q_7
\end{vmatrix}
$$

Note: $p_i = P_i/P$ for a P mm, D hour unit hydrograph; i.e. if $P_1 = 30$ mm and the unit hydrograph is a 10 mm, D hour unit hydrograph, then $p_1 = 30/10 = 3$.

Once a unit hydrograph has been derived for a catchment, it may be used to predict the surface runoff for any storm event by the process of convolution. This is shown diagramatically in Figure 10.7b, along with the unit hydrograph definition (Fig. 10.7a).

The process may also be expressed in matrix form as

$$\mathbf{P} \cdot \mathbf{U} = \mathbf{q} \tag{10.10}$$

where $\mathbf{P}$ is the matrix of net rainfalls, $\mathbf{U}$ is the matrix of unit hydrograph ordinates and $\mathbf{q}$ is the matrix of surface runoff ordinates.

Alternatively, the process may be laid out in tabular form as shown in Table 10.1. A computer program for the matrix method is included in Appendix B.

Derivation of unit hydrographs

For gauged catchments, unit hydrographs are derived by hydrograph analysis of measured storm events. In general, storm events are complex (i.e. varying intensity and duration), and the unit hydrograph must be 'unearthed' from such events. As surface runoff can be predicted by convoluting the unit hydrograph with net rainfall, the converse must be true.

The matrix equation (10.10) may be solved for $\mathbf{U}$ by inversion. First $\mathbf{P}$ is converted into a square matrix by pre-multiplying by its transpose $|\mathbf{P}^T|$, hence:

$$\mathbf{P}^T \cdot \mathbf{P} \cdot \mathbf{U} = \mathbf{P}^T \cdot \mathbf{q}$$

This equation may now be inverted to give

$$\mathbf{U} = [\mathbf{P}^T \cdot \mathbf{P}]^{-1} \cdot \mathbf{P}^T \cdot \mathbf{q}$$

In practice, the method suffers from complications due to instability which arises because the unit hydrograph is only an approximate model and because of data errors. Full details of this method and its application are given in the Institute of Hydrology Report No. 71 (1981).

Synthetic unit hydrographs

For ungauged catchments, unit hydrographs cannot be derived directly. However, measures of catchment characteristics may be used to estimate a unit hydrograph. Such unit hydrographs are termed 'synthetic', and examples of two are shown in Figure 10.8. They have a simple triangular form, whose shape is determined by three parameters – the time to peak (T_p), the peak runoff (Q_p) and the time base (T_B).

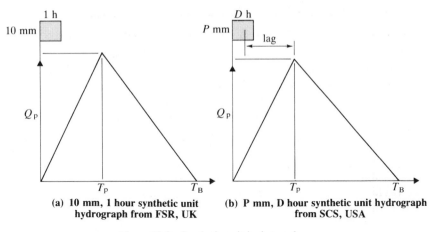

(a) **10 mm, 1 hour synthetic unit hydrograph from FSR, UK**
(b) **P mm, D hour synthetic unit hydrograph from SCS, USA**

Figure 10.8 Synthetic unit hydrographs.

The 10 mm, 1 hour synthetic unit hydrograph in the *Flood studies report* was derived using multiple regression techniques on data from gauged (rural) catchments, and is given by

$$T_p = 46.6\ L^{0.14}\ \mathrm{S}1085^{-0.38}\ (1 + \mathrm{URB})^{-1.99}\ \mathrm{RSMD}^{-0.4} \qquad (10.11)$$

$$Q_p = 2.2\ \mathrm{AREA}/T_p \qquad\qquad\qquad\qquad\qquad (10.12)$$

$$T_B = 2.52\ T_p \qquad\qquad\qquad\qquad\qquad\qquad (10.13)$$

where L is the mainstream length (km), URB is the fraction of catchment urbanised and S1085 and RSMD are as defined in (10.1), and where T_p and T_B are in hours and Q_p is in m^3/s.

The P mm, D hour synthetic unit hydrograph derived by the US Soil Conservation Service is given by

$$T_p = \mathrm{lag} + D/2 \qquad Q_p = 0.208AP/T_p \qquad T_B = 2.67T_p$$

where lag $\simeq 0.6\ t_c$, t_c is the time of concentration (h), A is the catchment area (km) and P is the net rainfall (mm), and where T_p and T_B are in hours and Q_p is in m^3/s.

These two unit hydrographs have a very similar shape. The main difference lies in the estimation of T_p, American practice being to calculate t_c rather than T_p from empirical formulae.

Design flood estimation using the unit hydrograph method

Figure 10.9 shows, in outline, the procedure for estimating a design flood event for any given return period. It is universally applicable. The central component of this procedure is the determination of the appropriate storm duration and associated rainfall depth and profile. Underestimation of the peak runoff will occur if the selected duration is either too short or too long, but the peak runoff is more sensitive to underestimation of the storm duration.

For the UK, full details of each component are given in the *Flood studies report,* Volumes I and II, and a summary is presented in the *Guide to the flood studies report.* For the USA, full details may be found in the *National engineering handbook* (US Soil Conservation Service 1972).

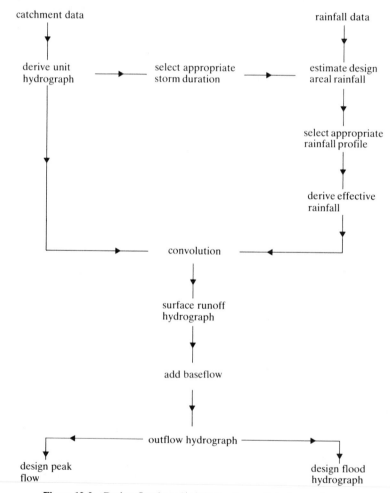

Figure 10.9 Design flood methodology using unit hydrographs.

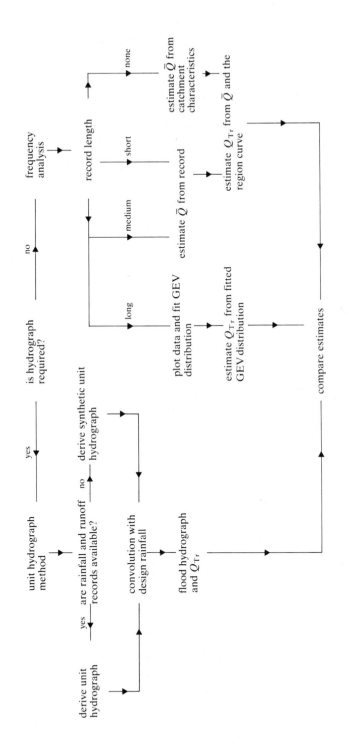

Figure 10.10 Summary of design flood estimation methods.

10.6   Summary of design flood procedures for rural catchments

This is illustrated in Figure 10.10. The two basic approaches of frequency analysis and unit hydrographs are complementary. The accuracy of each method depends on the amount and quality of available data. Estimates from gauged catchments are more accurate than those from ungauged catchments. It is prudent to estimate Q_{Tr} by both these methods. They should then be compared, in the light of the available data (and its influence on the accuracy of the result), before making a final choice of design flood magnitude.

Example 10.1   Design flood estimation

Estimate the 100 year flood event for the River Nidd at Hunsingore Weir using the following methods:

(a) frequency analysis of the runoff record;
(b) regional growth curves in conjunction with (i) $\bar{Q}$ from data and (ii) $\bar{Q}$ from catchment characteristics;
(c) synthetic unit hydrograph analysis.

Compare the results, and suggest a suitable design value.

Data. Table 10.2 lists the annual maximum discharges for water years 1934/5 to 1968/9. Table 10.3 lists the catchment characteristics.

Table 10.2   Annual maximum discharges for the River Nidd at Hunsingore weir.

Water year (Oct./Sept.)	Peak discharge (m^3/s)	Water year (Oct./Sept.)	Peak discharge (m^3/s)
1934/5	189.02	1952/3	65.60
1935/6	91.80	1953/4	213.70
1936/7	115.52	1954/5	86.73
1937/8	78.55	1955/6	111.74
1938/9	162.99	1956/7	75.06
1939/40	—	1957/8	100.40
1940/1	—	1958/9	81.27
1941/2	65.08	1959/60	92.82
1942/3	76.22	1960/1	151.79
1943/4	149.30	1961/2	88.89
1944/5	—	1962/3	111.54
1945/6	261.82	1963/4	148.63
1946/7	181.59	1964/5	87.76
1947/8	90.28	1965/6	251.96
1948/9	95.47	1966/7	138.72
1949/50	158.01	1967/8	305.75
1950/1	172.92	1968/9	226.48
1951/2	179.12		

Table 10.3 Catchment characteristics for the River Nidd at Hunsingore weir.

AREA (km^2)	484
L (km)	65.09
S1085 (m/km)	4.01
STMFRQ (no./km^2)	1.23
SAAR (mm)	993
r	0.33
M52D (mm)	61.0
SMDBAR (mm)	8.7
RSMD (mm)	34.5
SOIL	0.483
URB	0.02
LAKE	0.25
REGION	3
const	0.0213

Solution

(a) Table 10.4 lists the ranked discharges from highest to lowest, and associated expected return periods based on the Gringorten plotting position formula (Equation (10.3)). These data are plotted in Figure 10.11 using an EV1 distribution and the fitted distribution is shown. Table 10.4 also lists the mean and standard deviation of the series, the location and scale parameters (calculated from (10.8) and (10.9)) and the predicted flood magnitudes for various return periods (calculated from (10.7)). Figure 10.12 shows the histogram of the data series and the fitted pdf for comparison.

(b) (i) $\bar{Q}$ from the data = 137.7 m^3/s (see Table 10.4).

(ii) To find $\bar{Q}$ from catchment characteristics, use (10.1):

$$\bar{Q} = 0.0213 \times 484^{0.94} \times 1.23^{0.27} \times 4.01^{0.16} \times 0.438^{1.23} \times 34.5^{1.03} \times (1 + 0.25)^{-0.85}$$

$$= 108 \text{ m}^3/\text{s}$$

Using Figure 10.4a, the ratios of $Q_{Tr}/\bar{Q}$ for various return periods may be abstracted. They are summarised below, together with the corresponding Q_{Tr} estimates for (i) and (ii)

T_r	5	10	25	50	100
$Q_{Tr}/\bar{Q}$	1.25	1.45	1.7	1.9	2.08
$\bar{Q}$ data	172.1	199.7	234.1	261.6	286.4
$\bar{Q}$ characteristics	135.0	156.6	183.6	205.2	224.6

These results have been plotted in Figure 10.11.

(c) Full details of this method can be found in the *Guide to the flood studies report,* and only a summary is given here.

Table 10.4 Frequency analysis for the River Nidd at Hunsingore weir.

Annual maximum discharge (m^3/s)	Year	Return period (years)
305.8	1967/8	57.4
261.8	1945/6	20.6
252.0	1965/6	12.5
226.5	1968/9	9.0
213.7	1953/4	7.0
189.0	1934/5	5.8
181.6	1946/7	4.9
179.1	1951/2	4.2
172.9	1950/1	3.8
163.0	1938/9	3.4
158.0	1949/50	3.0
151.8	1960/1	2.8
149.3	1943/4	2.6
148.6	1963/4	2.4
137.8	1966/7	2.2
115.5	1936/7	2.1
111.7	1955/6	1.9
111.5	1962/3	1.8
100.4	1957/8	1.7
95.5	1948/9	1.6
92.8	1959/60	1.6
91.8	1935/6	1.5
90.3	1947/8	1.4
88.9	1961/2	1.4
87.8	1964/5	1.3
86.7	1954/5	1.3
81.3	1958/9	1.2
78.6	1937/8	1.2
76.2	1942/3	1.1
75.1	1956/7	1.1
65.6	1952/3	1.1
65.1	1941/2	1.1

Statistics

Mean = 137.70

Standard deviation = 63.02

$u = 109.35$

$\alpha = 49.15$

Fitted distribution $Q = 109.35 + 49.15\,y$

Predicted values from the fitted distribution.

Return period (years)	Discharge (m^3/s)
100	335.5
50	301.1
25	266.6
10	220.0
5	183.1

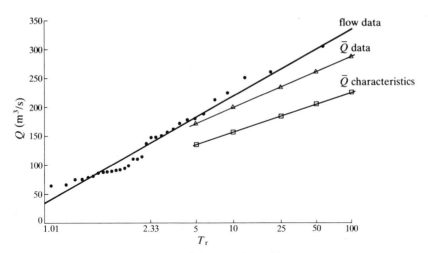

Figure 10.11   Frequency analysis for the River Nidd at Hunsingore weir.

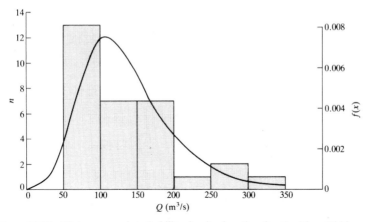

Figure 10.12   Histogram and probability density function for the River Nidd.

First calculate the parameters of the synthetic 10 mm, 1 hour unit hydrograph:

$$T_p = 46.6 \times 65.09^{0.14} \times 4.01^{-0.38} \times (1 + 0.02)^{-1.99} \times 34.5^{-0.4} \quad \text{(from (10.11))}$$

$$T_p = 11.5 \text{ h}$$

$$Q_p = 2.2 \times 484/11.5 \qquad \text{(from (10.12))}$$

$$Q_p = 92.6 \text{ m}^3/\text{s}$$

$$T_B = 2.52T_p \qquad \text{(from (10.13))}$$

$$T_B = 29 \text{ h}$$

To obtain a unit hydrograph ordinate corresponding to Q_p and to reduce calculation a new T-hour unit hydrograph is formed such that

$$T_p' = 5T$$

and

$$T_p' = T_p + (T - 1)/2$$

Hence

$$T = 2.44 \text{ h} \qquad T_p' = 12.2 \text{ h} \qquad Q_p' = 87.1 \text{ m}^3/\text{s} \qquad T_B' = 30.8 \text{ h}$$

Table 10.5 Synthetic unit hydrograph analysis for the River Nidd at Hunsingore weir.
Unit hydrograph ordinates ($m^3/s/10$ mm net rain).

u (m^3/s)	17.4	34.8	52.3	69.7	87.1	75.4	64.1	52.4	41.2	30.0	18.3
time (h)	2.4	4.9	7.3	9.8	12.2	14.6	17.1	19.5	22.0	24.4	26.8

Results of convolution.

Time (h)	Net rain (mm)	Surface runoff (m^3/s)	Runoff + baseflow (m^3/s)
2.4	0.7	1.2	14.9
4.9	1.8	5.6	19.3
7.3	2.9	14.9	28.6
9.8	4.3	31.6	45.3
12.2	8.5	63.1	76.8
14.6	10.6	111.1	124.8
17.1	8.5	168.6	182.3
19.5	4.3	225.4	239.1
22.0	2.9	274.8	288.5
24.4	1.8	302.6	316.3
26.8	0.7	301.1	314.8
29.3		274.7	288.4
31.7		236.1	249.8
34.2		190.8	204.5
36.6		142.6	156.3
39.0		94.2	107.9
41.5		54.0	67.7
43.9		27.4	41.1
46.4		13.5	27.2
48.8		5.4	19.1
51.2		1.3	15.0

Next, a storm duration (D) must be selected. The *Flood studies report* recommends

$$D = T_p(1 + \text{SAAR}/1000)$$

and

$$D = \text{odd multiple integer of } T$$

In this case,

$$D = 12.2(1 + 993/1000) = 24.3 \text{ h}$$

Take $D = 11 \times T = 26.8$ h. Using this duration, an areal rainfall depth may be calculated which will produce a 100 year flood event:

$$P = 93.4 \text{ mm}$$

This rainfall depth is distributed according to a selected profile (75% winter profile from *Flood studies report* Vol. II), and the net rainfall is found by multiplying by the percentage runoff (PR) equation (from *Flood studies report* Vol. I).

Finally the net rainfall profile is convoluted with the unit hydrograph and a baseflow allowance added. This is summarised in Table 10.5 and plotted in Figure 10.13.

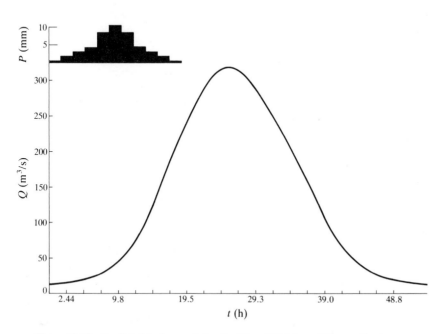

Figure 10.13 Predicted hydrograph for the River Nidd for a 100 year flood event.

Summary of results

Method	Q_{100} (m^3/s)
(a) EV1	336
(b) (i) $\bar{Q}$ data + region curve	286
(ii) $\bar{Q}$ characteristics + region curve	225
(c) synthetic unit hydrograph	316

Method (a) is generally reliable for long records (32 years in this case). However, its use should be restricted to predicting Q_{Tr} for $T_r < 2N$. The data are not a perfect fit to an EV1, as illustrated in the pdf and histogram of Figure 10.12. The largest recorded flood in the record is 306 m³/s, and the probability that a 100 year event will occur in a 32 year period is 27.5%. This suggests that the 100 year event is likely to be in excess of 300 m³/s. (Note: for region 3, $k = 0$ and hence the GEV corresponds to an EV1.)

For larger return periods, method (b) is considered to give a result with a smaller margin of error than method (a), since the region curves were based on a pooled data set containing several hundred station years. However, in this particular case it does not appear to be superior to method (a) when plotted (refer to Fig. 10.11).

Method (b) (ii). This is primarily for use on ungauged catchments. It illustrates the wide differences in estimates that can occur with limited data.

Method (c). Because the unit hydrograph was derived from catchment characteristics, it is subject to the same criticisms as method (b) (ii). However, it does give a design flood hydrograph, which is necessary when flood routing calculations are required. The runoff ordinates may be scaled up or down, according to the chosen value of Q_{Tr}. If a unit hydrograph had been derived from hydrograph analysis, the method should have given a more reliable estimate.

In conclusion, the most likely value of the 100 year flood event is 336 m³/s, as given by method (a). An alternative would be to average methods (a) and (b) (i) to give 311 m³/s.

10.7 Flood routing

General principles

So far, the discussion has centred on methods of estimating flood events at a given location. However, the engineer requires estimates of both the stage and discharge along a watercourse resulting from the passage of a flood wave. The technique of flood routing is used for this purpose.

There are two distinct kinds of problem:

(a) reservoir routing: to find the outflow hydrograph over the spillway from the inflow hydrograph.

(b) channel routing: to find the outflow hydrograph from a river reach from the inflow hydrograph.

In each case the peak flow of the outflow hydrograph is less than and later than that of the inflow hydrograph. These processes are referred to as 'attenuation' and 'translation', respectively.

To determine the outflow hydrograph from the inflow hydrograph requires the application of the continuity equation in the form:

$$I - O = dV/dt \tag{10.14}$$

where I is the inflow rate, O the outflow rate, V is the volume and t is time.

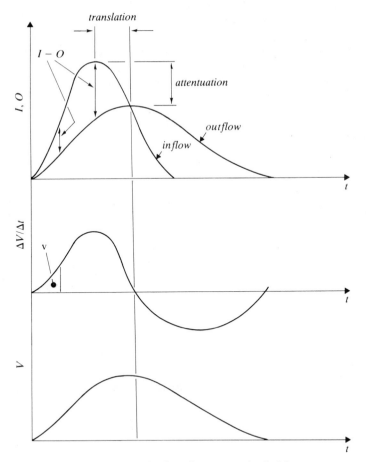

Figure 10.14 Flood routing – general principles.

This is shown diagramatically in Figure 10.14. Expressing (10.14) in a finite difference form gives

$$\frac{(I_t + I_{t+\Delta t})}{2}\Delta t - \frac{(O_t + O_{t+\Delta t})}{2}\Delta t = V_{t+\Delta t} - V_t \qquad (10.15)$$

where I_t and O_t are inflow and outflow rates at time t and $I_{t+\Delta t}$ and $O_{t+\Delta t}$ are inflow and outflow rates at time $t + \Delta t$. This equation may be solved successively through time for a known inflow hydrograph if the storage volume can be related to outflow (reservoir case) or channel properties.

Reservoir routing

This case is shown in Figure 10.15. The outflow is governed by the height (stage h) of water above the spillway crest level, and the volume of live storage is also governed by this height. Hence, for a given reservoir, both the volume and ouflow can be expressed as functions of stage. This is achieved by a topographical survey and application of a suitable weir equation, respectively.

Rearranging (10.15) in terms of unknown and known values,

$$\frac{2V_{t+\Delta t}}{\Delta t} + O_{t+\Delta t} = I_t + I_{t+\Delta t} + \frac{2V_t}{\Delta t} - O_t$$

or

$$\boxed{\frac{2V_{t+\Delta t}}{\Delta t} + O_{t+\Delta t}} = I_t + I_{t+\Delta t} + \boxed{\frac{2V_t}{\Delta t} + O_t} - 2O_t \qquad (10.16)$$

i.e. 　　　　　　　　　 unknown　　　　　　　 known

Given the stage/storage and stage/discharge characteristics, a relationship between $2V/\Delta t + O$ and O can be derived, which leads to a simple tabular solution of (10.16). The method is best illustrated by an example.

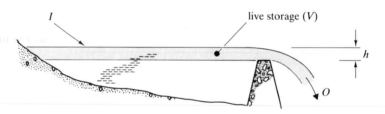

Figure 10.15　Reservoir routing.

Example 10.2 Reservoir flood routing

Determine the outflow hydrograph resulting from the probable maximum flood (PMF) at Ardingly Reservoir from the following data:

reservoir plan area at spillway crest level = 0.8 km^2;
reservoir plan area 3 m above spillway crest level = 1.0 km^2;
spillway type = circular shaft;
discharge equation: $Q = 64h^{3/2}$;
discharge preceding occurrence of PMF = $5 \text{ m}^3/\text{s}$.

PMF

time (h)	0	2	4	6	8	10	12	14	16	18	20
inflow (m³/s)	5	8	15	30	85	160	140	95	45	15	10

Solution

First the relationship between O and $2V/\Delta t + O$ must be established. An approximate equation for V in terms of h may be derived by assuming a linear variation of area with stage:

$$A_h = \left(0.8 + h\frac{(1 - 0.8)}{3}\right) \times 10^6$$

where A_h is the reservoir area at stage h. Hence

$$V_h = \left(\frac{0.8 + A_h}{2}\right)h \times 10^6$$

$$= (0.8 + 0.033h)h \times 10^6 \text{ m}^3$$

Also, the outflow is related to stage through the weir equation.
Hence Table 10.6 may be prepared. (Note $\Delta t = 2$ h, i.e. 7200 s to correspond with the inflow data.)
 A graph of O against $2V/\Delta t + O$ is shown in Figure 10.16a.

Table 10.6 Relationships between stage, volume and outflow for Example 10.2.

h (m)	V (m³ × 10⁶)	O (m³/s)	$2V/\Delta t + O$ (m³/s)
0.0	0.000	0.0	0
0.5	0.408	22.6	136
1.0	0.833	64.0	295
1.5	1.274	117.6	471
2.0	1.732	181.0	662

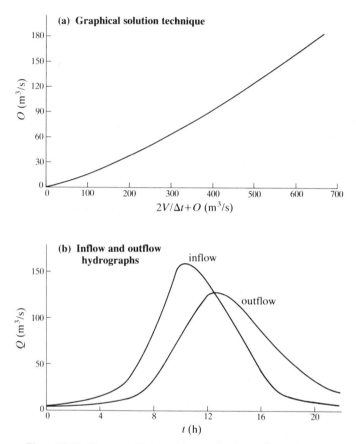

Figure 10.16 Reservoir flood routing – solution to Example 10.2.

Table 10.7 Tabular solution for reservoir routing.

Time (h)	Inflow (m^3/s)	$I_t + I_{t+\Delta t}$ (m^3/s)	$2V/\Delta t + O$ (m^3/s)	$(2V/\Delta t + O) - 2O$ (m^3/s)	Outflow (m^3/s)
0.00	5.00		46.71	36.62	5.00
2.00	8.00	13.00	49.62	38.58	5.52
4.00	15.00	23.00	61.58	46.46	7.56
6.00	30.00	45.00	91.46	65.24	13.11
8.00	85.00	115.00	180.24	114.06	33.09
10.00	160.00	245.00	359.06	194.03	82.52
12.00	140.00	300.00	494.03	244.46	124.79
14.00	95.00	235.00	479.46	239.20	120.13
16.00	45.00	140.00	379.20	201.99	88.60
18.00	15.00	60.00	261.99	152.94	54.53
20.00	10.00	25.00	177.94	112.90	32.52

The solution may now proceed by using the tabular method shown in Table 10.7, once the initial conditions have been specified. At time zero, the outflow is 5 m³/s, the corresponding stage is 0.183 m (from the weir equation) and the storage volume is 0.148×10^6 m³. Hence, the first row of Table 10.6 may be completed. The outflow at all succeeding times is calculated from (10.16) as follows:

$$\left(\frac{2V}{\Delta t} + O\right)_{t+\Delta t} = (I_t + I_{t+\Delta t}) + \left(\frac{2V}{\Delta t} + O - 2O\right)_t$$

$$\text{column (4)} = \text{column (3)} + \text{column (5)}$$

Then estimate

$$O_{t+\Delta t} \qquad \text{from} \quad \left(\frac{2V}{\Delta t} + O\right)_{t+\Delta t}$$

$$\begin{array}{cc} \text{column} & \text{using} \\ \text{(6)} & \text{Figure 10.16a} \end{array}$$

Finally,

$$\left(\frac{2V}{\Delta t} + O - 2O\right)_{t+\Delta t} = \left(\frac{2V}{\Delta t} + O\right)_{t+\Delta t} - 2O_{t+\Delta t}$$

$$\text{column (5)} \qquad \text{column (4)} \quad -2 \times \text{column (6)}$$

The complete inflow and outflow hydrographs are shown in Figure 10.16b.

Channel routing

In this case the storage volume is not a simple function of stage, and therefore solution of the continuity equation is more complex. It may be solved using the full equations of gradually varied unsteady flow as described in Chapter 5. However, there are simpler techniques which can be applied if previous inflow and outflow hydrographs have been recorded. These are referred to as hydrological methods, and are based on the so-called Muskingum method (after McCarthy in 1938).

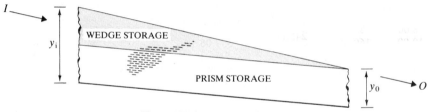

Figure 10.17 Channel routing.

Channel storage may be considered to consist of two parts, prism and wedge storage, as shown in Figure 10.17. Assuming no sudden change of cross section within the reach, then approximate expressions for inflow, outflow and storage are

$$I = a y_i^n \qquad O = a y_o^n$$

where a and n are constants. Now

$$\text{Prism storage} = b y_o^m$$

$$\text{wedge storage} = c(y_i^m - y_o^m)$$

where b and c are constants. So

$$\text{Total storage } V = \text{prism storage} + \text{wedge storage}$$

$$= b y_o^m - c y_o^m + c y_i^m$$

Substituting for y_i and y_o, and assuming m = n,

$$V = \frac{b}{a} O - \frac{c}{a} O + \frac{c}{a} I$$

or

$$V = \frac{b}{a} \left(\frac{c}{b} I + O - \frac{c}{b} O \right)$$

Taking $K = b/a$ and $X = c/b$

$$V = K(XI + O - XO)$$

$$= K[O + X(I - O)] \tag{10.17}$$

where K is called the storage constant and has dimensions of time and X is a dimensionless weighting factor between 0 and 0.5 (but normally between 0.2 and 0.4).

Equation (10.17) is the Muskingum equation. It is obviously only an approximation, but has been used widely with reasonable results.

Substitution of this equation into the continuity equation yields

$$O_{t+\Delta t} = C_0 I_{t+\Delta t} + C_1 I_t + C_2 O_t \tag{10.18}$$

where

$$C_0 = -\frac{KX - 0.5\,\Delta t}{K - KX + 0.5\,\Delta t}$$

$$C_1 = \frac{KX + 0.5\,\Delta t}{K - KX + 0.5\,\Delta t}$$

$$C_2 = \frac{K - KX - 0.5\,\Delta t}{K - KX + 0.5\,\Delta t}$$

Hence, the outflow may be determined through time from the inflow hydrograph, the known values of K and X and a starting value for the outflow.

In the Muskingum method, both K and X may be determined by a graphical technique from a previously recorded event. A value for X is assumed, and a plot of V is drawn, derived from the known inflows and outflows against $(O + X(I - O))$. If the assumed value of X is correct, then a straight-line plot with gradient K should result (cf. (10.17)). If this is not the case, a new value of X is chosen and the procedure repeated.

This graphical technique is tedious. Cunge (1969) presented a simpler alternative to this approach, known as the Muskingum–Cunge method. He demonstrated that K is approximately equal to the time of travel of the flood wave, i.e.

$$K \simeq \Delta L/c \qquad\qquad (10.19)$$

where ΔL is the length of the river reach, and c is the flood wave celerity $(c \simeq \sqrt{gy})$

$$X \simeq 0.5 - \frac{\bar{Q}_p}{2 S_0 \bar{B} c \Delta L} \qquad\qquad (10.20)$$

where $\bar{Q}_p$ is the mean flood peak and $\bar{B}$ is the mean surface width of the channel.

Using these equations allows rapid calculation of K and X, and the Muskingum–Cunge method may also be applied to rivers without recorded outflow hydrographs.

Volume III of the *Flood studies report* is devoted to the subject of flood routing, and includes a review of methods and their application to British rivers. The subject is currently being actively researched.

10.8 Design floods for reservoir safety

The safe design of reservoirs entails, amongst other things, the provision of a spillway which will prevent flood waters overtopping the dam, causing subsequent collapse and the release of a flood wave. In the past, many such failures have occurred, causing loss of life and property on a disastrous scale (for details, see Binnie 1981). As a result of such failures in the UK, the Reservoirs (Safety Provisions) Act of 1930 was introduced at the instigation of the Institution of Civil Engineers (ICE). Subsequently, the Reservoirs (Safety Provisions) Act 1975 was introduced, though this is only just being implemented at the time of writing (1984).

The main purpose of these acts is to ensure that new reservoirs are designed by competent engineers (known as Panel engineers, appointed through ICE) and that existing reservoirs are regularly inspected and repaired in accordance with the recommendations of the Panel engineers.

The design philosophy behind spillway sizing is to make them sufficiently large to pass the 'probable maximum flood' (PMF) safely. This ensures that a dam will never fail due to overtopping.

The estimation of a PMF is problematic, because such events should never occur. The interim report on reservoir safety (ICE 1933), re-published with additions in 1960, tackled this problem by gathering information on the largest recorded historical floods. Using this information, they produced an 'envelope curve' of specific peak discharge (Q/A) against catchment area (A). Floods lying on the envelope curve were termed 'Normal Maximum Floods', and it was suggested that reservoirs be designed for a 'Catastrophic Flood' equal to twice the Normal Maximum Flood.

With the advent of the *Flood studies report,* the Insitution of Civil Engineers produced a new guide, *Floods and reservoir safety* (1978). In this guide, the concept of the PMF was introduced, and details were given of how it may be calculated. The essence of the method is to estimate an inflow hydrograph based on the unit hydrograph model. The storm rainfall (depth and profile) to be used is the estimated maximum rainfall, as calculated by the Meteorological Office. This is convoluted with a unit hydrograph in which all the parameters are set to maximise runoff. Finally, the PMF is routed through the reservoir to find the outflow hydrograph.

Since the introduction of the 1930 Act, there have been no major dam failures in the UK due to overtopping. However, many reservoirs are now over 100 years old, and are suffering from the effects of age. Some of these reservoirs have had their spillway capacities increased as a result of the 1978 guide. It is to be expected that many more will be similarly treated, and that the inspection and repair of old reservoirs will assume more prominence than the construction of new ones (in the UK) for the foreseeable future.

10.9   Methods of flood prediction for urban catchments

The runoff response of urban catchments to rainfall is different from that of rural catchments. For a given rainfall, flood rise times are quicker, flood peaks higher and flood volumes larger. This is due to two main physical differences between urban and rural catchments; the large proportion of impermeable areas (roads, roofs, pavements, etc.) and the existence of man-made drainage systems. The percentage runoff from impermeable areas is typically 60–90%. The drainage systems convey this increased volume to the outflow point of the catchment much more quickly than natural drainage would. The design of drainage systems calls for the application of both hydrological and hydraulic principles.

Historically, such designs have been based on the rational method (attributed to Lloyd-Davies 1906). Some of the shortcomings of this method were overcome in the Transport and Road Research Laboratory (TRRL) hydrograph method introduced in 1963 (refer to TRRL 1976). The most up-to-date method is known as the Wallingford Procedure (1981), which incorporates both design and simulation methods. The procedure embodies a modified rational method and a completely new method based on a conceptual model of rainfall–runoff processes. These models were developed jointly by the Hydraulics Research Station (now Hydraulics Research Limited), the Institute of Hydrology and the Meteorological Office, under the aegis of the National Water Council (NWC). Much of the hydrological work is derived from the methods in the *Flood studies report* with additional research carried out specifically for urban catchments. The Wallingford Procedure is described in five volumes, and is only suitable for computer solution. To this end, both mainframe (WASSP) and micro (Micro-WASSP) versions are available from Hydraulics Research Limited.

The following two sections describe, in outline, the rational method and the Wallingford procedure. Complete details are available in the references.

The rational method

The basis of this method is a simplistic relationship between rainfall and runoff of the form

$$Q_P = CiA \tag{10.21}$$

where Q_P is the peak runoff rate, i is the rainfall intensity, A is the catchment area (normally impermeable area only) and C is the runoff coefficient. In the modified rational method this equation becomes

$$Q_P = \frac{C_V C_R i A}{0.36} \tag{10.22}$$

with the units: Q_P in l/s, i in mm/h, A in hectares, and where C_V is the volumetric runoff coefficient (0.6–0.9), C_R is the routing coefficient ($=1.3$) and A is the impermeable area only.

To apply these equations to urban storm water drainage design requires a knowledge of the critical storm duration to estimate i. The assumption made is that this storm duration is equal to the time of concentration of the catchment (t_c), given by

$$t_c = t_e + t_f \qquad (10.23)$$

where t_e is the time of entry into the drainage system (between 3 and 8 min) and t_f is the time of flow through the drainage system.

To calculate t_f, a pipe size (D), length (L) and gradient (S_0) must be chosen, and the full bore velocity (V) calculated (using the HRS tables or charts). Thus, for a single pipe $t_f = L/V$, t_c is found from (10.23) and Q_p from (10.22) (i being found from tables of rainfall intensity for various durations). If this value of Q_P exceeds the pipefull discharge for the pipe initially chosen, then the procedure is repeated with a larger pipe size. For a complete drainage system this analysis is carried out sequentially in the downstream direction. t_c and A will, of course, increase, and consequently i will decrease, in the downstream direction.

Whilst the rational method is often satisfactory from a design viewpoint, it can be criticised for its inability to simulate a real storm event for the following reasons:

(a) the inaccuracy of the runoff coefficient C_V – note that this is improved in the modified rational method;
(b) the rainfall intensity varies continuously down the system: this does not correspond to a real storm event;
(c) the assumption of full bore conditions to calculate t_f – in reality, each pipe will be running partially full most of the time.

For these reasons the modified rational method is only recommended for the design of drainage systems in catchments which do not exceed 150 ha in area.

The Wallingford hydrograph method

This is a completely new approach based on a conceptual model of the physical processes occurring in an urban catchment. It consists of two parts; a surface runoff model (illustrated in Fig. 10.18) and a routing model.

In the surface runoff model, the net rainfall depth is calculated from a percentage runoff equation derived from regression analysis, and an allowance is made for depression storage. The time distribution of this net rainfall is calculated according to whether it falls on pitched roofs or paved

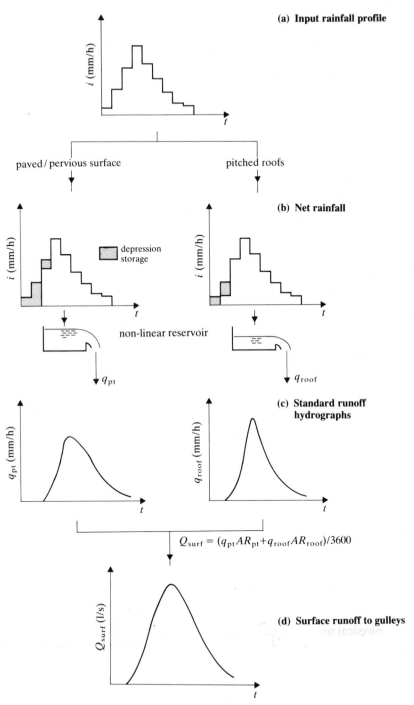

Figure 10.18 Surface runoff model for WASSP.

Table 10.8 Methods and models included in the WASSP package.

Method	Rainfall models	Overland flow models	Pipe flow models	Sewer ancillaries models	Construction cost model	Flood alleviation benefit model
modified rational method	intensity–duration–frequency relationship	percentage runoff model + time of entry	pipe-full velocity	storm overflow	—	—
hydrograph method	rainfall profiles	complete surface runoff model	Muskingum–Cunge	storm overflow, storage tank, pumping station	TRRL resource cost model	—
		sewered sub-area model may be used for selected sub-areas				
optimising method	as for modified rational method	as for modified rational method	pipe-full velocity	—	—	—
simulation method	as for hydrograph method	complete surface runoff model	Muskingum–Cunge and surcharged flow	as for hydrograph method plus tailwater level		Middlesex Polytechnic flood Hazard Research Project model (not included in programs)
		sewered sub-area model (without surcharging) may be used for selected sub-areas				

or pervious areas. The runoff response of these areas is derived by routing the net rainfall through a non-linear reservoir to produce a hydrograph of surface runoff. Nine such standard hydrographs are used for paved or pervious areas (depending on catchment slope and the paved area per gulley) and one for pitched roofs. These standard hydrographs are then multiplied by the relevant contributing areas and combined to form the surface runoff hydrograph.

In the routing model, the surface runoff hydrographs entering each pipe length are then routed through the drainage network using a version of the Muskingum–Cunge method (refer to Section 10.7).

In this manner, the flood hydrographs at each pipe junction are calculated, giving a complete picture of the performance of the drainage system to any given rainfall input. This method may be used for design or for simulation of unsurcharged systems (i.e. systems in which pressurised flow does not occur due to backup in the manholes).

The Wallingford simulation method

This method uses the same surface runoff and routing models as the hydrograph method. In addition, the effects of surcharging and surface flooding are incorporated. Thus, the response to flood events considerably in excess of the design flood may be realistically studied for an existing drainage system. Proposed flood alleviation measures, such as intercepting sewers or on/off line storage tanks, may be tested to gauge their effect on the existing system.

A summary of the WASSP package methods and models is given in Table 10.8.

References and further reading

Binnie, G. M. 1981. *Early Victorian water engineers*. London: Thomas Telford.
Chatfield, C. 1983. *Statistics for technologists*. London: Chapman and Hall.
Cunge, J. A. 1969. On the subject of a flood propagation method. *J. Hydraulics Res. IAHR* **7**, 205–30.
Fleming, G. 1975. *Computer simulation techniques in hydrology*. London: Elsevier.
Hall, M. J. 1984. *Urban hydrology*. Barking: Elsevier Applied Science.
Institute of Hydrology 1981. *Derivation of a catchment average unit hydrograph*. Rep. No. 71, NERC.
Institution of Civil Engineers 1978. *Floods and reservoir safety*. London.
Institution of Civil Engineers 1960. *Floods in relation to reservoir practice*. London.
Linsley, R. K., M. A. Kohler and J. L. H. Paulhus 1982. *Hydrology for engineers*, 3rd edn. London: McGraw-Hill.
Lloyd-Davies, D. E. 1906. The elimination of storm water from sewerage systems. *Proc. Instn Civ. Engrs* **164**(2), 41–67.

National Water Council 1981. *Design and analysis of urban storm drainage. The Wallingford procedure.* Vol. 1: *Principles, methods and practice.* Vol. 2: *Program user's guide.* Vol. 3: *Maps.* Vol. 4: *Modified rational method.* Vol. 5: *Programmer's manual.* Wallingford: Hydraulics Research Limited.

Natural Environment Research Council 1975. *Flood studies report.* Vol. I: *Hydrological studies.* Vol. II: *Meteorological studies.* Vol. III: *Flood routing studies.* Vol. IV: *Hydrological data.* Vol. V: *Maps.* London.

Shaw, E. M. 1983. *Hydrology in practice.* Wokingham: Van Nostrand Reinhold.

Sutcliffe, J. V. 1978. *Methods of flood estimation: A guide to the flood studies report.* Wallingford: Institute of Hydrology.

Transport and Road Research Laboratory 1976. *A guide for engineers to the design of storm sewer systems.* Road Note 35, 2nd edn.

US Soil Conservation Service 1972. *National engineering handbook* (Section 4: *Hydrology*). Washington, DC: US Government Printing Office.

Wilson, E. M. 1983. *Engineering hydrology,* 3rd edn. London: Macmillan.

11

Dimensional analysis and the theory of physical models

11.1   Introduction

The preceding chapters have explored various aspects of hydraulics. In each case a fundamental concept has been described and, where possible, that concept has been translated into an algebraic expression which has then been used as the basis of a mathematical model. Mathematical models are functions which represent the behaviour of a physical system, and which can be solved on a computer or calculator. A mathematical model is very convenient, since it is available whenever the engineer needs to use it. However, it may have occurred to the reader that some problems could be so complex that no adequate mathematical model could be formulated. If such a problem is encountered, what is the engineer to do? To deal with such problems it is necessary to find an alternative to mathematical models. One alternative which is frequently adopted is the use of scale model experiments. However, this approach also raises questions. For example, even when the experimental results have been obtained, there may be no self-evident (e.g. geometrical) relationship between the model behaviour and the behaviour of the full scale prototype. Thus, if an engineer wishes to employ model tests, two problems must be faced:

(a) the design of the model and of the experimental procedure;
(b) the correct interpretation of the results.

To this end it is necessary to identify physical laws which apply equally to the behaviour of model and prototype. Our understanding of such laws has developed progressively over the last century or so.

11.2   The idea of 'similarity'

Ideas about basic forms of similarity are often gained early in life from simple shapes. These ideas are formalised mathematically as 'geometrical similarity'. Geometrical similarity requires:

(a) that all the corresponding lengths of two figures or objects are in one ratio;
(b) that all corresponding angles are the same for both figures or objects.

Now, (a) may be expressed algebraically as

$$L''/L' = \lambda_L$$

where λ_L is a 'scale factor' of length (Fig. 11.1a). Similar statements may be made about other characteristics of two systems, where such systems exhibit some form of 'similarity', i.e. where they have certain features in common. For example, measurements of the velocity patterns in two systems may reveal that at corresponding co-ordinates there is a relationship between the velocity U' in one system and the velocity U'' in the other system. If that relationship is in the form $U''/U' = \lambda_U$, then the two systems are 'kinematically' similar (Fig. 11.1b). This could equally well be restated in terms of the fundamental dimensions as

$$\lambda_T = T''/T' \qquad \lambda_L = L''/L'$$

since velocity has dimensions of LT^{-1}.

There may also be systems which exhibit similarity in their force patterns (Fig. 11.1c) so that

$$\lambda_F = F''/F'$$

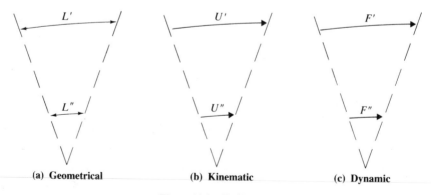

(a) **Geometrical**   (b) **Kinematic**   (c) **Dynamic**

Figure 11.1   Similarity.

and since force = mass × acceleration, force may be said to possess the dimensions MLT^{-2}, therefore, for systems with similar forces,

$$\lambda_T = T''/T' \qquad \lambda_L = L''/L' \qquad \lambda_M = M''/M'$$

must all be true. Systems exhibiting similarity of this nature are referred to as being 'dynamically' similar.

These statements indicate the sorts of scales which may be of interest. They do not, however, help us to determine scale magnitudes. For this purpose, we must develop equations which reflect the appropriate dimensional considerations.

11.3 Dimensional homogeneity and its implications

Consider the relationship

$$3 \text{ eggs} + 2 \text{ bananas} = 4 \text{ Rolls–Royces}$$

This will immediately be rejected as nonsense. There is no link, numerical, dimensional or of any other sort, between the left-hand and the right-hand sides. The point of this statement of the obvious is to impress the fact that most people intuitively accept the idea of homogeneity or harmony within an equation. Dimensional homogeneity must be true for any equation which purports to describe a set of physical events.

If any physical quantity, J, is considered, it will be possible to reduce it to some function of the three fundamental dimensions, mass, length and time: i.e.

$$J = f[M, L, T]$$

Furthermore, if the magnitude of J is compared for two similar systems, then

$$\frac{J''}{J'} = \frac{f[M, L, T]''}{f[M, L, T]'} = \lambda_J$$

Self-evidently, this ratio must be dimensionless. This is true if the function is in the form of a product, and therefore

$$J = K[M^a L^b T^c] \tag{11.1}$$

where K is a numeric, and a, b and c are powers or indices whose magnitudes have to be determined. As an example, if J is a velocity, then in (11.1), $a = 0$, $b = 1$ and $c = -1$, since velocity has dimensions LT^{-1}.

In one sense it can be argued that dimensional relationships are arbitrary, since magnitudes depend on the choice of units (feet, metres, pounds, kilogrammes, etc.). For this reason, an equation which is a statement of a physical law is often used in a dimensionless form. Dimensionless equations are completely general, and are therefore frequently the basis for the representation of experimental data.

11.4 Dimensional analysis

Dimensional analysis is a powerful tool for deriving the dimensionless relationships referred to above. The methodology has long been used by engineers and scientists, and the techniques have been progressively refined over the years. Three approaches will be mentioned here. They are (1) the indicial method, (2) Buckingham's method and (3) the matrix method. However, it must be emphasised that all methods are absolutely dependent on the correct identification of all the factors which govern the physical events being analysed. The omission of a single factor may give quite misleading results. The procedure is best explained through the medium of a worked example. To illustrate the reason for using this procedure, the reader should imagine that we have been presented with a set of data relating to some unfamiliar or new type of engineering system. The indicial method will be used here.

Example 11.1 Dimensional analysis procedure

As part of a development programme, scale model tests have been carried out on a new hydraulic machine. The experimental team have presented the following data. Thrust force F, the flow velocity u, the viscosity μ and density ρ of the fluid. A typical size of the system, L, is also given. Two questions must be posed, namely (a) how to analyse or plot the data in the most informative way and (b) how to relate the performance of the model to that of the working prototype.

Solution

It seems reasonable to postulate that the force F is related to the other given quantities, so one might say

$$F = f[\rho, u, \mu, L]$$

The form of the function is completely unknown, but it has been proposed above:

(a) that the function must be in the form of a power product;
(b) that there must be a dimensional balance between both sides of the equation.

From (a), the equation may be rewritten as

$$F = K[\rho^a u^b \mu^c L^d] \tag{11.2}$$

To meet the requirement of (b), each quantity must be reduced to its fundamental dimensions.

Force, F, is measured in newtons, but newtons are not fundamental dimensions. However, by definition, force is equated to the product of mass (M) and acceleration (which has dimensions LT^{-2}). The fundamental dimensions of F are therefore MLT^{-2}. The dimensions of the other terms are: $\rho = ML^{-3}$, $u = LT^{-1}$, $\mu = ML^{-1}T^{-1}$ and $L = L$. These dimensional relationships may be conveniently tabulated as a matrix:

$$
\begin{array}{c}
 \\ M \\ L \\ T
\end{array}
\begin{array}{cccccc}
F & \rho & u & \mu & L \\
\end{array}
\left[
\begin{array}{ccccc}
1 & 1 & 0 & 1 & 0 \\
1 & -3 & 1 & -1 & 1 \\
-2 & 0 & -1 & -1 & 0
\end{array}
\right]
$$

Expressing each quantity in (11.2) in terms of its dimensions,

$$MLT^{-2} = K[(ML^{-3})^a (LT^{-1})^b (ML^{-1}T^{-1})^c L^d] \tag{11.3}$$

For dimensional homogeneity, the function has the dimensions MLT^{-2}, and since a product relationship exists, the sum of the indices (or powers) of each dimension in the function may be equated to the index of the same dimension on the left-hand side of the equation.

Thus, on the left-hand side, M has the index 1. In the function bracket, M has the indices a and c. Therefore,

$$1 = a + c \tag{11.4}$$

Similarly, T has the index -2 on the left-hand side, and the indices $-b$ and $-c$ in the function, therefore

$$2 = b + c \tag{11.5}$$

The corresponding indicial equation for L is

$$1 = -3a + b - c + d \tag{11.6}$$

There are thus three equations (one for each fundamental dimension) but four unknown indices, so that a complete solution is unattainable. However, a partial solution is worthwhile. To obtain the partial solution, it is necessary to select (or guess) three 'governing variables'. Suppose that ρ, u and L are selected. The indices of ρ, u and L are a, b and d, respectively. So (11.4), (11.5) and (11.6) must be rearranged in terms of a, b and d.

From (11.4),

$$a = 1 - c \tag{11.7}$$

From (11.5),

$$b = 2 - c \tag{11.8}$$

Substituting (11.7) and (11.8) into (11.6)

$$1 = -3(1 - c) + (2 - c) - c + d$$

and therefore

$$d = 2 - c \tag{11.9}$$

Substituting for a, b and d in (11.2),

$$F = K[\rho^{1-c} u^{2-c} \mu^c L^{2-c}]$$

$$= K\left[\rho u^2 L^2 \left(\frac{\mu}{\rho u L}\right)^c\right]$$

Since the function represents a product, it may be restated as

$$F = \rho u^2 L^2 K\left(\frac{\rho u L}{\mu}\right)^{-c}$$

or as

$$\frac{F}{\rho u^2 L^2} = K\left(\frac{\rho u L}{\mu}\right)^{-c} \tag{11.10}$$

where K and c are unknown.

There are a number of important points to be made about (11.10).

(a) Two groups have emerged from the analysis, $F/\rho u^2 L^2$ and $\rho u L/\mu$. If the reader cares to check, it will be found that both groups are dimensionless. For conciseness, dimensionless groups are referred to as 'Π' groups. Thus we might state

$$\Pi_1 = F/\rho u^2 L^2$$

and

$$\Pi_2 = \rho u L/\mu$$

Π_2 is, of course, the Reynolds Number, which has been encountered before.

(b) Dimensionless groups are independent of units and of scale. Π_1 and Π_2 are therefore equally applicable to the model or to the prototype.

(c) Both Π groups represent ratios of forces, as will now be shown:

$$\text{the 'inertia force' of a body, } F = \text{mass} \times \text{acceleration}$$

$$\text{the mass of a body} = \rho \times \text{volume} = \rho L^3$$

$$\text{acceleration} = du/dt, \text{ which has the dimensions of (velocity/time)} = LT^{-2}$$

Therefore

$$\text{mass} \times \text{acceleration} = \rho L^3 L T^{-2} = \rho L^4 T^{-2} (= \rho u^2 L^2)$$

Therefore Π_1 represents the ratio of the thrust force to the inertia force:

$$\text{the inertia force per unit area} = \rho L^4 T^{-2} L^{-2} = \rho L^2 T^{-2}$$

Now $\rho L^2 T^{-2}$ may be rewritten as ρu^2, so looking now at Π_2, the ratio (inertia force/viscous force) can be written as

$$\frac{\rho u^2}{\text{viscous force}}$$

Viscous force is represented by the viscous shear stress, $\tau = \mu\, du/dy$. Dimensionally, this is the same as $\mu u/y$. Therefore,

$$\frac{\rho u^2}{\mu(u/y)} = \frac{\rho u y}{\mu}$$

which is Π_2.

(d) All three fundamental dimensions are present in (11.10). Therefore, if the model is to truly represent the prototype, then both model and prototype must conform to the law of dynamic similarity. For this to be so, the magnitude of each dimensionless group must be the same for the model as for the prototype:

$$\Pi_1' = \Pi_1''$$

$$\Pi_2' = \Pi_2''$$

where Π' refers to the prototype and Π'' to the model.

If the above statement (regarding equality of Π groups) is not fulfilled, it must follow that ratios of forces in the model do not correspond to the ratios of forces in the prototype, and hence that the model and prototype are not dynamically similar.

(e) It would, of course, be possible to choose alternatives to ρ, u and L as governing variables. Different Π groups might then emerge. However, because the product $\rho u L$ represents the inertia force (which is a parameter relevant to any flow), these three quantities are very frequently selected. The groups which are ultimately the most helpful for experimental analysis have emerged largely by trial and error.

The reader is strongly advised to review Example 11.1 carefully, since it contains all the main ideas and processes which form the basis of dimensional analysis.

The indicial method of dimensional analysis is perfectly satisfactory as long as the number of variables involved is small. For problems involving larger numbers of variables, a more orderly process is helpful. This leads us to the use of the Buckingham or the matrix methods.

11.5   Dimensional analysis involving more variables

Buckingham's method

Buckingham's ideas were published in a paper in 1915. In outline, he proposed:

(a) that if a physical phenomenon was a function of m quantities and n fundamental dimensions, dimensional analysis would produce $(m - n)$ Π groups;
(b) that each Π should be a function of n governing variables plus one more quantity;
(c) the governing quantities must include all fundamental dimensions;
(d) the governing quantities must not combine among themselves to form a dimensionless group;
(e) as each Π is dimensionless, the final function must be dimensionless, and therefore dimensionally

$$f[\Pi_1, \Pi_2, \ldots, \Pi_{(m-n)}] = M^0 L^0 T^0$$

Referring back to Example 11.1, there were five quantities (F, ρ, u, μ, L) and three dimensions (M, L, T), from which we derived two groups. To obtain Π_1, using Buckingham's approach, with ρ, u, and L as governing variables,

$$\Pi_1 = \rho^a u^b L^c F = M^0 L^0 T^0$$

Therefore

$$(ML^{-3})^a (LT^{-1})^b L^c (MLT^{-2}) = M^0 L^0 T^0$$

Equating indices,

$$M \quad a + 1 = 0 \qquad \text{therefore } a = -1$$

$$T \quad -b - 2 = 0 \qquad \text{therefore } b = -2$$

$$L \quad -3a + b + c + 1 = 0$$

substituting for a and b,

$$+3 - 2 + c + 1 = 0 \qquad \text{therefore } c = -2$$

so

$$\Pi_1 = F/\rho u^2 L^2$$

Any remaining Π groups are produced in the same way, and thus $\Pi_2 = \mu/\rho u L$, as before.

The matrix method

This is based upon the work of a number of investigators, notably Langhaar (1980) and Barr (1983). The method used here is described in detail in Appendix A, and is a development of the previous work. It lends itself to computer or manual manipulation. The procedure entails:

(1) setting up a dimensional matrix with the governing variables represented by the first n columns;
(2) partitioning the matrix to form an '$n \times n$' matrix and an additional matrix;
(3) applying Gauss–Jordan elimination to produce a leading diagonal of unity in the $n \times n$ matrix;
(4) abstracting the requisite Π groups from the final index matrix.

All of the remaining examples in this chapter employ this method.

11.6 Applications of dynamic similarity

Pipe flow

The history of the development of the various pipe flow equations has already been outlined in Chapter 4. It is noteworthy that dimensional procedures underlie a number of these developments, such as Nikuradse's diagram. An example illustrating the use of these procedures follows.

Example 11.2   Analysis of pipe flow

The hydraulic head loss, h_f, in a simple pipeline is assumed to depend on the following quantities:

the density (ρ) and viscosity (μ) of the fluid;
the diameter (D), length (L) and roughness (k_S) of the pipe;
a typical flow velocity (usually the mean velocity) V.

    (a) Develop the appropriate dimensionless groups to describe the flow.
    (b) A 10 km pipeline, 750 mm diameter, is to be used to convey oil ($\rho = 850$ kg/m^3, $\mu = 0.008$ kg/m s). The design discharge is 450 l/s. The pipeline will incorporate booster pumping stations at suitable intervals. As part of the design procedure, a model study is to be carried out. A model scale $\lambda_L = 1/50$ has been selected, and air is to be used as the model fluid. The air has a density of 1.2 kg/m^3 and viscosity of 1.8×10^{-5} kg/m s.
    At what mean air velocity will the model be correctly simulating the flow of oil?
    If the head loss in the model is 10 m for the full pipe length, what will be the head loss in the prototype?

Solution

(a) If the matrix method is employed, the dimensional matrix is first produced. The principal quantities of interest are:

(i)   the characteristics of the fluids: ρ, μ;
(ii)   the characteristics of the pipeline: length L, diameter D and roughness k_S;
(iii) the characteristics of the flow: typical (usually mean) velocity V, head loss h_f;

The dimensional matrix ($\mathbf{M_D}$) is then

$$
\begin{array}{c}
M \\ L \\ T
\end{array}
\begin{array}{cccccccc}
V & D & \rho & \vert & h_f & \mu & k_S & L \\
0 & 0 & 1 & \vert & 0 & 1 & 0 & 0 \\
1 & 1 & -3 & \vert & 1 & -1 & 1 & 1 \\
-1 & 0 & 0 & \vert & 0 & -1 & 0 & 0
\end{array}
\qquad (11.11)
$$

The governing variables are assumed to be V, D and ρ.
    Applying the Gauss–Jordan elimination procedure,

$$
\begin{array}{c}
\rho \\ D \\ V
\end{array}
\begin{array}{cccccccc}
\rho & D & V & \vert & h_f & \mu & k_S & L \\
1 & 0 & 0 & \vert & 0 & 1 & 0 & 0 \\
0 & 1 & 0 & \vert & 1 & 1 & 1 & 1 \\
0 & 0 & 1 & \vert & 0 & 1 & 0 & 0
\end{array}
\qquad (11.12)
$$

and hence $\mathbf{M_k^*}$, the index matrix is

$$
\begin{array}{c}
\begin{array}{cccc} \Pi_1 & \Pi_2 & \Pi_3 & \Pi_4 \end{array} \\
\begin{array}{c} \rho \\ D \\ V \\ h_f \\ \mu \\ k_S \\ L \end{array}
\left[
\begin{array}{cccc}
0 & -1 & 0 & 0 \\
-1 & -1 & -1 & -1 \\
0 & -1 & 0 & 0 \\
1 & 0 & 0 & 0 \\
0 & 1 & 0 & 0 \\
0 & 0 & 1 & 0 \\
0 & 0 & 0 & 1
\end{array}
\right]
\end{array}
\qquad (11.13)
$$

Therefore

$$
\Pi_1 = \frac{h_f}{D} \qquad \Pi_2 = \left(\frac{\rho D V}{\mu}\right)^{-1}
$$

$$
\Pi_3 = \frac{k_S}{D} \qquad \Pi_4 = \frac{L}{D}
$$

Because D was used as a governing variable, Π_1 emerged as $h_f D^{-1}$. It is conventional to use $h_f L^{-1}$ in preference to $h_f D^{-1}$, since $h_f L^{-1}$ represents the hydraulic gradient. This makes no practical difference, providing the model is geometrically similar to the prototype, i.e.

$$
\frac{D''}{D'} = \frac{L''}{L'}
$$

in which case,

$$
\frac{(h_f D^{-1})''}{(h_f D^{-1})'} = \frac{(h_f L^{-1})''}{(h_f L^{-1})'}
$$

(b) Proceeding to the numerical solution, the prototype velocity is obtained, as usual:

$$
V' = \frac{0.45}{(\pi/4) \times 0.75^2} = 1.019 \text{ m/s}
$$

For dynamic similarity, the ratios of forces in the model must equal the corresponding ratios in the prototype. In pipe flow the force ratio is represented by the Reynolds Number, so

$$
\Pi_2'' = \Pi_2'
$$

For the prototype,

$$
\Pi_2' = \frac{\rho' V' D'}{\mu'} = \frac{850 \times 1.019 \times 0.75}{0.008} = 81\,202
$$

Therefore

$$\Pi_2'' = 81\ 202 = \frac{1.2 \times V'' \times (0.75/50)}{1.8 \times 10^{-5}}$$

Therefore the velocity in the model $V'' = 81.2$ m/s. The velocity which is required if the model is to truly represent the prototype is known as the 'corresponding velocity'.

To estimate the head loss in the prototype, the modified Π_1 equation ($\Pi_1 = h_f L^{-1}$) is used:

$$\Pi_1' = \Pi_1''$$

i.e.

$$\left(\frac{h_f}{L}\right)' = \left(\frac{h_f}{L}\right)''$$

In the model, $h_f'' = 10$ m and $L'' = 200$ m. Therefore

$$\frac{h_f'}{10\,000} = \frac{10}{200}$$

So $h_f' = 500$ m, which will be the total head required from the booster pumps.

Free surface flows

In Example 11.2, the evaluation of V'' presented no problem. This was so due to the dominance of one pair of forces (momentum and shear), which are both represented by the Reynolds Number. Hydraulics problems are by no means always so straightforward. A case in point is the family of free surface flows. These flows are controlled by momentum, shear and gravity forces, and this combination highlights a difficulty which has to be overcome through a fundamental grasp of the mechanics of fluids.

Example 11.3   Open channel flow

State the physical quantities which relate to steady open channel flow. Hence derive the dimensionless groups which describe such a flow. Why is it theoretically impossible to produce a valid scale model? How is this problem usually circumvented?

Solution

The principal characteristics of a steady channel flow are:

a typical velocity (usually the mean velocity), V;
the frictional resistance, F;

the fluid characteristics, i.e. density (ρ) and viscosity (μ);
the geometrical characteristics of the channel, i.e. the depth, y (or hydraulic radius, R), bed slope S_0, and surface roughness, k_S;
the gravitational acceleration, g.

The dimensional matrix is

$$
\begin{array}{c}
 \\
M \\
L \\
T
\end{array}
\begin{array}{c}
\begin{array}{cccccccc}
V & R & \rho & \mu & g & k_S & S_0 & F
\end{array} \\
\left[
\begin{array}{cccccccc}
0 & 0 & 1 & 1 & 0 & 0 & 0 & 1 \\
1 & 1 & -3 & -1 & 1 & 1 & 0 & 1 \\
-1 & 0 & 0 & -1 & -2 & 0 & 0 & -2
\end{array}
\right]
\end{array}
\qquad (11.14)
$$

Following Gauss–Jordan elimination,

$$
\begin{array}{c}
 \\
\rho \\
R \\
V
\end{array}
\begin{array}{c}
\begin{array}{cccccccc}
\rho & R & V & \mu & g & k_S & S_0 & F
\end{array} \\
\left[
\begin{array}{cccccccc}
1 & 0 & 0 & 1 & 0 & 0 & 0 & 1 \\
0 & 1 & 0 & 1 & -1 & 1 & 0 & 2 \\
0 & 0 & 1 & 1 & 2 & 0 & 0 & 2
\end{array}
\right]
\end{array}
\qquad (11.15)
$$

The index matrix is then

$$
\begin{array}{c}
 \\
\rho \\
R \\
V \\
\mu \\
g \\
k_S \\
S_0 \\
F
\end{array}
\begin{array}{c}
\begin{array}{ccccc}
\Pi_1 & \Pi_2 & \Pi_3 & \Pi_4 & \Pi_5
\end{array} \\
\left[
\begin{array}{ccccc}
-1 & 0 & 0 & 0 & -1 \\
-1 & 1 & -1 & 0 & -2 \\
-1 & -2 & 0 & 0 & -2 \\
1 & 0 & 0 & 0 & 0 \\
0 & 1 & 0 & 0 & 0 \\
0 & 0 & 1 & 0 & 0 \\
0 & 0 & 0 & 1 & 0 \\
0 & 0 & 0 & 0 & 1
\end{array}
\right]
\end{array}
\qquad (11.16)
$$

Therefore

$$
\Pi_1 = \frac{\rho R V}{\mu} \qquad \Pi_2 = \frac{V^2}{gR}
$$

$$
\Pi_3 = \frac{k_S}{R} \qquad \Pi_4 = S_0 \qquad \Pi_5 = \frac{F}{\rho R^2 V^2}
$$

In Example 11.2, the Reynolds Number was the sole criterion for dynamic similarity. In channel flow there are two criteria: Reynolds Number (Π_1) and Froude Number ($\Pi_2 = \mathrm{Fr}^2$). Furthermore, in modelling channel flows, it is almost universal practice to use water as the 'model fluid'. This means that $\rho' = \rho''$ and $\mu' = \mu''$.

From Π_1

$$
\frac{\rho V' R'}{\mu} = \frac{\rho V'' R''}{\mu}
$$

therefore $V'R' = V''R''$, or

$$\frac{V''}{V'}(=\lambda_V) = \frac{R'}{R''} \tag{11.17}$$

From Π_2,

$$\left(\frac{V^2}{Rg}\right)' = \left(\frac{V^2}{Rg}\right)''$$

Therefore, since g must be the same for model as for prototype,

$$\frac{V''^2}{V'^2} = \frac{R''}{R'}$$

Therefore

$$\frac{V''}{V'}(=\lambda_V) = \left(\frac{R''}{R'}\right)^{\frac{1}{2}} \tag{11.18}$$

It is clear that (11.17) and (11.18) are incompatible unless $R'' = R'$, i.e. unless the model and the prototype are identical in size. This is impracticable. It is possible to overcome this problem by considering the nature of the flow. In channels, as in pipes, frictional resistance is a function of the viscosity (and therefore of the Reynolds Number) for laminar and transitional turbulent flows. Once the flow is completely turbulent, resistance is independent of viscosity. Therefore, if the magnitude of the Reynolds Number representing the model flow is sufficiently great to indicate 'complete turbulence', (11.17) may be ignored. Model scaling is then based on (11.18) only. In practice, the design of hydraulic models involves rather more complexities than have emerged here, as will be seen in Section 11.7.

Hydraulic machines

The quantities which are usually considered in a dimensional analysis of hydraulic machines are:

(a) the power (P) and rotational speed (N) of the machine;
(b) the pressure head (H) generated by the machine;
(c) the corresponding discharge (Q);
(d) the typical machine size (D) and roughness (k_S);
(e) the fluid characteristics $(\rho$ and $\mu)$.

(Note that it is conventional for the pressure head to be represented as gH rather than simply as H.)

If ρ, N and D are used as governing variables, the following groups emerge:

$$\Pi_1 = \frac{P}{\rho N^3 D^5} \qquad \Pi_2 = \frac{Q}{ND^3} \qquad \Pi_3 = \frac{gH}{N^2 D^2}$$

$$\Pi_4 = \frac{\rho ND^2}{\mu} \qquad \Pi_5 = \frac{k_s}{D}$$

Most hydraulic machines operate in the completely turbulent zone of flow, so that the Reynolds Number term (Π_4) is neglected. Since geometrical similarity must apply if two machines are to be compared, the geometrical term (Π_5) is also usually ignored. Thus, Π_1, Π_2 and Π_3 are the groups used in analysis. However, one further group, known as the 'specific speed' of a machine, is derived by combining groups as follows:

Pumps. For pumps, the interest centres on the discharge and head of a particular machine. The product

$$\Pi_2^{1/2} \Pi_3^{-3/4} = \frac{NQ^{1/2}}{(gH)^{3/4}} = K_N \qquad (11.19)$$

is a dimensionless specific speed. Historically the 'g' term was disregarded, and so conventionally the specific speed becomes

$$N_S = NQ^{1/2}/H^{3/4} \qquad (11.20)$$

This is not dimensionless.

Turbines. For turbines the emphasis lies with the power output and head terms. The product

$$\Pi_1^{1/2} \Pi_3^{-5/4} = \frac{NP^{1/2}}{\rho^{1/2}(gH)^{5/4}} = K_N \qquad (11.21)$$

is again a dimensionless specific speed. However, since both the ρ and g terms are constant, they are usually ignored, and therefore

$$N_S = NP^{1/2}/H^{5/4} \qquad (11.22)$$

Again this is not dimensionless.

The name 'specific speed' is an unfortunate term since neither K_N nor N_S represent a meaningful 'speed' anywhere in the machine. The name owes its origins to a concept which arose early in the development of hydraulic machines. It is best to think of it as a numeric whose value is related to the geometrical form and designed duty of a given machine. This value will be a constant for a particular 'family' of similar machines, and will be calculated for the designed optimum operating conditions.

Example 11.4   Pump similarity

A 100 mm diameter pump, which is driven by a synchronous electric motor, has been tested with the following results:

  power supplied to pump = 3.2 kW;
  pump speed = 1450 rev/min, head = 12.9 m, discharge = 20 l/s;
  efficiency = 79%; fluid: water at $18°C$.

A geometrically similar pump is required to deliver 100 l/s against a 20 m head. The new pump will be driven by a synchronous motor. The speeds may be either 960 rev/min or 1450 rev/min. Calculate the principal characteristics (size, speed and power) of the new machine. Calculate also the specific speed, N_S.

Solution

(Note that in the solution Π'' terms refer to the 100 mm pump).
Π_3 is constant, i.e. $\Pi_3'' = \Pi_3'$. Therefore

$$\frac{9.81 \times 12.9}{1450^2 \times 0.1^2} = \frac{9.81 \times 20}{N'^2 D'^2}$$

so

$$N'D' = 180.5$$

or

$$N' = 180.5/D'$$

Π_2 is constant, i.e. $\Pi_2'' = \Pi_2'$. Therefore

$$\frac{0.02}{1450 \times 0.1^3} = \frac{0.1}{N'D'^3}$$

so

$$N'D'^3 = 7.25$$

or

$$N' = 7.25/D'^3$$

Thus, combining these two equations

$$N' = \frac{180.5}{D'} = \frac{7.25}{D'^3}$$

therefore $D'^2 = 0.0402$, i.e. $D' = 0.2$ m. Hence

$$N' = 180.5/0.2 = 902.5 \text{ rev/min.}$$

Therefore, the speed must be 960 rev/min.

Check diameter:

$$960 = 180.5/D'$$

therefore $D' = 0.188$ m. In practice, D' would probably be rounded up to 0.2 m.

Check discharge from Π_2:

$$\frac{0.02}{1450 \times 0.1^3} = \frac{Q'}{960 \times 0.2^3}$$

therefore $Q' = 0.106 \text{ m}^3/\text{s}$.

To estimate the power of the new pump, use Π_1:

$$\frac{3.2}{1000 \times 1450^3 \times 0.1^5} = \frac{P'}{1000 \times 960^3 \times 0.2^5}$$

therefore $P' = 29.72$ kW.

The specific speed is obtained from (11.20), so for the new pump

$$N_s' = \frac{960 \times 0.1^{\frac{1}{2}}}{20^{\frac{3}{4}}} = 32.1$$

Check against the original pump:

$$N_s'' = \frac{1450 \times 0.02^{\frac{1}{2}}}{12.9^{\frac{3}{4}}} = 30.1$$

The difference between N_s'' and N_s' arises due to rounding up the speed of the new pump from 902.5 to 960 rev/min.

11.7 Hydraulic models

The design of a major new hydraulic system (e.g. a hydroelectric scheme, a dock or a hydraulic structure) may be approached in one of three ways:

(a) by theoretical reasoning and the use of numerical models;
(b) on the basis of previous experience of similar systems;
(c) by scale model experiments.

It has already been pointed out that, even with modern computing facilities, many complex problems still defy complete theoretical analysis. A combination of past experience, theory and dimensional analysis will provide partial or complete solutions to a number of problems (e.g. design of pumps or turbines, design of conventional hydraulic structures). However, there still remain many problems which are tractable only through experimentation. Such problems will be exemplified here by reference to river flows, tidal flows and flows through hydraulic structures.

River models

For present purposes, a river will be defined as a channel in which flow is unidirectional. River models may be constructed:

(a) to study flow patterns only. The model then has a 'rigid' or 'fixed' bed (e.g. of wood or concrete) or
(b) to study flow and sediment movement. The model must then be constructed with a particulate 'mobile' bed.

River models may be 'undistorted' or 'distorted'. An undistorted model is geometrically similar to the prototype, and therefore has the same scale (λ_L) for the vertical and horizontal dimensions.

A distorted model has differing horizontal and vertical scales (λ_x and λ_y, respectively, with $\lambda_x < \lambda_y$) to save space and cost (but note that distorted mobile bed models may not simulate prototype sediment transport behaviour accurately). The possibility of distortion is based on studies of river flows. The flow pattern may be envisaged as two distinct zones (Fig. 11.2a):

(1) The zone near to the bank in which velocity must vary in the x-direction (due to frictional shear at the sides) and in the y-direction (due to friction at the bed). According to Keulegan (1938) this zone extends approximately $2.5y$ from each bank (and therefore occupies $2 \times 2.5y = 5y$ of the total width of the river).

(2) The central zone, in which velocity varies in the y-direction, but varies very little in the x-direction. The flow is therefore practically two-dimensional. The properties of a two-dimensional flow may be represented

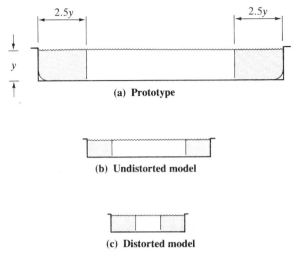

(a) Prototype

(b) Undistorted model

(c) Distorted model

Figure 11.2   River models.

by a 'sample', it is unnecessary to reproduce the whole flow. Therefore if the model channel width $B'' > 5y''$, a sample of the central zone will be reproduced (Fig. 11.2c).

As a first step towards selecting a model scale, it is useful to estimate the scale of the smallest model which will ensure a completely turbulent flow. Complete turbulence exists only if the Reynolds Number exceeds a certain minimum value. Unfortunately, the minimum value is not a constant, but is a function of the relative roughness (a glance at Fig. 4.4 will confirm this). A modified version of the Moody diagram has been produced for channel flows (Henderson 1966, p. 23) so if k_S is known, the minimum Re may be obtained. This is not the usual approach. Modern investigators (e.g. Yalin 1971) use the modified version of the Reynolds Number based on friction velocity:

$$\mathrm{Re}_* = \rho u_* k_S / \mu$$

It has been found that complete turbulence occurs if $\mathrm{Re}_* > 70$.

Example 11.5   Model of a large river

Explain the difference between distorted and undistorted models, and justify the use of distortion. Illustrate the application of distortion by reference to a model study of a lowland river having the following prototype dimensions: depth = 5 m, width = 190 m, slope = 0.0001, mean velocity = 1.15 m/s, $k_S = 0.02$ m.

Solution

The first part of the question has been covered in the preceding text.

The preliminary stage in finding the model scale is to calculate the value of Re_*. To calculate u_* recall that

$$u_* = \left(\frac{\tau_0}{\rho}\right)^{1/2}$$

$$\text{friction factor, } \lambda = \frac{8\tau_0}{\rho V^2} = 8\left(\frac{u_*}{V}\right)^2$$

$$V = C(RS_0)^{1/2}$$

where

$$C = \left(\frac{8g}{\lambda}\right)^{1/2} = \left(\frac{V}{u_*}\right)g^{1/2}$$

Therefore, substituting for C in the Chézy formula and rearranging,

$$u_* = (gRS_0)^{1/2} \simeq (gyS_0)^{1/2}$$

The data for the lowland river will therefore lead to

$$u_*' = (9.81 \times 5 \times 0.0001)^{1/2} = 0.07 \text{ m/s}$$

$$Re_*' = \frac{\rho u_*' k_s'}{\mu} = \frac{1000 \times 0.07 \times 0.02}{1.14 \times 10^{-3}} = 1228$$

Now $u_*'' = \lambda_V u_*'$, and from (11.18) $\lambda_V = \lambda_R^{1/2} \simeq \lambda_y^{1/2}$. Also $k_s'' = \lambda_L k_s' = \lambda_y k_s'$. Therefore

$$Re_*'' = \rho(u_*'\lambda_y^{1/2})(k_s'\lambda_y)/\mu = Re_*'\lambda_y^{3/2} = 1228\lambda_y^{3/2}$$

Since $Re_*'' \geqslant 70$ for fully turbulent flow,

$$1228\lambda_y^{3/2} \geqslant 70$$

i.e.

$$\lambda_y \geqslant \left(\frac{70}{1228}\right)^{2/3} \geqslant \frac{1}{6.752}$$

For an undistorted model, horizontal length scale and vertical length scale are the same.

Therefore $B'' = B' \times \lambda_x$, i.e.

$$B'' = 190 \times \frac{1}{6.752} = 28.14 \text{ m}$$

and $y'' = y' \times \lambda_y$, i.e.

$$y'' = 5 \times \frac{1}{6.752} = 0.74 \text{ m}$$

The model is thus the size of a substantial river! This is completely impracticable in terms of space and cost.

Turning to the concept of the 'distorted' model, $y'' = 0.74$. Therefore the zones of three-dimensional flow near the two banks account for $2 \times (2.5 \times 0.74) = 3.7$ m. Therefore, if $B > 3.7$ m, a sample of the central zone will be reproduced, say $\lambda_x = 1/40$, which gives

$$B'' = 190 \times (1/40) = 4.75 \text{ m}$$

Other scales are determined as follows:

$$S_0 \propto y/x,$$

so

$$\lambda_S = \frac{\lambda_y}{\lambda_x} = \frac{1/6.75}{1/40} = 5.93$$

therefore

$$S_0'' = S_0' \lambda_S = 0.0001 \times 5.93 = 0.000593$$

$Q = V \times \text{area} = VBy$, therefore

$$\lambda_Q = \lambda_V \lambda_B \lambda_y = \lambda_y^{1/2} \lambda_x \lambda_y = \lambda_x \lambda_y^{3/2}$$

so

$$\lambda_Q = \frac{1}{40} \left(\frac{1}{6.75}\right)^{3/2} = \frac{1}{701}$$

Now $Q' = 1.15 \times 190 \times 5 = 1092.5 \text{ m}^3/\text{s}$, therefore

$$Q'' = 1092.5 \times \frac{1}{701} = 1.56 \text{ m}^3/\text{s}$$

(which would still strain the pumping resources of most laboratories!)

Now that the model has been distorted, the variation of depth with distance may not accurately reproduce the prototype characteristics. It is usual to carry out adjustments to the surface roughness in the model (k_S'') until an acceptable degree of accuracy is attained. This is known as 'calibration' of the model.

Models of tidal and wave phenomena

Problems relating to tides and waves are encountered in the design of docks, coastal defences, outfalls (discharge pipes) into estuarial or coastal waters, maritime structures, etc. Physical models (Photograph 5) are still widely used for these studies, though recent progress in the application of finite element analysis is having an increasing impact. Since tides are a particular type of wave, this section is, in effect, all about waves.

Waves may be simply (and roughly) divided into two categories, namely 'long' waves and 'short' waves. The parameter of wave length is L/y, and it is generally held that for long waves $L/y > 20$ and for short waves $L/y < 2$. Short waves are primarily wind-generated in deep waters. Long waves may be tidal or due to swell action. The principal characteristics of a wave (see Fig. 11.3) are the geometrical terms (height H, length L and mean water depth y) and the temporal terms (periodic time T and celerity c). Dimensional analysis reveals that the main parameters are the Froude and Strouhal groups:

$$\Pi_1 = c/(gy)^{1/2} \quad \text{(for long waves)}$$

or

$$\Pi_1 = c/(gL)^{1/2} \quad \text{(for short waves)}$$

$$\Pi_2 = cT/L$$

Photograph 5   Hydraulic model of Port Belawan, Sumatra.

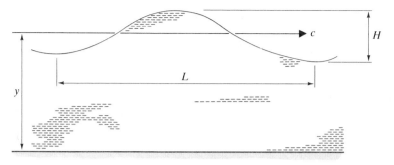

Figure 11.3 Wave profile.

None of the above equations inherently excludes the possibility of some distortion. It must be recalled, however, that waves are a function of gravitation and of the fluid. A wave shape cannot be arbitrarily distorted. On the other hand, current patterns can be reproduced by distorted models. Thus, for example, wave diffraction patterns around a breakwater are not accurately reproduced in a distorted model, though in practice experienced researchers can often interpret the results obtained from distorted models with sufficient accuracy.

Example 11.6 Harbour model

A major and developing port is subject to excessive wave action during storms. This leads on occasion to damage to vessels at their moorings. A model test is to be carried out to optimise the design of a new breakwater. The harbour occupies 0.7×2 km, and the maximum space in the hydraulics laboratory is 10×20 m. Select the model scales.

Solution

The absolute maximum length scale would be

$$\lambda_L = \frac{20}{2000} = \frac{1}{100}$$

However, allowances must be made for access and for the wave- and tide-producing mechanisms, so say $\lambda_L = 1/120$. (It is being assumed here the model will be undistorted, since wave diffraction is the centre of interest.)

Other scales are as follows:

$$\text{wave height} \quad \lambda_H (= H''/H') = 1/120$$

$$\text{wave period } T_W = L/c_W,$$

therefore

$$\lambda_{T_W} = \frac{\lambda_L}{\lambda_{c_W}} = \frac{1/120}{\sqrt{1/120}} = \frac{1}{10.95}$$

say, $\lambda_{T_W} = 1/11$;

$$\text{tidal period } T_T = L/c_T,$$

therefore

$$\lambda_{T_T} = \frac{\lambda_L}{\lambda_{c_T}} = \frac{\lambda_L}{\sqrt{\lambda_L}} \approx \frac{1}{11}$$

Therefore tidal period in model = 12/11 = 1.09 h.

Models of estuaries are more complex to design and operate, since all of the following quantities may be of interest: tidal currents, waves, incoming river flow, differential density effects at the interface between salt and fresh water, and sediment transport. It is often necessary to use two models to provide sufficient data. For example:

Stage I – construct small distorted model (λ_x might be 1/500). After calibration, current patterns may be investigated.

Stage II – construct larger, undistorted model of the environment in the immediate vicinity of the area of interest in the estuary. This would be designed to reproduce the current patterns observed during Stage I. Events such as local scour can be more readily observed on the larger scale.

Sediment transport may be caused by a unidirectional flow (such as the current in a river or estuary) or by wave action. Where the sediment is carried by a unidirectional flow, a dimensional analysis produces the following dimensionless groups:

$$\Pi_1 = \frac{\rho u_* D}{\mu} \qquad \Pi_2 = \frac{\rho u_*^2}{(\rho s - \rho)gD}$$

$$\Pi_3 = \frac{y}{D}\left(\text{or } \frac{R}{D}\right) \qquad \Pi_4 = \frac{\rho s}{\rho}$$

In the above equations, D is the representative sediment diameter (usually D_{50}), ρs is the sediment density, y is the depth of flow and R is the hydraulic radius. The other terms have their usual meaning. The Π_2 group is a form of the Froude Number (u_*^2/gD) multiplied by a density ratio, and is often known as a 'densimetric Froude Number'. If an hydraulic model is being designed to investigate this type of sediment transport problem, the criterion $\Pi_1 > 70$ must be met to ensure fully turbulent flow. The principal similarity criterion is then based on the Froude Number, i.e. the model must be designed and operated such that $\Pi_2' = \Pi_2''$. It is not always easy to construct a model in which $\Pi_1 > 70$, $\Pi_2' = \Pi_2''$ and $\rho_S' = \rho_S''$. For this reason, models quite often use a sediment which has a different density to that of the prototype. A range of granular materials is available for this purpose, for example coal dust ($\rho_S = 1300 \text{ kg/m}^3$), Araldite ($\rho_S = 1120 \text{ kg/m}^3$) or polystyrene ($\rho_S = 1040 \text{ kg/m}^3$).

Where sediment transport is primarily the result of wave action rather than current action, the problems of the designer of the model are multiplied. Sediment transport by waves occurs only in transitional or shallow waters, where the wave motion is transmitted to the sea bed (see Section 8.2) or the shore. The fluid velocity induced at the bed alternates in direction such that a graph of velocity (or displacement) against time is sinusoidal in form. The direction of the shear stress at the bed must therefore also alternate. In effect, the wave motion sets up a boundary layer, in which the velocities are a function both of y/δ and time (δ is the boundary layer thickness). The effects of wave refraction and shoaling are also significant in this type of problem. One way of approaching the dimensional analysis is to adopt a 'representative velocity', say the maximum x-direction velocity (u_δ) for $y = \delta$. Four groups emerge:

$$\Pi_1 = \frac{\rho u_\delta D}{\mu} \qquad \Pi_2 = \frac{\rho u_\delta^2}{(\rho_S - \rho)gD} \qquad \Pi_3 = \frac{x_\delta}{D} \qquad \Pi_4 = \frac{\rho_S}{\rho}$$

where x_δ is the maximum displacement of a fluid particle at $y = \delta$. Most of these quantities can be measured, using modern instrumentation techniques. However, prototype sites are not always easily instrumented (e.g. at the entry to a busy port instruments are prone to damage by passing ships). Further discussion of this difficult aspect of modelling is beyond the scope of this text. The interested reader is referred to Yalin's (1971) extensive treatment.

Models of hydraulic structures

Hydraulic structures are small compared with rivers or estuaries. In consequence, hydraulic models of such structures can usually be constructed to a relatively large scale (say $1/10 < \lambda_L < 1/50$) and need not be distorted. Velocity scales are usually based on the Froude Number (for flows with a free surface) or the Reynolds Number (for ducted flows). However, there are some exceptions to this general rule. For example, to simulate cavitation, the pressures in the model should be the same as those for the prototype. This is not always possible. Taking the case of a spillway model, the velocities in the model will be lower than those in the prototype, so the lowest model pressures will probably not be as low as those in the prototype. Nevertheless, an experienced researcher will usually be able to use the results from the model to predict conditions on the prototype. Models can also be used to investigate local scour problems, or the current patterns set up by the structure.

Example 11.7 Spillway model

A spillway system is to be designed for a dam (Fig. 11.4). The design discharge over the spillway is 15 m³/s per metre width at a design head of 3 m. An hydraulic model is to be used to confirm the estimates of the spillway performance, and also to help in the design of the scour bed just downstream of the stilling basin. Estimate the scales of such a model if the scour bed is armoured with stones having a representative size $D_{50} = 35$ mm.

Solution

For fully turbulent flow, $Re_* > 70$. To obtain Re_*, u_* must first be evaluated. From (3.7),

$$\frac{V}{u_*} = \frac{1}{0.4} \ln \left(\frac{y}{k_s} \right)$$

where k_s represents the roughness, which will be taken as being equal to $D_{50}/2$. The magnitude of the velocity, V, over the scour bed is obtained from the continuity equation, $V = Q/A = 15/6 = 2.5$ m/s. Therefore

$$\frac{2.5}{u_*} = \frac{1}{0.4} \ln \left(\frac{6}{0.035/2} \right)$$

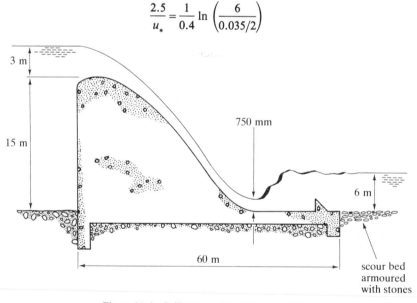

Figure 11.4   Spillway section (Example 11.7).

Therefore $u_* = 0.171$ m/s. Hence,

$$Re_*' = \frac{1000 \times 0.171 \times 0.035}{0.001} = 5985$$

and $\lambda_L^{1.5} = 70/5985$, so $\lambda_L = 1/19.4$ (say scale $= 1/20$). Now, $\lambda_Q = \lambda_L^{2.5} = 1/1789$, so $Q'' = 8.385$ l/s and $\lambda_V = \lambda_L^{\frac{1}{2}} = 1/4.472$. The densimetric Froude Numbers are, for the prototype,

$$\Pi_2' = \frac{1000 \times 0.171^2}{(2650 - 1000) \times 9.81 \times 0.035} = 0.0516$$

and for the model (assuming the same sediment density),

$$\Pi_2'' = \frac{1000 \times (0.171/4.472)^2}{(2650 - 1000) \times 9.81 \times (0.035/20)} = 0.0516$$

so the sediment would be of 1.75 mm diameter in the model. The sediment in the model can be of the same density as that in the prototype.

References and further reading

Barr, D. I. H. 1983. A survey of procedures for dimensional analysis. *Int. J. Mech. Engng Education* **11**(3), 147–59.

Henderson, F. M. 1966. *Open channel flow*. New York: Macmillan.

Keulegan, G. H. 1938. Laws of turbulent flows in open channels. *J. Res.* **21**, Paper No. 1151. Washington, DC: US National Bureau of Standards.

Langhaar, H. L. 1980. *Dimensional analysis and theory of models*. Florida: Robert E. Krieger.

Novak, P. and J. Cabelka 1981. *Models in hydraulic engineering*. London: Pitman.

Sharp, J. J. 1981. *Hydraulic modelling*. London: Butterworth.

Yalin, M. S. 1971. *Theory of hydraulic models*. London: Macmillan.

PART II
Aspects of hydraulic engineering

12

Pipeline systems

12.1 Introduction

In Part I, the principles of hydraulics were explained. Part II considers how these principles may be applied to practical design problems. In this chapter, various aspects of pipelines are considered. The starting point is the design of simple (one pipe) pipelines. This is followed by a discussion of series, parallel and branched pipelines, leading to the analysis of distribution systems. Finally, the steady flow design of pumping mains is discussed, and the important topic of surge protection for pumping mains and turbine installations is introduced.

12.2 Design of a simple pipe system

Aspects of design

Consider Figure 12.1, which shows a typical pipeline between two storage tanks. Broadly, there are two aspects to the design of this system:

(a) hydraulic calculations;
(b) detail design.

Under (a), a suitable pipe diameter must be determined for the available head and required discharge. This aspect has already been covered in Chapter 4. However, it should be noted that to estimate the local head losses, the proposed valves and fittings must be known. In addition, the maximum and minimum pressures within the pipeline must be found, to ensure that the pressure rating of the pipe is sufficient and to check that sub-atmospheric pressures do not occur. This is most easily visualised by drawing the energy line and hydraulic gradient on the plan of the pipe longitudinal section (refer to the following section).

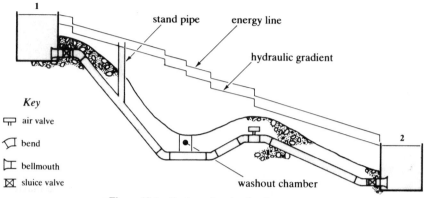

Figure 12.1   Design of a simple pipe system.

Under (b), the following must be considered:

 (i) pipeline material, pressure rating and jointing system;
 (ii) provision of valves, bends, fittings and thrust blocks;
(iii) locations of any necessary air valves and washouts.

The energy line and hydraulic gradient

Applying Bernoulli's energy equation in Figure 12.1,

$$\frac{p_1}{\rho g} + \frac{V_1^2}{2g} + z_1 = \frac{p_2}{\rho g} + \frac{V_2^2}{2g} + z_2 + h_f + h_L$$

where h_f is the frictional head loss and h_L is the local head loss. As

$$p_1 = p_2 = 0 \qquad V_1 = V_2 = 0$$

then

$$z_1 - z_2 = H = h_f + h_L$$

This equation is used to find the required diameter for the given discharge (cf. Example 4.5).

At any point between the reservoirs,

$$\frac{p}{\rho g} + \frac{V^2}{2g} + z = \text{height of the energy line}$$

and

$$\frac{p}{\rho g} + z = \text{height of the hydraulic gradient}$$

Both of these lines have a slope of $S_f (h_f/L)$, and local head losses are represented by a step change. They are both shown in Figure 12.1. The energy line begins and ends at the water level in the upper and lower tanks, and the hydraulic gradient is always a distance $V^2/2g$ below the energy line.

The usefulness of the hydraulic gradient lies in the fact that it represents the height to which water would rise in a standpipe (piezometer). Hence, the location of the maximum and minimum pressures may be found by finding the maximum and minimum heights between the pipe and the hydraulic gradient.

If the hydraulic gradient is below the pipe, then there is sub-atmospheric pressure at that point. This condition is to be avoided since cavitation may occur (if $p/\rho g < -7.0$ m), and if there are any leaks in the pipeline matter will be sucked into the pipe, possibly causing pollution of the water supply.

Pipe materials and jointing systems

Table 12.1 lists typical pipe materials, associated linings and jointing systems for water supply pipelines.

The choice of material will depend on relative cost and ground conditions. All of the materials in Table 12.1 are in common usage. More detailed guidance is given in Twort *et al.* (1985). Pipe joints, as shown in Figure 12.2, are normally of the spigot and socket type for underground pipes, where there are no lateral forces to resist. Flanged joints are used in pumping stations, service reservoirs, etc., where lateral forces must be resisted and where easy removal of pipe sections is required.

Thrust blocks

Thrust blocks should be provided at all fittings where a change of velocity or flow direction occurs. The forces acting may be calculated by applying

Table 12.1 Pipe materials and joints for water supply pipes.

Material	Protective lining	Joint type(s)
cast iron (old water mains)	bitumen	spigot and socket with run lead sealer, flanged
ductile iron	bitumen, spun concrete	spigot and socket with rubber ring sealer, flanged
steel	epoxy resin	welded (large pipes), screwed (small pipes)
uPVC	none	spigot and socket with rubber ring sealer, sleeved with chemical sealer, flanged
asbestos cement	bitumen	sleeved with rubber ring sealer

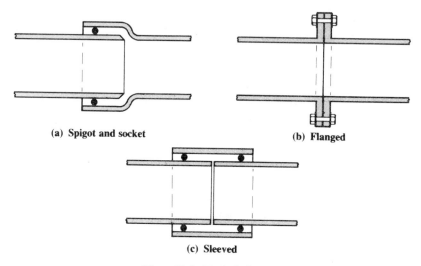

Figure 12.2   Typical pipe joints.

the momentum equation (refer to Ch. 2). Thrust blocks are normally designed to withstand either the static head or the pipe test pressure, whichever is greater.

Air valves and washouts

Air valves should be provided at all high points in a water main, so that entrained air is removed during normal operation and air is evacuated during filling. To fulfil this dual role, double orifice air valves are commonly used. It should be noted, however, that air valves should not be placed in any regions of sub-atmospheric pressure, because air would then enter the pipe, rather than be expelled from it.

Washouts are normally placed at all low points so that the water main may be completely emptied for repair or inspection.

12.3   Series, parallel and branched pipe systems

Introduction

Figure 12.3 shows all three cases. To determine the heads and discharges is more complex than in simple pipe problems, and requires the use of the continuity equation in addition to the energy and frictional head loss equations.

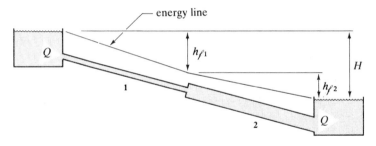

(a) Series

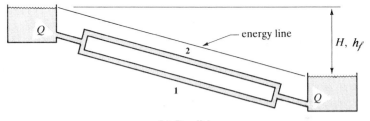

(b) Parallel

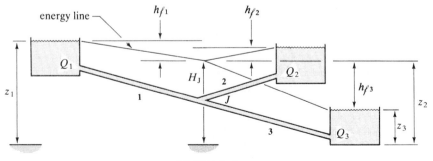

(c) Branched

Figure 12.3 Series, parallel and branched pipes.

Series solution

Energy $H = h_{f1} + h_{f2}$
Continuity $Q = Q_1 = Q_2$

As h_{f1}, h_{f2} are initially unknown, a solution method is as follows:

(1) guess h_f;   (3) if $Q_1 = Q_2$, then the solution is correct;
(2) calculate Q_1 and Q_2;   (4) if $Q_1 \neq Q_2$, then return to (1).

Parallel solution

Energy $\quad H = h_{f1} = h_{f2}$
Continuity $\quad Q = Q_1 + Q_2$

This problem can be solved directly for Q_1 and Q_2.

Branched solution

Energy $\quad h_{f1} = z_1 - H_J$
$\qquad\quad h_{f2} = z_2 - H_J$
$\qquad\quad h_{f3} = H_J - z_3$

where H_J is the energy head at junction J.

Continuity $\quad Q_3 = Q_1 + Q_2$

As H_J is initially unknown, a method of solution is as follows:

(1) guess H_J;
(2) calculate Q_1, Q_2 and Q_3;
(3) if $Q_1 + Q_2 = Q_3$, then the solution is correct;
(4) if $Q_1 + Q_2 \neq Q_3$, then return to (1).

Example 12.1 Branched pipe system

Given the following data for the system shown in Figure 12.3c, calculate the discharge and the pressure head at the junction J:

Pipe	Length (km)	Diameter (mm)	Roughness, k_S (mm)
1	5	300	0.03
2	2	150	0.03
3	4	350	0.03

Item	Elevation (m above datum)
Reservoir 1	800
Reservoir 2	780
Reservoir 3	700
Junction J	720

Solution

The method is that given in the previous section, using Figure 4.5 (HRS chart) to find the discharges. Try $H_J = 750$ m above datum.
Then

$$h_{f1} = 800 - 750 = 50 \text{ m}$$

$$\frac{100 h_{f1}}{L_1} = \frac{50 \times 100}{5000} = 1$$

therefore

$$Q_1 = 140 \text{ l/s}$$

and

$$h_{f2} = 780 - 750 = 30 \text{ m}$$

$$\frac{100 \, h_{f2}}{L_2} = \frac{100 \times 30}{2000} = 1.5$$

therefore

$$Q_2 = 30 \text{ l/s}$$

and

$$h_{f3} = 750 - 700 = 50 \text{ m}$$

$$\frac{100 \, h_{f3}}{L_3} = \frac{100 \times 50}{400} = 1.25$$

therefore

$$Q_3 = 250 \text{ l/s}$$

Hence, for

$$H_J = 750 \text{ m}$$

$$Q_1 + Q_2 = 170 \text{ l/s}$$

$$Q_3 = 250 \text{ l/s}$$

This is not the correct solution. A better guess would be achieved by reducing H_J, hence reducing Q_3 and increasing Q_1 and Q_2. For $H_J = 740$ m above datum,

$$h_{f1} = 60 \text{ m} \qquad 100 \, S_{f1} = 1.2 \qquad Q_1 = 160 \text{ l/s}$$

$$h_{/2} = 40 \text{ m} \qquad 100\,S_{/2} = 2.0 \qquad Q_2 = 35\,\text{l/s} \qquad Q_1 + Q_2 = 195\,\text{l/s}$$

$$h_{/3} = 40 \text{ m} \qquad 100\,S_{/3} = 1.0 \qquad Q_3 = 210\,\text{l/s}$$

For $H_J = 735$ m above datum,

$$h_{/1} = 65 \text{ m} \qquad 100\,S_{/1} = 1.3 \qquad Q_1 = 165\,\text{l/s}$$

$$h_{/2} = 45 \text{ m} \qquad 100\,S_{/2} = 2.25 \qquad Q_2 = 37\,\text{l/s} \qquad Q_1 + Q_2 = 202\,\text{l/s}$$

$$h_{/3} = 35 \text{ m} \qquad 100\,S_{/3} - 0.875 \qquad Q_3 = 200\,\text{l/s}$$

This is an acceptable solution. The pressure head at J (neglecting velocity head) is given by

$$p_J/\rho g = H_J - z_J$$

$$= 735 - 720 = 15 \text{ m}$$

It is worth noting that if the junction elevation were greater than 735 m above datum, then negative pressures would result. This problem could be overcome by increasing the diameter of pipe No. 2.

12.4 Distribution systems

General design considerations

A water supply distribution system consists of a complex network of inter-connected pipes, service reservoirs and pumps which deliver water from the treatment plant to the consumer. Water demand is highly variable, both by day and season. Supply, by contrast, is normally constant. Consequently, the distribution system must include storage elements, and must be capable of flexible operation. Water pressures within the system are normally kept between a maximum (about 70 m head) and a minimum (about 20 m head) value. This ensures that consumer demand is met, and that undue leakage due to excessive pressure does not occur. The topography of the demand area plays an important part in the design of the distribution system, particularly if there are large variations of ground levels. In this case, several independent networks may be required to keep within pressure limitations. For greater operational flexibility, however, they are usually interconnected through booster pumps or pressure reducing valves.

In addition to new distribution systems a common need is for improve-ment to existing (often ageing) systems. It is good practice to use a ring main system in preference to a branching system. This prevents the occurrence of 'dead ends' with the consequent risk of stagnant water, and permits more flexible operation, particularly when repairs must be carried out. Many existing systems have very high leakage rates (30–50%). Leaks are often very difficult to locate, and have an important bearing on the accuracy of any hydraulic analysis of the system.

An essential prerequisite to the improvement of an existing system is to have a clear understanding of how that system operates. This is often quite difficult to achieve. Plans of the pipe network, together with elevations, diameters, water levels, etc., are required. In addition, the demands must be estimated on a *per capita* consumption basis, or preferably by simultaneous field measurements of pressures and flows at key points in the system.

The analysis of such systems is generally carried out by computer simulation, in which a numerical model of the system is initially calibrated to the field data before being used in a predictive mode. The model is a simplified version of the real system and, in particular, demands from the system are assumed to be concentrated at pipe ends or junctions. This allows relatively simple models to be used without great loss of accuracy, providing that a judicious use of pipe junctions is made.

Such models have been succesfully used to locate areas of leakage within a system. However, if leakage rates are high and of unknown location, then a computer simulation may give misleading results, and therefore be of little value.

Many computer models have been developed with varying degrees of success and applicability. One successful model developed by the Water Research Centre is called WATNET (see Creasy 1982), and is currently in use in all of the English Regional Water Authorities.

Hydraulic analysis

The solution methods described for the analysis of series, parallel and branched pipes are not very suitable for the more complex case of networks. A network consists of loops and nodes as shown in Figure 12.4. Applying the continuity equation to a node,

$$\sum_{i=1}^{n} q_i = 0 \tag{12.1}$$

where n is the number of pipes joined at the node. The sign convention used here sets flows into a junction as positive.

Applying the energy equation to a loop, then

$$\sum_{i=1}^{m} h_{fi} = 0 \tag{12.2}$$

where m is the number of pipes in a loop. The sign convention sets flow and head loss as positive clockwise.

In addition,

$$h_{fi} = f(q_i) \tag{12.3}$$

where $f(q_i)$ represents the Darcy–Weisbach/Colebrook–White equation. Equations (12.1)–(12.3) comprise a set of simultaneous non-linear equations, and an iterative solution is generally adopted. The two standard solution techniques, the loop method and the nodal method, are now discussed.

The loop method

This method, originally proposed by Hardy-Cross in 1936, essentially consists of eliminating the head losses from (12.2) and (12.3) to give a set of equations in discharge only. It may be applied to loops where the external discharges are known and the flows within the loop are required. The basis of the method is as follows:

(1) assume values for q_i to satisfy $\sum q_i = 0$;
(2) calculate h_{fi} from q_i;
(3) if $\sum h_{fi} = 0$, then the solution is correct;
(4) if $\sum h_{fi} \neq 0$, then apply a correction factor δq to all q_i and return to (2).

A reasonably efficient value of δq for rapid convergence is given by

$$\delta q = -\frac{\sum h_{fi}}{2\sum h_{fi}/q_i} \tag{12.4}$$

Step (2) may be carried out using the HRS charts or tables (for hand calculations) or using Barr's explicit formula for λ (for computer solution). Due account must be taken of the sign of q_i and h_{fi}, and of the use of an appropriate convention (i.e. clockwise positive).

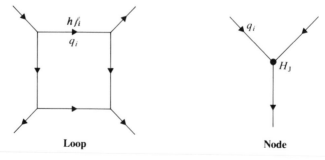

Figure 12.4   Networks.

Equation (12.4) may be derived as follows. Using the Darcy–Weisbach formula,

$$h_f = \frac{\lambda L V^2}{2gD}$$

or, for a given pipe and assuming that λ is constant,

$$h_f = kQ^2$$

Taking the true flow to be Q, then

$$Q = (q_i + \delta q)$$

and taking the true head loss to be H_f

$$H_f = k(q_i + \delta q)^2$$

Expanding by the binomial theorem,

$$H_f = kq_i^2 \left[1 + 2\frac{\delta q}{q_i} + \frac{2(2-1)}{2!} \left(\frac{\delta q}{q_i}\right)^2 + \cdots \right]$$

ignoring second-order terms and above (for $\delta q \ll q_i$). Then

$$H_f = kq_i^2 \ [1 + (2\delta q/q_i)]$$

For a loop,

$$\Sigma H_{fi} = 0 = \Sigma kq_i^2 + 2\delta q \Sigma \ kq_i^2/q_i$$

or

$$0 = \Sigma h_{fi} + 2\delta q \ \Sigma h_{fi}/q_i$$

Hence

$$\delta q = - \frac{\Sigma \ h_{fi}}{2 \ \Sigma h_{fi}/q_i}$$

Example 12.2   Flows in a pipe loop

For the square pipe loop shown in Figure 12.5, find:

(a) the discharges in the loop;
(b) the pressure heads at points B, C and D, if the pressure head at A is 70 m and A, B, C and D have the same elevations.

All pipes are 1 km long and 300 mm in diameter, with roughness 0.03 mm.

Solution

(a) It is convenient for hand solution to use a tabular layout in conjunction with the HRS charts or tables for finding h_{fi} from q_i and d.

Initial trial (assume values for q_i)

Pipe	q_i(l/s)	h_{fi}(m)	h_{fi}/q_i
A–B	+60	+2.00	0.0333
B–C	+40	+0.93	0.0233
C–D	0	0	0
A–D	−40	−0.93	0.0233
Σ		+2.00	0.0799

Note the 'positive clockwise' sign convention for q_i and h_{fi}.

Apply correction factor,

$$\delta q = -\frac{2}{2(0.0799)} = -12.5 \text{ l/s}$$

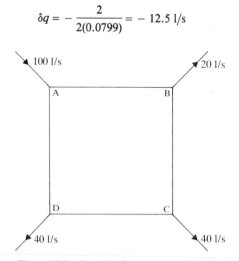

Figure 12.5   Example flow in a pipe loop.

Second trial (new discharges $= q_i - 12.5$)

Pipe	$q_i(\text{l/s})$	$h_{fi}(\text{m})$	h_{fi}/q_i
A–B	+47.5	+1.3	0.0274
B–C	+27.5	+0.48	0.0175
C–D	−12.5	−0.12	0.0096
A–D	−52.5	−1.58	0.0301
Σ		+0.08	0.0846

$$\delta q = - \frac{0.08}{2(0.0846)} = -0.5 \text{ l/s}$$

As $\delta q = -0.5$ l/s, this solution is sufficiently accurate for most practical purposes.

(b) To find the pressure heads $(p/\rho g)$ at B, C and D, apply the energy equation. Ignoring velocity heads and recalling that the elevations are the same at A, B, C and D,

$$\frac{p_B}{\rho g} = \frac{p_A}{\rho g} - h_{fA-B} = 70 - 1.3 = 68.7 \text{ m}$$

$$\frac{p_C}{\rho g} = \frac{p_B}{\rho g} - h_{fB-C} = 68.7 - 0.48 = 68.22 \text{ m}$$

$$\frac{p_D}{\rho g} = \frac{p_A}{\rho g} - h_{fA-D} = 70 - 1.58 = 68.42 \text{ m}$$

As a check

$$\frac{p_C}{\rho g} = \frac{p_D}{\rho g} - h_{fD-C} = 68.42 - 0.12 = 68.3$$

Comparing with the previous estimate $(p_C/\rho g = 68.22 \text{ m})$, the difference is equal to Σh_{fi} (the closing error).

The nodal method

This method, originally proposed by Cornish in 1939, consists of eliminating the discharges from (12.1) and (12.3) to give a set of equations in head losses only. It may be applied to loops or branches where the external heads are known and the heads within the networks are required. The basis of the method is as follows:

(1) assume values for the head (H_j) at each junction;
(2) calculate q_i from H_j;

(3) if $\Sigma\, q_i = 0$, then the solution is correct;
(4) if $\Sigma\, q_i \neq 0$, then apply a correction factor δH to H_j and return to (2), where

$$\delta H = \frac{2\,\Sigma\, q_i}{\Sigma q_i/h_{fi}} \qquad (12.5)$$

The derivation of (12.5) is similar to that for (12.4).

Step (2) may be carried out using the HRS charts or tables (for hand calculations) or using the Colebrook–White/Darcy–Weisbach equation for q in terms of h_f (for computer solution). An appropriate sign convention must be used for q_i and h_{fi}, e.g. q_i and h_{fi} positive entering a node.

Example 12.3   Flows in a branched pipe network

Resolve Example 12.1 using the nodal method.

Solution

Assume a trial value of $H_J = 750$ m

Pipe	h_{fi}(m)	q_i(l/s)	q_i/h_{fi}
1	+ 50	+ 140	+ 2.8
2	+ 30	+ 30	+ 1.0
3	− 50	− 250	+ 5.0
Σ		− 80	8.8

Apply correction factor,

$$\delta H = \frac{2(-80)}{8.8} = -18.18 \text{ m}$$

Second trial $H_J = 750 - 18.18 = 731.82$ m

Pipe	h_{fi}(m)	q_i(l/s)	q_i/h_{fi}
1	+ 68.18	+ 168	+ 2.5
2	+ 48.18	+ 37	+ 0.8
3	− 31.82	− 191	+ 6.0
Σ		+ 14	9.3

$$\delta H = \frac{2(14)}{9.3} = + 3.01$$

Third trial $H_J = 731.82 + 3.01 = 734.83$ m

As the solution to Example 12.1 gave $H_J = 735$ m, then convergence has been achieved and the solution is as given in Example 12.1.

Complex networks

For more complex networks (e.g. more than one loop or junction) the loop and nodal methods may both be applied with minor modifications. In the case of the loop method, the correction factor δq at each iteration must be carried over from one loop to the next through any common pipes. For the nodal method, the correction factor δH is applied to successive nodes through the network at each iteration.

The choice of loop or nodal method for the analysis of complex networks depends on the available data. In terms of efficiency of solution and required computer storage space, the loop method requires more storage but converges more quickly, whereas the nodal method requires less storage but convergence is slower and not always achieved. The WATNET program uses a hybrid combined loop–nodal method, which gives the benefits of both methods. For details of other available methods refer to Novak (1983).

12.5 Design of pumping mains

Introduction

Typical applications of pumping mains include river abstractions (low level supply to high level demand), borehole supplies from groundwater and surface water and foul water drainage from low-lying land. In all of these cases there are at least three elements to consider:

(a) hydraulic design;
(b) economic matching of pump and pipeline;
(c) detail design.

These three elements are now considered in turn.

Hydraulic design

The primary requirement is to determine a suitable pump and pipe combination for the required design discharge (Q). Consider Figure 12.6, which shows a simple pumping main. At start-up, the pump is required to deliver

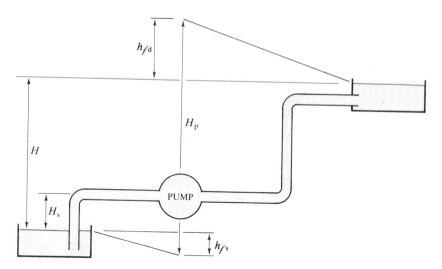

Figure 12.6 Simple pumping main.

the design discharge (Q) against the static head (H). However, as soon as flow commences, frictional losses are introduced (h_{fs} and h_{fd}) which vary with discharge. To attain the design discharge (Q), the head provided by the pump (H_p) must exactly match the static plus friction heads at Q. Hence the discharge is a function of both the pump and the pipeline. For a given system, the head–discharge characteristic curves for the pump may be superimposed on that for the pipeline, as shown in Figure 12.7. The point of intersection of the two characteristic curves locates the one possible combination of head and discharge for the system under steady flow conditions. The intersection point is referred to as the operating point.

To obtain the required discharge (Q), it may be necessary to investigate several pump and pipe combinations. In addition, it is obviously desirable that the pump should be running at or near peak efficiency at the design discharge (Q). This condition may be checked by drawing the pump efficiency curve (as shown in Fig. 12.7).

Pumps are often arranged in series (for large static heads) or parallel (for varying discharge requirements). To obtain the discharge in these cases requires the estimation of their combined characteristic curves. These are obtained from the characteristic for the single pump case as follows:

For series operation (of n pumps)

$$H_{np} = nH_p \qquad Q_{np} = Q_p$$

where subscript np denotes n pumps and subscript p denotes one pump.

For parallel operation (of n pumps)

$$H_{np} = H_p \qquad Q_{np} = nQ_p$$

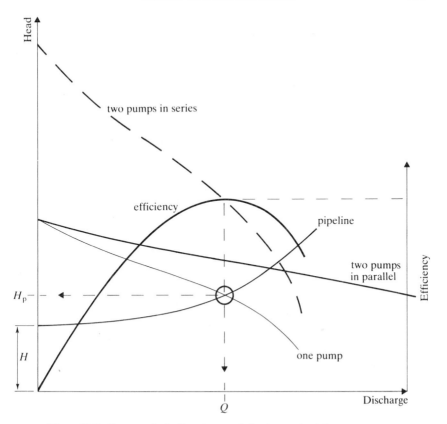

Figure 12.7 Pump and pipeline characteristics for a mixed flow pump.

The resulting characteristic curves for $n = 2$ are shown in Figure 12.7.

A secondary requirement is to prevent cavitation occurring in the pump. This is likely to occur if the water level in the suction well is several metres below the pump. As cavitation in pumps has already been described in Section 7.7, it is not discussed further here.

Example 12.4   Design of a simple pumping main

Given the pump characteristics below, determine the pump efficiency and power requirement for a pipeline of diameter 300 mm, length 5 km, roughness 0.03 mm and a static lift of 10 m. Comment on the suitability of this pump and pipeline combination.

Pump characteristics

Discharge (l/s)	0	10	20	30	40	50	60	70
Head (m)	26.25	24.00	21.75	19.50	17.50	15.00	11.75	6.75
Efficiency (%)	0	28	51	68	80	85	80	64

Solution

(a) Determine the pipeline characteristic by finding the frictional losses at the discharges given in the table above. These are most readily found using the HRS charts or tables.

Pipeline characteristic

Discharge (l/s)	20	30	40	50	60	70
Friction head (m)	1.35	2.8	4.7	7.25	9.75	13.5
Friction + static head (m)	11.35	12.8	14.7	17.25	19.75	23.5

(b) Draw the characteristic curves as shown in Figure 12.8.

(c) The operating point is given by:

$$Q_d = 45.5 \text{ l/s} \qquad H_p = 16 \text{ m} \qquad \eta = 84\%$$

The power consumption is given by:

$$P = \rho g Q H / \eta$$
$$= 10^3 \times 9.81 \times 45.5 \times 10^{-3} \times 16/0.84 \text{ W}$$
$$= 8.5 \text{ kW}$$

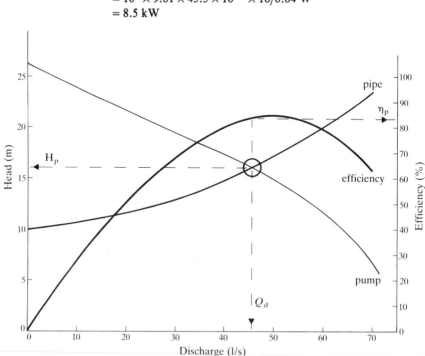

Figure 12.8 Characteristic curves for Example 12.4.

The operating point is satisfactory since the system is operating very close to the peak efficiency point of the pump.

Economics of pumping mains

For the required design discharge, a variety of possible pump and pipe combinations are possible. The final choice is not determined purely on the basis of hydraulic considerations. It is necessary to determine the least-cost solution. There are three basic elements involved:

(a) the pipe cost;
(b) the pump cost;
(c) the running costs.

Pipe costs increase with diameter but, conversely, pump and running costs will decrease with pipe diameter (because of smaller frictional losses). Hence a least cost solution is possible, as is shown in Figure 12.9.

The normal method for finding this solution is:

(1) estimate the capital cost of pipes (material and laying) and the associated pumps for various pump diameters;
(2) convert capital cost to annual cost by assessing interest charges or by discounted cash flow analyses;

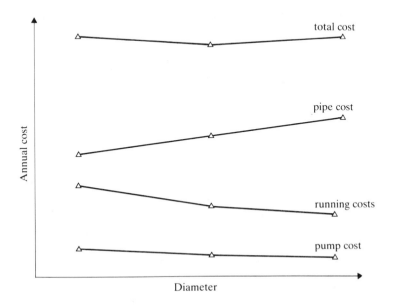

Figure 12.9 Variation of pumped pipeline costs with pipe diameter.

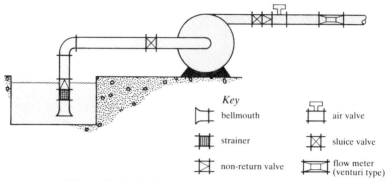

Figure 12.10   Typical centrifugal pump installation.

(3) estimate the annual power charges and other running costs (maintenance, etc.);

(4) tabulate the results and compare.

Details of such calculations are given in Twort *et al.* (1985).

Detail design

A typical centrifugal pump installation is shown in Figure 12.10. The pump is installed between two valves for easy removal in case of repair or maintenance. On the suction side, a combined bellmouth entry and strainer are necessary, together with a non-return valve to ensure self-priming. On the delivery side, a second non-return valve is necessary to prevent damage from possible surge pressures. In addition, an air valve and flow meter (venturi type) are desirable.

12.6   Surge protection

General description

The main thrust of this chapter has been the application of principles associated with steady full-pipe flows. However, it would be a serious omission in any design procedure if the possibility of an unsteady flow was not considered. It was asserted in Chapter 6 that changes in conditions at a controlling boundary could produce such flows. The first step, therefore, is to check this aspect of the system for all possible operating conditions, especially those associated with an emergency (pump breakdown, emergency shutdown, etc.). The techniques introduced in Chapter 6 would be suitable for this purpose. If a surge problem is predicted, clearly some counter-measures must be taken. These measures are given the general title of surge protection measures.

The crux of the problem is how to modify the system economically in such a way as to reduce the effects of the unsteady flow to an acceptable level. Two approaches are possible:

(a) Surge protection by mechanical means – this might entail limiting the rate of valve closure, for example. However, for some systems, the scale of the system or the nature of emergency shutdown procedures would preclude this approach.
(b) Surge protection by hydraulic means – this implies the incorporation of an hydraulic device which automatically limits the pressure rise in the system.

Since this is an hydraulics text, we shall concentrate on the second group of methods. This is not in any way to discount the methods outlined under (a) where they are appropriate.

The simple surge tower

This is one of the simplest devices to analyse and to design. It consists of a large vertical tube connected at its base to the pipeline (Photograph 6 and Fig. 12.11). The top is open to the atmosphere. It is usually sited as close as possible to the controlling boundary which is causing the surge. With the

Photograph 6 Construction of the surge tower for Dinorwig pumped storage scheme, Wales.

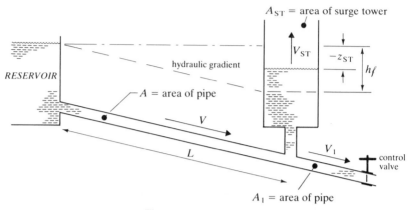

Figure 12.11 Surge tower.

valve open and steady flow in the pipe, the equilibrium water level in the surge tower will correspond to the pressure in the pipeline (i.e. it will lie on the hydraulic gradient). If the control valve is suddenly and completely closed, a rise in pressure is generated as the water decelerates. This causes a rise in the surge tower level. Since the valve is shut, the discharge in the pipe is all diverted into the surge tower. Thus, at time t after valve closure, if V is the mean velocity in the pipeline and V_{ST} the mean velocity in the surge tower, the continuity equation may be written

$$VA = V_{ST}A_{ST} \tag{12.6}$$

The rise in the water level in the surge tower means that the difference between reservoir level and surge tower level is now z_{ST}. The corresponding pressure difference, δp, is given by

$$\delta p = \rho g z_{ST}$$

The resultant force acting on the fluid in the pipeline is obtained by taking the sum of the forces due to the pressure difference δp, the frictional resistance and any local losses. Taking each force in turn:

$$\text{force due to pressure difference} = \delta p \, A = \rho g z_{ST} A$$

$$\text{resistance due to pipe friction} = \rho g h_f A = \frac{\rho g \lambda L V^2 A}{2gD}$$

$$\begin{array}{l}\text{resistance due to local losses} \\ \text{(e.g. at the entry to the surge tower)}\end{array} = \rho g h_L A = \frac{\rho g K_L V^2 A}{2g}$$

Since L and D are constant, and λ will be assumed to remain constant, the pipe friction resistance term may be written

$$\rho g h_f A = \rho g K_f V^2 A / 2g$$

where $K_f = \lambda L / D$. The total resistance will retard the flow of water in the pipeline upstream of the surge tower. The mass of water in the pipe is ρAL, and its rate of change of momentum is $\rho AL \, dV/dt$. The momentum equation may therefore be written as

$$-\left(-\rho g z_{ST} A + \frac{\rho g K_f V^2 A}{2g} + \frac{\rho g K_L V^2 A}{2g} \right) = \rho AL \frac{dV}{dt}$$

Dividing by ρA throughout,

$$-g\left(-z_{ST} + \frac{K_f V^2}{2g} + \frac{K_L V^2}{2g} \right) = L \frac{dV}{dt} \tag{12.7}$$

As it stands, this is not a standard differential equation, but it can be converted into an homogeneous equation as follows. The surge tower velocity, $V_{ST} = dz_{ST}/dt$, so (12.6) may be written as

$$V = \frac{dz_{ST}}{dt} \frac{A_{ST}}{A} \tag{12.6a}$$

Therefore

$$\frac{dV}{dt} = \frac{d^2 z_{ST}}{dt^2} \frac{A_{ST}}{A}$$

Equation (12.7) may therefore be rewritten as

$$-g\left[-z_{ST} + \frac{K_f}{2g}\left(\frac{dz_{ST}}{dt} \frac{A_{ST}}{A} \right)^2 + \frac{K_L}{2g}\left(\frac{dz_{ST}}{dt} \frac{A_{ST}}{A} \right)^2 \right] = L \frac{d^2 z_{ST}}{dt^2} \frac{A_{ST}}{A} \tag{12.8}$$

This is a second-order differential equation, and readers who are familiar with these equations will recognise that it represents a 'damped' harmonic motion. A graph of z_{ST} against t would therefore take the form of a sinusoidal curve whose amplitude decreases with time. Unfortunately, (12.8) is not quite in the standard mathematical form, due to the fact that the turbulent flow leads to resistance in terms of V^2 (or $(dz/dt)^2$). It will be shown below that a numerical solution is available. However, some useful information may be obtained if the friction and local loss terms are ignored.

This yields

$$-gz_{ST} = L \frac{d^2 z_{ST}}{dt^2} \frac{A_{ST}}{A} \qquad (12.9)$$

Now this is a standard second-order equation, which represents an un-damped harmonic motion. A graph of z_{ST} against t would take the form of a sine curve. It is easily shown that the maximum amplitude of the curve is

$$z_{ST\,max} = \pm V_0 \sqrt{\frac{L}{g} \frac{A}{A_{ST}}} \qquad (12.10)$$

where V_0 is the original velocity of flow in the pipeline before valve closure. The periodic time T_p may also be deduced:

$$T_p = 2\pi \sqrt{\frac{L}{g} \frac{A_{ST}}{A}} \qquad (12.11)$$

Equations (12.8) and (12.9) both draw broadly the same picture of events. When the valve is shut, the pipe flow is diverted into the surge tower. The water level in the tower accelerates upwards, overshoots the reservoir water level and comes to rest at elevation $z_{ST\,max}$. The difference in head between the surge tower and the reservoir cannot be sustained, so the surge tower level begins to fall and water flows in the negative direction in the pipeline. The surge tower level falls until it comes to rest at some level ($z_{ST\,min}$, say) below reservoir level. The flow then reverses and the surge tower level begins to rise again, and so on.

Numerical solution for the equations. A numerical description of the motion in the surge tower may be obtained by solving (12.6) and (12.7) simultaneously. A finite difference method is often used, and for this purpose (12.7) is rearranged slightly. The flow in the pipeline may be positive (i.e. in the original direction) or negative (in the reverse direction). The total resistance will always oppose motion, and accordingly the sign of each resistance term must be negative (for a 'positive' flow) or positive (for a 'negative' flow). To achieve this end it is usual to write '$\pm V |V|$' instead of 'V^2'. Thus, the friction loss term becomes $\pm K_f V |V| / 2g$, where $|V|$ is the modulus (i.e. the positive numerical value) while V may be positive or negative, depending on flow direction. Equation (12.7) now becomes

$$\delta V = \frac{g}{L} \left(\bar{+} z_{ST} \pm \frac{K_f V |V|}{2g} \pm \frac{K_L V |V|}{2g} \right) \delta t \qquad (12.12)$$

and (12.6a) becomes

$$\delta z_{ST} = \frac{VA\delta t}{A_{ST}} \qquad (12.13)$$

Providing that conditions (V, z_{ST}) are known at time t, these values are substituted into (12.12) and (12.13). The equations are then evaluated to obtain corresponding conditions at time $(t + \delta t)$. A constant time interval (δt) is usually used, and it should be noted that the magnitude of δt will have an effect on the accuracy of the solution. For high accuracy, $\delta t \ll T_p$. The calculations may be performed manually using a calculator, in which case it is advantageous to use a tabular layout. It is also very easy to produce a program for use on a microcomputer.

This numerical technique has the advantage of being directly related to the physical pattern of events taking place. Other techniques are now available for the computer-based approach (e.g. those based on the Runge–Kutta method), but these are beyond the scope of this text. A worked example follows to illustrate the above technique.

Example 12.5   Surge tower

A hydroelectric scheme is served by a tunnel 3 m in diameter, 1.5 km long. A surge tower of diameter 10 m is sited at the downstream end of the tunnel at the entry to the penstocks. The local loss coefficient for the entry to the surge tower is 0.785 (referred to flow velocity in the tunnel). The initial discharge is 32 m^3/s. The friction factor for the tunnel is 0.04. Estimate the peak upsurge height, and the periodic time, for a sudden and complete shutdown.

Solution

For $Q = 32$ m^3/s, $V_0 = 4.53$ m/s and $h_{f0} = 20.92$ m,

$$K_f = \lambda L/D = 20$$

$$A = 7.068 \text{ m}^2 \qquad A_{ST} = 78.54 \text{ m}^2$$

δt must now be selected. Its magnitude will have an effect on the accuracy of the final result. For the present example, a value of $\delta t = 2$ s has been selected. The first three steps of the finite difference solution are laid out below.

Step 1   (2 s after closure. The data from $t = 0$ are used to predict conditions at $t = 2$ s. From (12.12),

$$\delta V = \frac{9.81}{1500}\left(-20.92 + \frac{20 \times 4.53 \,|\, 4.53 \,|}{2 \times 9.81} + \frac{0.785 \times 4.53 \,|\, 4.53 \,|}{2 \times 9.81}\right) \times 2$$

$$= 0.01 \text{ m/s}$$

and from (12.13)(with $V = 4.53$ m/s),

$$\delta z_{ST} = \left(4.53 \times \frac{7.068}{78.54}\right) \times 2 = 0.815 \text{ m}$$

Step 2 $(t = 4 \text{ s})$. $z_{ST} = -20.92 + 0.815 = 20.105$ m; $V = 4.53 - 0.01 = 4.52$ m/s

$$\delta V = \frac{9.81}{1500} \left(-20.105 + \frac{20 \times 4.52 \,|\, 4.52\,|}{2 \times 9.81} + \frac{0.785 \times 4.52 \,|\, 4.52\,|}{2 \times 9.81}\right) \times 2$$

$$= 0.02 \text{ m/s}$$

$$\delta z_{ST} = \left(4.52 \times \frac{7.068}{78.54}\right) \times 2 = 0.814 \text{ m}$$

Step 3 $(t = 6 \text{ s})$ $z_{ST} = -20.105 + 0.814 = -19.291$ m; $V = 4.52 - 0.02 = 4.5$ m/s

$$\delta V = \frac{9.81}{1500} \left(-19.291 + \frac{20 \times 4.50 \,|\, 4.50\,|}{2 \times 9.81} + \frac{0.785 \times 4.50 \,|\, 4.50\,|}{2 \times 9.81}\right) \times 2$$

$$= 0.028 \text{ m/s}$$

$$\delta z_{ST} = \left(4.50 \times \frac{7.068}{78.54}\right) \times 2 = 0.81 \text{ m}$$

The solution has been completed on a microcomputer. The results are presented in graphical form in Figure 12.12. The peak upsurge is 7 m above zero (reservoir) datum. The periodic time is about 254 s.

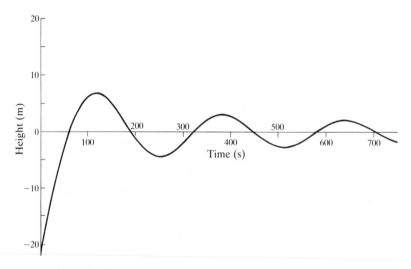

Figure 12.12 Graph of surge height versus time (Example 12.5).

A time interval $\delta t = 5$ s was also used for comparison, and the differences were

$$
\left.
\begin{array}{l}
\text{peak upsurge } +5\% \\
\text{periodic time } +1.7\%
\end{array}
\right\}
\begin{array}{l}
\text{compared with} \\
\text{results for } \delta t = 2 \text{ s}
\end{array}
$$

Surge protection for hydroelectric schemes

A typical high head hydroelectric scheme (Fig. 12.13) might consist of a reservoir feeding a low pressure tunnel or pipeline. The tunnel might be of considerable length and have a small gradient. At the end of the tunnel the system would be divided into a series of high pressure penstocks laid down the hillside to the turbine house. It might well be convenient to construct the surge tower as a vertical excavation at the downstream end of the tunnel, as shown. The tunnel is then protected from excessive surge pressures, which could otherwise cause damage (possibly even a tunnel collapse) that would be extremely expensive to repair. Another important aspect of the surge tower would be the periodic time T_p. The flow to the turbine has to be continuously regulated (or 'governed') so that the output from the turbines matches the electrical power requirement. In a badly designed system, the governor and surge tower can interact in such a way that the system is continuously 'hunting', i.e. a steady flow to the turbines is hardly ever achieved.

A wide variety of surge tower designs has been developed to match the differing requirements of designers. Pickford (1969) gives an excellent summary of these, together with an extensive survey of numerical methods.

The whole of this argument has been based on the assumption that the control valve has been completely closed. For any valve movement other

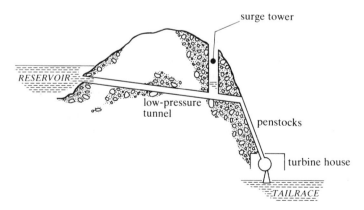

Figure 12.13 Typical hydroelectric scheme.

than complete closure, there must be a flow in the pipe between the surge tower and the control valve (Fig. 12.11). The continuity equation thus becomes

$$VA - V_1 A_1 = V_{ST} A_{ST} = \frac{dz_{ST}}{dt} A_{ST}$$

which may be converted into finite difference form:

$$\delta z_{ST} = \left(\frac{VA - V_1 A_1}{A_{ST}} \right) \delta t \qquad (12.14)$$

V_1 has to be determined from the hydraulic characteristic of the controlling boundary.

The emphasis here has been placed on the ability of the surge tower to limit the effects of unsteady flows. The surge tower will additionally provide a small local 'reservoir' of liquid which assists in preventing negative pressures by feeding water towards the valve when the valve is opened and the water in the pipeline from the reservoir is still accelerating to meet demand.

The surge tower should, theoretically, be placed as near to the controlling boundary as possible, though different systems may impose a variety of requirements.

Surge protection in pumped mains

Unsteady flow can be generated by pumps when they are started or stopped. During start-up, the pump operating point moves from the zero discharge point to its normal steady-state position. The transient pressures generated during start-up are not usually serious, though they should be checked. If a pump is switched off (or if a power failure occurs), then a negative pressure surge is generated downstream of the pump and a positive one upstream. The initial negative surge may generate a vapour cavity at any high point along the pipeline. It has already been stated (Section 6.7) that extremely high surge pressures may be generated when the cavity subsequently closes. This condition must be avoided.

A variety of approaches are used, including the following:

Mechanical methods. A typical mechanical method would be the use of a flywheel attached to the pump shaft. The increase in the pump inertia means that the rate of deceleration is reduced. A drawback of this method is the increase in starting power, so it is not suitable for all pumps.

Bypassing. This entails the installation of an additional pipeline (Fig. 12.14) around the pump. This pipeline incorporates a non-return valve. During normal operation the pressure downstream of the pipe is higher than that

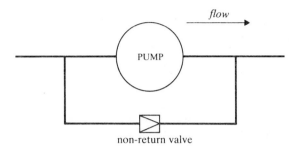

Figure 12.14 By-pass system.

upstream, so the valve remains closed. If the pump fails and the pressure downstream falls below the upstream pressure, then the valve opens and flow is admitted, limiting the fall in pressure. The bypass is also useful in long pipelines, where there may be a number of pumping stations along the pipeline. At times of low demand some of the pumps may be turned off, and the flow will then proceed through the bypass. Bypassing will not work in pipelines where there is negative pressure upstream of the pump.

Air chambers. An air chamber is, in effect, a very compact surge tower with its upper end capped so as to form an air-tight cylinder (Fig. 12.15). Under normal operating conditions (V_0, p_0) the chamber is partially filled with liquid. The pressure of the air trapped above the liquid must be the same as that of the liquid. Consequently, under surge conditions, the liquid level in the chamber will rise or fall to correspond with pressure in the pipeline, in much the same way as for the surge tower. Air chambers can usually be designed to be much more compact than surge towers.

Equations of motion can be developed for a pipeline with an air chamber. These are comparable in form with (12.12) and (12.14). In the momentum equation, an additional term is required to account for the air pressure in

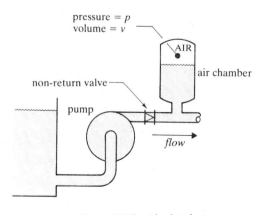

Figure 12.15 Air chamber.

the chamber. To estimate the air pressure it is usually assumed that the pressure and volume of the air are related by

$$pv^{1.2} = \text{constant}$$

(readers who are familiar with the gas laws will realise that this equation lies between the equations for isothermal and adiabatic compression). The implication of these remarks is that the instantaneous pressure change in the pipe will almost certainly be greater for an air chamber than for a surge tower. The volume of liquid in the chamber must be adequate to prevent the pressure in the pipe falling to vapour pressure. Conversely, the air volume must be adequate to 'cushion' positive surge pressures. It will be necessary to check the volume of air regularly (air is soluble in water), and to pump air in from time to time to maintain the required level.

A wealth of information is available on various aspects of surge control. The reader is referred in the first instance to Fok (1978), Pickford (1969), Parmakian (1963) and Wood (1970).

References and further reading

Creasy, J. D. (ed.) 1982. *A guide to water network analysis and the WRC computer program WATNET.* Tech. Rep. 177. Medmenham: Water Research Centre.

Fok, A. T. K. 1978. Design charts for air chambers on pump pipe lines. *Am. Soc. Civ. Engrs, J. Hydraulics Divn* **104**(HY9), 1289–303.

Novak, P. (ed.) 1983. *Developments in hydraulic engineering—1.* London: Applied Science Publishers.

Parmakian, J. 1963. *Waterhammer analysis.* New York: Dover.

Pickford, J. 1969. *Analysis of surge.* London: Macmillan.

Twort, A. C., F. M. Law and F. W. Crowley 1985. *Water supply,* 3rd edn. London: Edward Arnold.

Wood, D. J. 1970. Pressure surge attenuation utilizing an air chamber. *Am. Soc. Civ. Engrs, J. Hydraulics Divn* **96**(HY5), 1143–55.

13

Hydraulic structures

13.1 Introduction

Hydraulic structures are devices which are used to regulate or measure flow. Some are of fixed geometrical form, while others may be mechanically adjusted. Hydraulic structures form part of most major water engineering schemes, for irrigation, water supply, drainage, sewage treatment, hydropower, etc. It is convenient to group the structures under three headings:

(a) flow measuring structures, e.g. weirs and flumes;
(b) regulation structures, e.g. gates or valves;
(c) discharge structures, e.g. spillways.

As the reader progresses through the chapter, it will be observed that for most of these structures the depth–discharge relationship is based on the Bernoulli (or specific energy) equation. However, some modifications have to be incorporated to account for the losses of energy which are inevitably incurred in real flows.

An immense amount of experimental and theoretical information has been accumulated during this century, which is reflected in the body of literature now available. The most succinct statements of present knowledge are usually to be found in the relevant British, International or American standards. However, standards may not necessarily be found to cover the more esoteric structures.

13.2 Thin plate (sharp-crested) weirs

This type of device is formed from plastic or metal plate of a suitable gauge. The plate is set vertically and spans the full width of the channel. The weir itself is incorporated into the top of the plate. The geometry of the weir depends on the precise nature of the application. In this section we will concentrate on two basic forms, the rectangular weir and the vee (or triangular) weir. However, other forms are available, such as the compound weir.

It has already been stated that the primary purpose of a weir is to measure discharge. Once the upstream water level exceeds the crest height P_S, water will flow over the weir. As the depth of water above the weir (h_1) increases, the discharge over the weir increases correspondingly. Thus, if there is a known relationship between h_1 and Q, we need only to measure h_1 in order to deduce Q. The 'ideal' relationship between h_1 and Q may readily be derived for each weir shape on the basis of the Bernoulli equation. If these relationships are compared, it is evident that the triangular weir possesses greater sensitivity at low flows, whereas the rectangular weir can be designed to pass a higher flow for a given head and channel width.

Rectangular weirs

There are two types of rectangular weir (see Fig. 13.1):

(a) 'uncontracted' or full width weirs comprise a plate with a horizontal crest extending from one side of the channel to the other − the crest section is as illustrated in the figure.
(b) a 'contracted' weir, by contrast, has a crest width which is less than the channel width.

Since the operation of a weir is based on the use of a gauged depth to estimate the discharge, we must know how these two quantities are related. The actual flow over a weir is quite complex, involving a three-dimensional velocity pattern as well as viscous effects. The simplest method of developing a numerical model which represents a weir is to use the Bernoulli equation as a starting point. An idealised relationship between depth and discharge is obtained. This relationship can then be modified to take account of the differences between ideal and real flows.

The rectangular weir equation. Before developing this equation it should be recalled that, even with an ideal fluid, the velocity distribution at the weir is not uniform (see Section 2.8) due to the variation in static pressure with depth. For this reason, the Bernoulli equation is first applied along one streamline, as shown in Figure 13.2. Thus, the total energy on the streamline A−A at Station 1 is

$$H_1 = \frac{p_1}{\rho g} + z_1 + \frac{u_1^2}{2g}$$

Writing $z_1 + p_1/\rho g = y_1$, we obtain $H_1 = y_1 + u_1^2/2g$. At Station 2 the liquid passes over the weir and forms an overspilling 'jet' whose underside (or

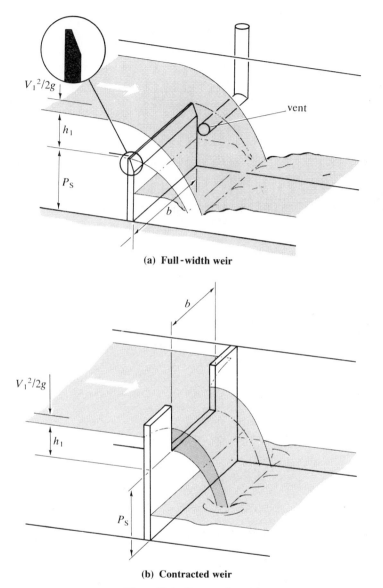

(a) **Full-width weir**

(b) **Contracted weir**

Figure 13.1 Rectangular weirs.

'lower nappe') is exposed to the atmosphere. The pressure distribution here cannot be of the linear hydrostatic form, and it is usual (after Weisbach) to assume that p_2 is atmospheric pressure. Therefore $H_2 = z_2 + u_2^2/2g$ and, assuming that no losses occur,

$$H_1 = H_2$$

i.e.

$$y_1 + u_1^2/2g = z_2 + u_2^2/2g$$

Therefore

$$u_2^2/2g = y_1 - z_2 + u_1^2/2g$$

and therefore

$$u_2 = [2g(y_1 - z_2 + u_1^2/2g)]^{1/2}$$

u_2 is thus a function of y_1 and z_2, i.e. u_2 varies with elevation above the weir crest. At Station 2, the ideal discharge through an elemental strip of width b and depth δz is

$$\delta Q_{ideal} = u_2 b\, \delta z = [2g(y_1 - z_2 + u_1^2/2g)]^{1/2} b\, \delta z$$

In order to integrate the equation it is necessary to know the limiting values of z_2. If the datum is now raised to the same elevation as the weir crest, the lower limit becomes $z_2 = 0$. A value for the upper limit may be obtained by making the rather drastic assumption that the elevation of the water surface at Station 2 is the same as that at Station 1, and therefore that $y_1 = z_2 = h_1$. (This implies a horizontal water surface, and therefore that streamlines at Station 2 are parallel and horizontal. This is physically impossible). With these assumptions,

$$Q_{ideal} = \int_0^{h_1} u_2 b\, dz = \int_0^{h_1} [2g(h_1 - z_2 + u_1^2/2g)]^{1/2} b\, dz$$

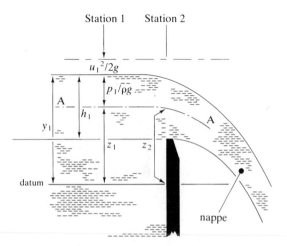

Figure 13.2 Flow over a thin plate weir.

$$= \tfrac{2}{3}b\sqrt{2g}\left[\left(h_1 + \frac{u_1^2}{2g}\right)^{3/2} - \left(\frac{u_1^2}{2g}\right)^{3/2}\right] \qquad (13.1)$$

as b is constant for a rectangular weir.

If $u_1^2/2g$ is negligible compared with h_1, then (13.1) reduces to

$$Q_{\text{ideal}} = \tfrac{2}{3}b\sqrt{2g}\, h_1^{3/2} \qquad (13.2)$$

Modifications to the rectangular weir equation. Due to the converging pattern of streamlines approaching the weir, the cross-sectional area of the jet just downstream of the weir will be significantly less than the cross section of the flow at the weir itself. This 'contraction' or 'vena contracta' in the flow implies that the actual discharge, Q, will be less than Q_{ideal}.

There are two further discrepancies in the theory:

(a) the pressure distribution in the flow passing over the weir is not atmospheric, but is distributed as shown in Figure 13.3;
(b) the flow approaching the weir will be subject to viscous forces: this produces two effects – (a) a non-uniform velocity distribution in the channel and (b) a loss of energy between Stations 1 and 2.

It was seen in Chapter 2 that it is conventional for Q and Q_{ideal} to be linked through the experimentally derived coefficient, C_d, i.e. $C_d = Q/Q_{\text{ideal}}$. C_d is not strictly a constant. It can be shown by dimensional analysis that $C_d = f\{\text{Re}, \text{We}, h_1/P_S\}$. However, provided that the overspilling liquid forms a 'jet' whose lower nappe springs clear of the weir plate, the value of C_d will not vary greatly with Q (or Re). The problems of calibration (i.e.

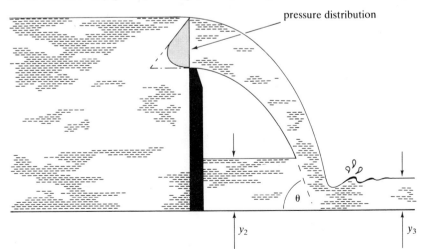

pressure distribution

Figure 13.3 Section through nappe.

evaluating C_d experimentally) become far greater if the nappe clings to the downstream face of the weir plate. Such conditions tend to arise when the discharge is small. Under these conditions, the effects of viscosity and surface tension combine to bring about unstable, fluctuating flow conditions. Since the value of h_1 is small at low flows, an accurate determination of C_d becomes impracticable. It will be seen later that these considerations lead to alternative weir shapes.

A vertical section through the overspilling water passing through any weir would clearly show that a vertical contraction occurred. However, in the case of a contracted weir, contraction additionally occurs at the sides of the jet. The jet of an uncontracted weir clings to the channel sides, so air cannot readily gain access to the underside of the nappe. Any air initially trapped beneath the nappe will tend to be entrained in the flow of water. This, in turn, implies the possibility of sub-atmospheric pressure beneath the nappe. For this reason the nappe for an uncontracted weir may tend to be drawn towards the weir more readily than the jet of a contracted weir. In order to offset this, some uncontracted weirs incorporate a vent pipe to admit air to the underside of the nappe (Fig. 13.1a). As the falling jet strikes the channel bed downstream of the weir, it must deflect away from the weir and along the channel. This means that there must be a net force in the direction of the acceleration (see Fig. 13.3). To produce this force, the depth y_2 (between the weir and the falling jet) must exceed y_3 downstream of the point of impact, such that

$$(\rho g b/2)(y_2^2 - y_3^2) = \rho Q V_3 - \rho Q V_3 \cos \theta \qquad (13.3)$$

$(V = Q/A = Q/by)$. This mechanism also tends to reduce the volume of air under the nappe, especially for an uncontracted weir.

A number of empirical discharge formulae have been developed which incorporate C_d. For uncontracted weirs under free discharge conditions two typical equations are:

(a) the Rehbock formula:

$$Q = \tfrac{2}{3}\sqrt{2g}\,(0.602 + 0.083\,h_1/P_S)b(h_1 + 0.0012)^{3/2}\ \text{m}^3/\text{s} \qquad (13.4)$$

which is valid for 30 mm $< h_1 <$ 750 mm, $b >$ 300 mm, $P_S >$ 100 mm, $h_1 < P_S$ (in foot-sec units $Q = \tfrac{2}{3}\sqrt{2g}(0.602 + 0.083\,h_1/P_S)b(h_1 + 0.004)^{3/2}$ ft^3/s)

(b) White's formula:

$$Q = 0.562(1 + 0.153\,h_1/P_S)b\sqrt{g}(h_1 + 0.001)^{3/2}\ \text{m}^3/\text{s} \qquad (13.5)$$

which is valid for $h_1 > 20$ mm, $P_S > 150$ mm, $h_1 < 2.2P_S$ (in foot-sec units $Q = 0.562 \ (1 + 0.153 \ h_1/P_S)b\sqrt{g}(h_1 + 0.003)^{\frac{3}{2}} \ \text{ft}^3/\text{s}$)

Ackers *et al.* (1978) suggest that the accuracy of the formulae depends, to a discernible extent, on the siting of the gauging station for measuring the upstream head (h_1). They recommend that this should be at least $2.67P_S$ upstream of the weir, to avoid undue drawdown effects. The British Standard (BS 3680) recommendation is that the station should be between $4h_1$ and $5h_1$ upstream of the weir. For American practice, see the *Water measurement manual* (US Bureau of Reclamation 1967).

For contracted rectangular weirs, (13.1) for Q_{ideal} is applicable. However, the magnitude of the actual discharge, Q, will be affected by the vertical and lateral contraction of the jet. A number of empirical equations have been developed to enable estimates of C_d to be made. Two such formulae are given below.

(a) The Hamilton–Smith formula is

$$C_d = 0.616(1 - h_1/P_S) \tag{13.6}$$

which is valid for $B > (b + 4h_1)$, $h_1/B < 0.5$, $h_1/P_S < 0.5$, $75 \text{ mm} < h_1 < 600$ mm, $P_S > 300$ mm, $b > 300$ mm. This has the advantage of simplicity, and therefore ease and speed in use. ($B = $ channel width)

(b) The Kindsvater-Carter formula makes use of the concept of the 'effective' head and width, h_e and b_e, where

$$h_e = (h + k_h) \text{ and } b_e = (b + k_b)$$

k_h and k_b are experimentally determined quantities which allow for the effects of viscosity and surface tension. It has been found that $k_h = \text{constant} = 0.001$ m.

The formula is therefore

$$Q = \tfrac{2}{3}C_e\sqrt{2g}b_e h_e^{\frac{3}{2}} \text{ m}^3/\text{s}$$

where

$$C_e = f\left\{\frac{b}{B}, \frac{h_1}{P_S}\right\}$$

and must be determined empirically. BS 3680 gives charts for the determination of C_e and k_b.

Example 13.1   Determination of discharge for a rectangular weir

A long channel 1.5 m wide terminates in a full width rectangular weir whose crest height is 300 mm above the stream bed. If the measuring station is recording a depth of 400 mm above the weir crest, estimate the discharge:

(a) assuming a flow with a negligible velocity of approach (take C_d as 0.7);
(b) including velocity of approach (taking the Coriolis coefficient, α, as 1.1 and C_d as 0.7);
(c) using White's formula.

Solution

(a) $Q = \frac{2}{3} C_d b \sqrt{2g} h_1^{3/2}$ (from (13.2))

$$= \frac{2}{3} \times 0.7 \times 1.5 \times \sqrt{2 \times 9.81} \times 0.4^{3/2} = 0.784 \text{ m}^3/\text{s}$$

(b) If the approach velocity is to be incorporated, (13.1) must be used. However, u_1 is initially unknown. The equation is therefore solved by successive approximations. For the sake of simplicity, it is usual to assume that the velocity at Station 1 is uniform, i.e. $u_1 = Q/A_1 = V_1$.

First trial.   If $Q = 0.784$ m^3/s (from (a)), then

$$V_1 = 0.784/[(0.3 + 0.4) \times 1.5] = 0.747 \text{ m/s}$$

Therefore

$$Q = \frac{2}{3} \times 0.7 \times 1.5 \times \sqrt{2 \times 9.81} \left[\left(0.4 + \frac{1.1 \times 0.747^2}{2 \times 9.81} \right)^{3/2} - \left(\frac{1.1 \times 0.747^2}{2 \times 9.81} \right)^{3/2} \right]$$

$$= 0.861 \text{ m}^3/\text{s}$$

Second trial. $V_1 = 0.861/[(0.3 + 0.4) \times 1.5] = 0.82$ m/s, therefore

$$Q = \frac{2}{3} \times 0.7 \times 1.5 \times \sqrt{2 \times 9.81} \left[\left(0.4 + \frac{1.1 \times 0.82^2}{2 \times 9.81} \right)^{3/2} - \left(\frac{1.1 \times 0.82^2}{2 \times 9.81} \right)^{3/2} \right]$$

$$= 0.875 \text{ m}^3/\text{s}$$

which is sufficiently accurate for practical purposes.

(c) Using White's formula (Equation (13.5)),

$$Q = 0.562 \left[1 + \left(0.153 \times \frac{0.4}{0.3} \right) \right] \times 1.5 \sqrt{9.81} \, (0.4 + 0.001)^{3/2}$$

$$= 0.807 \text{ m}^3/\text{s}$$

'Submergence' and the modular limit

All of the above equations have rested upon the assumption that the downstream depth, y_3, is too low to impede the free discharge over the weir. There is then a unique relationship between h_1 and Q. This is known as 'modular flow'. Weirs are designed for modular operation. In the field, however, circumstances may arise in which modular operation is not possible. A partial blockage of the channel downstream of a weir might be an example of such circumstances.

Consider what happens to Q if h_1 remains constant but y_3 increases until it rises above the crest. Up to a certain limiting value of y_3, h_1 and Q will be unaffected. The modular limit of operation occurs when this limiting y_3 has been reached. If y_3 increases further, h_1 will have to increase in order to maintain the discharge. The weir is then said to be 'submerged'. Thus, for a given upstream depth the discharge through a submerged weir is less than the free discharge. Furthermore, the discharge is now related to the depths upstream and downstream of the weir (i.e. to h_1 and y_3), and not just to h_1. The measurement of y_3 is problematical, since the surface of the water downstream of the weir is highly disturbed. If an estimate of the discharge Q_S of a submerged weir has to be made, Villemonte's formula may be used:

$$\frac{Q_S}{Q} = \left[1 - \left(\frac{h_1}{y_3} \right)^{3/2} \right]^{0.385} \tag{13.7}$$

where Q is the free discharge which corresponds to the upstream depth h_1. Much information is given in the British Standard (BS 3680) and in Ackers *et al.* (1978).

Vee weirs

It has been stated that rectangular weirs suffer from a loss of accuracy at low flows. The vee weir largely overcomes this problem. The variation of b with height, together with the narrow nappe width in the jet, means that for a given increase in Q, the increase in h_1 for a vee weir will be much greater than for a rectangular weir. Conversely, the greater sensitivity limits the range of discharge for which a vee weir can be economically applied. The underlying theory and assumptions are the same as for the rectangular weir excepting, of course, the fact that b is not a constant but a function of z. Thus, the discharge through an elementary strip across the weir (Fig 13.4) will be

$$\delta Q_{\text{ideal}} = u_2 b \, \delta z = \left[2g \left(h_1 - z_2 + \frac{u_1^2}{2g} \right) \right]^{1/2} b \, \delta z$$

However

$$b = 2 z_2 \tan(\theta/2)$$

so

$$Q_{\text{ideal}} = \int_0^{h_1} \left[2g\left(h_1 - z_2 + \frac{u_1^2}{2g}\right) \right]^{\frac{1}{2}} 2 z_2 \tan\left(\frac{\theta}{2}\right) \, dz \qquad (13.8)$$

The approach velocity, u_1, is almost always negligible for vee weirs in view of the smaller discharges for which they are designed. Therefore

$$Q_{\text{ideal}} = \int_0^{h_1} [2g(h_1 - z_2)]^{\frac{1}{2}} 2 z_2 \tan(\theta/2) \, dz$$
$$= \tfrac{8}{15}\sqrt{2g} \tan(\theta/2) h_1^{\frac{5}{2}} \qquad (13.9)$$

The value of C_d is a function of Re, We, θ and h_1, and magnitudes are given in BS 3680 for a wide range of weirs. The $90°$ weir is probably the most widely used. As a first approximation, a C_d of 0.59 may be used for this angle.

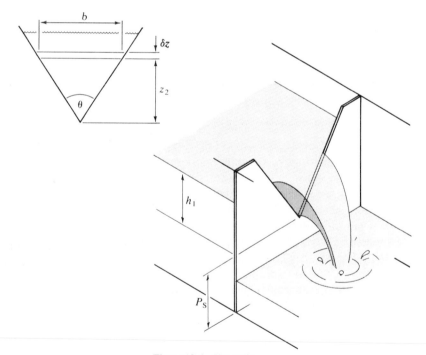

Figure 13.4   Vee weir.

Plate weirs of special form

A wide variety of weir shapes have emerged over the years. One example is the Sutro weir, in which the variation of b with z is such that Q is directly proportional to h_1. Other examples include the Trapezoidal weir and the Compound weir (a rectangular weir with a vee weir sunk in its crest). The reader is directed to Ackers *et al.* (1978) for further information.

13.3 Long-based weirs

In contrast to plate weirs, long-based weirs are larger and generally more heavily constructed (e.g. from concrete). They are usually designed for use in the field, and consequently may have to handle large discharges. An ideal long-based weir has the following characteristics:

(a) it is cheaply and easily fabricated (perhaps off-site);
(b) it is easily installed;
(c) it possesses a wide modular range;
(d) it produces a minimum afflux (i.e. increase in upstream depth due to the installation of the weir);
(e) it requires a minimum of maintenance.

There are a number of different designs, of which a selection is considered in detail below.

The rectangular ('broad-crested') weir

Rectangular weirs are solid weirs of rectangular cross section, which span the full width of a channel. They form part of the family of critical depth meters introduced in Chapter 5. However, it is worth recapitulating the following facts:

(a) A hump placed on the bed of the channel results in a local increase in the velocity of flow and a corresponding reduction in the elevation of the water surface.

(b) Given a hump of sufficient height, critical flow will be produced in the flow over the hump. There is then a direct relationship between Q and h_1, i.e. the flow is modular. Long-based weirs are designed for modular flow.

By definition, a rectangular weir is not streamlined. This, in turn, implies that the streamlines at the upstream end of the weir will not be parallel, since the flow will be accelerating. If frictional resistance is ignored, then the streamlines will become parallel and the flow become critical given a sufficient length of crest. It is then possible to derive a straightforward

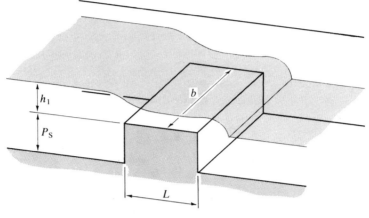

(a) **Flow over broadcrested weir**

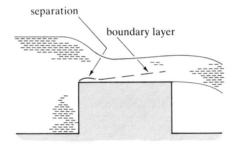

(b) **Effect of boundary layer on flow**

Figure 13.5   Broad-crested weir.

expression for Q_{ideal} in term of H_1 (see Fig. 13.5), as was seen in Chapter 5 (N.B. $H_1 = h_1 + V_1^2/2g$). Thus, as $y_c = \frac{2}{3}H_1$,

$$Q_{ideal} = \sqrt{gb(\tfrac{2}{3}H_1)^{3/2}}$$

In SI units this becomes

$$Q_{ideal} = 1.705 \, bH_1^{3/2} \text{ m}^3/\text{s} \tag{13.10}$$

(in foot-sec units, $Q_{ideal} = 3.09 \, bH_1^{3/2}$ ft^3/s). This equation may be modified to account for the differences between real and ideal fluids, using the relationship $Q/Q_{ideal} = C_d$:

$$Q = C_d \times 1.705 \, bH_1^{3/2} = CbH_1^{3/2} \text{ m}^3/\text{s} \tag{13.11}$$

In reality, the flow over a long-crested rectangular weir is more complex than ideal flow. Frictional shearing action promotes the growth of a boundary layer, which in turn implies that the water surface profile takes the form shown in Figure 13.5b. Additionally, as the flow approaches the critical condition, there is a tendency for ripples to form on the surface.

The value of C_d has to be derived empirically. Singer proposed an equation for C_d in the form

$$C_d = 0.848C_F \qquad (13.12)$$

The correction factor, C_F, is a function of h_1/L and $h_1/(h_1 + P_S)$ and is presented in graphical form in Ackers *et al*. However, as a first approximation, where $0.45 < h_1/L < 0.8$ and $0.35 < h_1/(h_1 + P_S) < 0.6$,

$$C_F \simeq 0.91 + 0.21\frac{h_1}{L} + 0.24 \left(\frac{h_1}{h_1 + P_S} - 0.35\right) \qquad (13.13)$$

These equations presuppose the establishment of critical flow over the crest. If the stage downstream of the weir rises above the stage which permits critical depth at the weir, then the weir will be submerged. An equation for Q_{ideal} for submerged flow can be developed, based on the energy and continuity equations. However, it is more usual to apply an empirical submergence correction factor ($= Q_S/Q$), which will be a function of both upstream depth (h_1) and depth over the crest (h_2). The modular equation (13.11) can be multiplied by this factor to give the actual discharge.

Round-nosed weirs

These are so named because of the radius formed on the leading corner. In other respects they resemble the broad-crested weir. The incorporation of the radius makes it the more robust of the two, and also means that the discharge coefficient is insensitive to minor damage. The magnitude of C_d is higher than that for a rectangular weir. However, very little is known about the performance of such weirs under submerged conditions.

It is worth noting that the region just downstream of any weir can be liable to scouring, especially in natural channels. For this reason, some form of 'stilling' arrangement may have to be incorporated.

Crump weirs

In 1952, E. S. Crump published details of a weir with a triangular profile, which had been developed at the Hydraulics Research Station. This was claimed to give a wide modular range, and also to give a more predictable

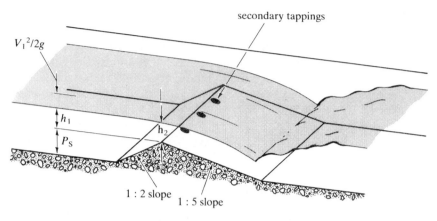

Figure 13.6 Crump weir.

performance under submerged conditions than other long-based weirs (Fig. 13.6).

Crump proposed upstream and downstream slopes of $1:2$ and $1:5$, respectively, which were based on sound principles. The upstream slope was designed so that sediment build-up would not reach the crest. The downstream slope was shallow enough to permit an hydraulic jump to form on the weir under modular flow conditions, thus providing an integral energy dissipator. Also, under submerged conditions, losses are not too high and the afflux is minimised. The primary gauging station is upstream of the weir. However, there is a second gauging point on the weir itself, just downstream of the crest. The second reading is used when the weir is submerged. The accuracy of the weir depends on the sharpness of the crest, so some weirs incorporate a metal insert at the crest. The secondary gauging tappings are drilled into the insert. The secondary tappings are sited near the crest in order to be clear of the disturbed flow further downstream.

The discharge equation for a Crump weir is in the form $Q = C_d C_v b g^{1/2} h_1^{3/2}$, which is clearly based on the same concepts as the corresponding equation for a rectangular weir. The value of C_d is about 0.63. The value of C_v varies with the ratio $h_1/(h_1 + P_S)$. For submerged flow,

$$\frac{Q_S}{Q} = 1.04 \left[0.945 - \left(\frac{h_p}{H_1} \right)^{3/2} \right]^{0.256} \tag{13.14}$$

where h_p is the head at the second gauging point. The onset of submergence occurs when

downstream depth $\geqslant 0.75 \times$ (upstream depth)

Since 1952, other variants have been developed, e.g. weirs with upstream and downstream slopes both $1:2$. However, some of these designs have a less stable C_d, and inferior submerged performance (see Ackers *et al.* 1978).

Example 13.2 Discharge over a broad-crested weir

A reservoir has a plan area of $100\,000$ m^2. The discharge from the reservoir passes over a rectangular broad-crested weir, having the following dimensions: $b = 15$ m, $P_S = 1$ m, $L = 2$ m. Estimate the discharge over the weir if the reservoir level is 1 m above the weir crest.

If there is no inflow into the reservoir, estimate the time taken for the reservoir level to fall from 1 m to 0.5 m above the weir crest. Assume that C_d remains constant.

Solution

From (13.13),

$$C_F = 0.91 + \left(0.21 \times \frac{1}{2}\right) + 0.24 \left(\frac{1}{1+1} - 0.35\right) = 1.051$$

Therefore

$$C_d = 0.848 \times 1.051 = 0.891$$

Substituting in (13.11),

$$Q = 0.891 \times 1.705 \times 15 \times 1^{\frac{3}{2}} = 22.79 \text{ m}^3/\text{s}$$

If the level in the reservoir falls from 1 m to 0.5 m, then let the upstream depth $= y$ when time $= t$ s after commencement of fall in reservoir level:

$$Q = 0.891 \times 1.705 \times 15 \times y^{\frac{3}{2}} = 22.79 y^{\frac{3}{2}} \text{ m}^3/\text{s}$$

Therefore, during a time interval δt, the outflow will be

$$-Q\,\delta t = -22.79 y^{\frac{3}{2}}\,\delta t \text{ m}^3$$

and the corresponding reduction in reservoir water level will be $-\delta y$. Thus, outflow will also be given by the product (plan area $\times \delta y$) $= -100\,000\,\delta y$. Therefore

$$100\,000\,\delta y = 22.79 y^{\frac{3}{2}}\,\delta t$$

i.e.

$$\delta t = \frac{100\,000\,\delta y}{22.79 y^{\frac{3}{2}}}$$

Integrating,

$$t = \frac{100\,000}{22.79} \int_{1.0}^{0.5} \frac{dy}{y^{3/2}} = \frac{100\,000}{22.79} \left[-\frac{2}{y^{1/2}} \right]_{1.0}^{0.5}$$

$$= 3635 \text{ s (or approx. 1 h)}$$

13.4 Flumes

This term is applied to devices in which the flow is locally accelerated due to:

(a) a streamlined lateral contraction in the channel sides;
(b) the combination of the lateral contraction, together with a streamlined hump in the invert (channel bed) (see Fig. 13.7).

The first type of flume is known as a venturi flume, and it has already been briefly introduced in Section 5.9. Flumes are usually designed to achieve critical flow in the narrowest (throat) section, together with a small afflux. Flumes are especially applicable where deposition of solids must be avoided (e.g. in sewage works or in irrigation canals traversing flat terrain).

A general equation for the ideal discharge through a flume may be developed on the basis of the energy and continuity principles.

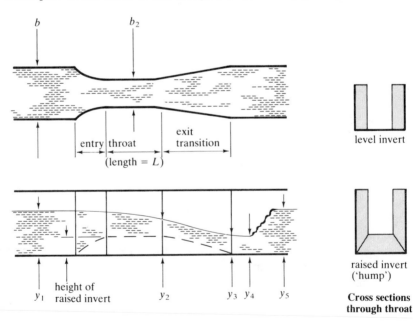

Figure 13.7   Venturi flume.

From the energy equation,

$$E_S = y_1 + \frac{V_1^2}{2g} = y_2 + \frac{V_2^2}{2g}$$

but

$$V_1 = Q/by_1 \text{ and } V_2 = Q/b_2y_2$$

Substituting for V_1 and V_2 and rearranging,

$$Q_{ideal} = \sqrt{\frac{2g(y_1 - y_2)(by_1)^2(b_2y_2)^2}{(by_1)^2 - (b_2y_2)^2}}$$

$$= by_1\sqrt{\frac{2g(y_1 - y_2)}{(by_1/b_2y_2)^2 - 1}} \tag{13.15}$$

If critical flow is attained in the throat, then $y_2 = 2/3E_S$. If this is substituted into (13.15), then (in SI units)

$$Q_{ideal} = 1.705b_2E_S^{\frac{3}{2}} = 1.705b_2C_vy_1^{\frac{3}{2}} \text{ m}^3/\text{s} \tag{13.16}$$

where C_v is the velocity of approach correction factor (in foot-sec units, $Q_{ideal} = 3.09b_2C_vy_1^{\frac{3}{2}}$ ft^3/s). The actual discharge is

$$Q = 1.705b_2C_dC_vy_1^{\frac{3}{2}} \text{ m}^3/\text{s} \tag{13.17}$$

(which is the same as (5.27)).

Although the flat bed venturi flume is simpler to construct, it is sometimes necessary to raise the invert in the throat to attain critical conditions. For either the flat or raised floor, the flume throat length should ideally be sufficient to ensure that the curvature of the water surface is small, so that in the throat the water surface is parallel with the invert.

A special group of 'short' flumes are less expensive and more compact than the 'ideal' throated flume. However, the water surface profile varies rapidly, and a theoretical analysis is not really possible. Empirical relationships must therefore be developed. For this reason, compact flumes tend to be of standard designs (e.g. the Parshall Flume) for which extensive calibration experiments have been carried out (see US Bureau of Reclamation 1967).

Example 13.3   Venturi flume

An open channel is 2 m wide and of rectangular cross section. A venturi flume with a level invert, having a throat width of 1 m is installed at one point. Estimate the discharge:

(a) if the upstream depth is 1.2 m and critical flow occurs in the flume;
(b) if the upstream depth is 1.2 m and the depth in the throat is 1.05 m.

Take $C_v = 1$ and $C_d = 0.95$.

Solution

(a) If the flow in the throat is critical, then using (13.17),

$$Q = 1.705 \times 0.95 \times 1.0 \times 1.2^{3/2} = 2.13 \text{ m}^3/\text{s}$$

(b) If $y_1 = 1.2$ m and $y_2 = 1.05$ m, we must use (13.15):

$$Q = 0.95 \times 2.0 \times 1.2 \sqrt{\frac{2 \times 9.81(1.2 - 1.05)}{(2 \times 1.2/1 \times 1.05)^2 - 1}} = 1.903 \text{ m}^3/\text{s}$$

The apparent ease of this approach (for flumes or for long-based weirs) to some extent disguises a number of difficulties which the designer must face. These are discussed below.

Flume design methodology

(1) The principal obstacle to modular operation is submergence due to high tailwater levels. To obtain modular conditions at low discharge, it is necessary to use a narrow throat combined with a small expansion angle in the exit transition. This approach suffers from two disadvantages:

(a) it is expensive;
(b) it will cause a large afflux at high discharges.

To illustrate the effect of installing a flume, Figure 13.8a shows how y_1 varies with Q for two different throat widths. The appropriate values of y_n are also given. The height of the banks upstream will impose a maximum limit on the afflux, which implies a corresponding lower limit on the value of b_2. Incorporating a raised throat invert in the form of a streamlined hump often assists in deferring submergence. Indeed, it may be possible to improve the matching of the flume to the channel by raising the invert and increasing b_2.

(2) The next problem is to determine the value of C_d. The flow in the main channel upstream probably approximates to a steady fully developed uniform flow. However, the flume is normally of a different material and

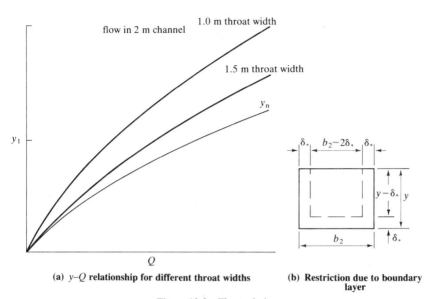

(a) y–Q relationship for different throat widths   (b) Restriction due to boundary
                                                                                                layer

Figure 13.8   Flume design.

surface roughness from that of the channel. Consequently, as the flow
passes through the flume, a boundary layer appropriate to the surface of the
flume will develop. The boundary layer will influence the values of C_d and
C_v. It may be assumed (a) that boundary layer growth commences at the
entry to the throat and (b) that the control is at the downstream end of the
throat where the boundary layer thickness is greatest (Fig. 13.8b). The
boundary layer has the effect of reducing the values of y_1 and b_2 to $(y_1 - \delta_*)$
and $(b_2 - 2\delta_*)$, respectively. So, for a rectangular flume and channel,

$$Q_{\text{ideal}} \propto b_2, \ y_1^{3/2}$$

and

$$Q \propto (b_2 - 2\delta_*), (y_1 - \delta_*)^{3/2}$$

Therefore

$$C_d = \frac{(b_2 - 2\delta_*)(y_1 - \delta_*)^{3/2}}{b_2 y_1^{3/2}} \tag{13.18}$$

and

$$C_v = \left(\frac{E_{S1} - \delta_*}{y_1 - \delta_*}\right)^{3/2} \tag{13.19}$$

(3) Matching flume geometry to a given channel must be determined by trial and error. For a flume and channel of rectangular cross section this is quite straightforward. A limiting value of y_1 must be related to the maximum flood discharge. A corresponding initial value of b_2 may then be obtained:

$$b_2 = Q/1.705 y_1^{3/2}$$

assuming modular flow, and with b_2 in metres.

In order to estimate C_d and C_v, δ_* must first be evaluated. For a first approximation,

$$\delta_* = 0.003L \tag{13.20}$$

which is valid for $\mathrm{Re} > 2 \times 10^5$ and $10^5 > L/k_S > 4000$ (N.B. A more detailed and accurate method of obtaining δ_* is also presented in the British Standard (BS 3680), together with illustrative worked examples).

The value of b_2 may then be checked:

$$b_2 = Q/1.705 C_d C_v y_1^{3/2} \tag{13.21}$$

and the conditions downstream of the flume should then be assessed to ascertain whether submergence is a potential problem.

For flumes and channels which are not of rectangular section the process remains essentially the same. However, the discharge function, $Q/1.705 y_1^{3/2}$, will now vary with depth.

The art of 'matching' lies in designing a flume for which the variation of discharge function with E_{S1} or y_1 approximates closely to the corresponding variation for the channel. For a more detailed treatment of this problem, Ackers et al. (1978) should be consulted.

(4) The designer needs to predict the range of discharge over which modular conditions apply. Two possible approaches to this will now be considered.

(a) The simplest method is that proposed in BS 3680. This suggests that, for modular flow in a flume with a full length exit transition, the total head upstream shall be at least 1.25 times the total head downstream (provided that flow downstream is subcritical). The corresponding relationship for a flume with a truncated exit transition is that total head upstream shall be at least 1.33 times that downstream.

(b) A more sophisticated method is suggested by ILRI (Bos 1978), which is as follows. Starting with the equation for modular flow,

$$Q = \text{constant} \times C_d C_v b_2 E_{S1}^{3/2}$$

or

$$Q = \text{constant} \times C_v b_2 E_{S2}^{3/2}$$

Hence

$$E_{S1} C_d^{2/3} = E_{S2} \tag{13.22}$$

In the exit transition, there will be losses of energy due both to friction and to 'separation'. This may be expressed as

$$E_{S2} - E_{S3} = h_{f2-3} + \frac{C_L(V_2 - V_3)^2}{2g} \tag{13.23}$$

The frictional head loss may be deduced from an appropriate equation (e.g. the Manning equation). The separation loss coefficient C_L has been evaluated empirically (Bos 1978): typical values are $C_L = 0.27$ for a $28°$ included exit angle, $C_L = 0.68$ for a $52°$ included exit angle.

If (13.22) is substituted into (13.23), then, with some rearrangement,

$$\frac{E_{S3}}{E_{S1}} = C_d^{2/3} - \frac{C_L(V_2 - V_3)^2}{2g E_{S1}} - \frac{h_{f2-3}}{E_{S1}}$$

This will apply as long as the hydraulic jump is downstream of the exit. Therefore, modular conditions will certainly exist if

$$\frac{E_{S3}}{E_{S1}} \leqslant C_d^{2/3} - \frac{C_L(V_2 - V_3)^2}{2g E_{S1}} - \frac{h_{f2-3}}{E_{S1}} \tag{13.24}$$

A similar approach may be applied to flow over long-based weirs.

It is not possible to predict the modular limit precisely:

(a) because C_d and C_L cannot be accurately estimated;
(b) because it is difficult to estimate the location of an hydraulic jump if it occurs in the exit transition.

In order to ensure that the streamlines in the throat are parallel, the length of the throat section should be at least twice E_{S1}. For 'short' flumes, of course, this criterion is not met.

The method outlined in this section will now be applied.

Example 13.4   Flume design using boundary layer theory

A flume is to be installed in a rectangular channel 2 m wide. The bed slope is 0.001, and Manning's n is 0.01. When the flume is installed, the increase in upstream depth must not be greater than 350 mm at the maximum discharge of 4 m^3/s. Take $C_L = 0.27$ (28° expansion).

Solution

For the maximum discharge (using Manning's equation),

$$4 = 2y_n \frac{R^{\frac{2}{3}}}{0.01} \times 0.001^{\frac{1}{2}}$$

therefore $y_n = 1.0$ m.
   For a first trial, take b_2 as 1.5 m. Then

$$E_{S1}^{\frac{3}{2}} \simeq \frac{4}{1.705 \times 1.5}$$

so $E_{S1} = 1.35$ m. Hence $y_1 = 1.2$ m. Assume throat length $= 2 \times 1.35 = 2.7$ m. Then

$$\delta_* = 0.003 \times 2.7 = 0.008 \text{ m}$$

$$C_d = \frac{[1.5 - (2 \times 0.008)](1.2 - 0.008)^{\frac{3}{2}}}{1.5 \times 1.2^{\frac{3}{2}}} = 0.98$$

$$C_v = \left(\frac{1.35 - 0.008}{1.2 - 0.008} \right)^{\frac{3}{2}} = 1.195$$

Therefore

$$y_1^{\frac{3}{2}} = 4/(1.705 \times 0.98 \times 1.195 \times 1.5)$$

so $y_1 = 1.213$ m. This is acceptable ($y_1 < y_n + 0.35$). If the first trial produces an unacceptable result, further trials must be undertaken.
   The normal depth for the channel has been calculated above. It is necessary for the engineer to ascertain:

(a) whether free discharge is attained with this depth;
(b) the downstream depth which corresponds to the modular limit. Here the ILRI method (Equation (13.24)) is used.

   To solve the equation, V_3 must be known. For a first approximation assume $E_{S3} = E_{S1}$. Hence, $y_3 = 0.48$ m and $V_3 = 4.17$ m/s. The hydraulic gradient is

$$S_{f2-3} = \frac{Q^2 n^2}{A^2 R^{\frac{4}{3}}}$$

In the exit transition both width and depth of flow are varying. As an approximation, the cross-sectional area and hydraulic radius are calculated for the throat. Critical depth in the throat is 0.9 m $(= \frac{2}{3}E_{S1})$. Thus

$$S_{f2-3} = \frac{4.17^2 \times 0.01^2}{(1.5 \times 0.9)^2 \left(\dfrac{1.5 \times 0.9}{1.5 + (2 \times 0.9)}\right)^{4/3}} = 0.0031$$

For a $28°$ exit angle, the length of the transition is 1 m for an expansion of 0.5 m, therefore $h_{f2-3} = 0.0031$ m

Also, for critical flow in the throat, $V_2 = 2.963$ m/s $(= \sqrt{gy_2})$. From (13.24),

$$\frac{E_{S3}}{1.35} = 0.98^{2/3} - \frac{0.27(2.963 - 4.17)^2}{2 \times 9.81 \times 1.35} - \frac{0.0031}{1.35}$$

Therefore $E_{S3} = 1.31$ m, and hence $y_3 = 0.5$ m, $V_3 = 4$ m/s, $Fr_3 = 1.806$. Assuming $y_4 \simeq y_3$, the hydraulic jump (conjugate depth) equation (Equation (5.21)) is used to estimate y_5. Hence, $y_5/y_4 = 2.10$, so at the modular limit $y_5 = 2.1 \times 0.5 = 1.05$ m.

Naturally, this calculation could be repeated for a series of discharges to cover the whole operational range.

13.5 Spillways

The majority of impounding reservoirs are formed as a result of the construction of a dam. By its very nature, the streamflow which supplies a reservoir is variable. It follows that there will be times when the reservoir is full and the streamflow exceeds the demand. The excess water must therefore be discharged safely from the reservoir. In many cases, to allow the water simply to overtop the dam would result in a catastrophic failure of the structure. For this reason, carefully designed overflow passages − known as 'spillways' (Photograph 7) − are incorporated as part of the dam design. The spillway capacity must be sufficient to accommodate the 'largest' flood discharge (the Probable Maximum Flood or 1 in 10 000 year flood) likely to occur in the life of the dam. Because of the high velocities of flow often attained on spillways, there is usually some form of energy dissipation and scour prevention system at the base of the spillway. This often takes the form of a stilling basin.

There are several spillway designs. The choice of design is a function of the nature of the site, the type of dam and the overall economics of the scheme. The following list gives a general outline of the various types and applications:

overfall and 'Ogee' spillways are by far the most widely adopted: they may be used on masonry or concrete dams which have sufficient crest length to obtain the required discharge;

Photograph 7   Spillway and control gates, Darwendale Dam, Zimbabwe.

chute and tunnel (shaft) spillways are often used on earthfill dams;
side channel and tunnel spillways are useful for dams sited in narrow
gorges;
siphon spillways maintain an almost constant headwater level over the
designed range of discharge.

Some typical designs are now considered.

Gravity ('Ogee') spillways

These are by far the most common type, being simple to construct and
applicable over a wide range of conditions. They essentially comprise a
steeply sloping open channel with a rounded crest at its entry. The crest
profile approximates to the trajectory of the nappe from a sharp crested
weir. The nappe trajectory varies with head, so the crest can be correct only
for one 'design head' H_d. Downstream of the crest region is the steeply
sloping 'face', followed by the 'toe', which is curved to form a tangent to
the apron or stilling basin at the base of the dam. The profile is thus in the
form of an elongated 'S' (Fig. 13.9). Profiles of spillways have been
developed for a wide range of dam heights and operating heads. A wealth
of information is available in references published by the US Bureau of
Reclamation (1964, 1977) and the US Army Waterways Experimental Station (1959). Spillways will here be considered mainly from the hydraulics
standpoint.

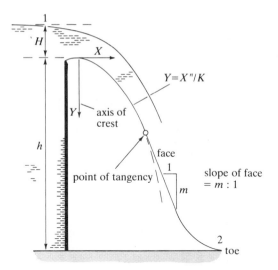

Figure 13.9   'Ogee' spillway.

The discharge relationship for a spillway is of the same form as for other weirs:

$$Q = \text{constant} \times C_d\sqrt{g}bH^{3/2}$$

In practice, this is usually written $Q = CbH^{3/2}$. C is not dimensionless, and its magnitude increases with increasing depth of flow. C usually lies within the range $1.6 < C < 2.3$ in metric units ($2.8 < C < 4.1$ in foot-sec units). The breadth, b, does not always comprise a single unbroken span. If control gates are incorporated in the scheme, the spillway crest is subdivided by piers into a number of 'bays' (Fig. 13.10). The piers form the supporting structure for the gates. The piers have the effect of inducing a lateral contraction in the flow. In order to allow for this effect in the discharge equation, the total span, b, is replaced by b_c, the contracted width:

$$b_c = b - knH$$

where n is the number of lateral contractions and k is the contraction coefficient, which is a function of H and of the shape of the pier.

Some other important aspects of spillway hydraulics are summarised below.

(a) If $H < H_d$, the natural trajectory of the nappe falls below the profile of the spillway crest, then there will therefore be positive gauge pressures over the crest. On the other hand, if $H > H_d$, then the nappe trajectory is higher than the crest profile, so negative pressure zones tend to arise. Frictional shear will accentuate this tendency, and cavitation may occur.

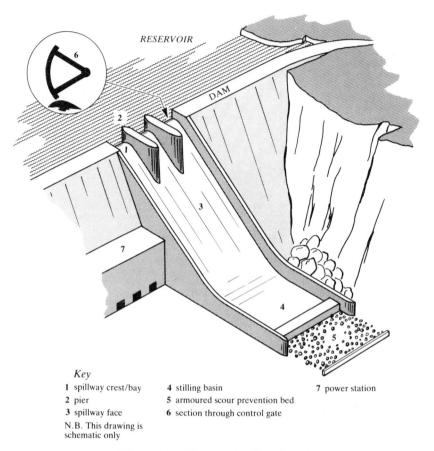

Key

1 spillway crest/bay   4 stilling basin   7 power station
2 pier   5 armoured scour prevention bed
3 spillway face   6 section through control gate
N.B. This drawing is
schematic only

Figure 13.10   Diagram of spillway layout.

However, in practice, this pressure reduction is not normally a serious problem unless $H > 1.5H_d$. Indeed, recent work suggests that separation will not occur until $H \rightarrow 3H_d$.

(b) Conditions in the flow down the spillway face may be quite complex, since:

(i)   the flow is accelerating rapidly, and may be 'expanding' as it leaves a bay–pier arrangement;
(ii) frictional shear promotes boundary layer growth;
(iii) the phenomenon of self-aeration of the flow may arise;
(iv) cavitation may occur.

For these reasons, the usual equations for non-uniform flow developed in Chapter 5 cannot really be applied. If it is necessary to make estimates of flow conditions on the spillway, then empirical data must be used. Each feature will now be examined in greater detail.

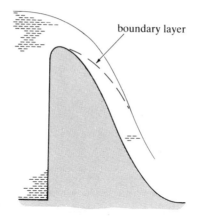

Figure 13.11 Growth of a boundary layer on spillway.

(i) In a region of rapidly accelerating flow, the specific energy equation (or Bernoulli's equation) is usually applied. It is possible to obtain very rough estimates of the variation of V and y down the spillway on this basis, accuracy will be slightly improved if a head loss term is incorporated. Nevertheless, in the light of (ii) and (iii), below, conditions on the spillway are far from those which underly the energy equations.

(ii) A boundary layer will form in the spillway flow, commencing at the leading edge of the crest. The depth of the boundary layer, δ_*, will grow with distance downstream of the crest. Provided that the spillway is of sufficient length, at some point the depth, δ_*, will meet the free surface of the water (Fig. 13.11).

The flow at the crest is analogous to the flow round any fairly streamlined body. This may imply flow separation, eddy shedding, or both. Such conditions may be instrumental in bringing about cavitation at the spillway face. There have been a number of reported incidents of cavitation in major dams.

(iii) Aeration has been observed on many spillways. It entails the entrainment of substantial quantities of air into the flow, which becomes white and foamy in appearance. The additional air causes an increase in the gross volume of the flow. Observations of aeration have led to the suggestion that the point at which aeration commences coincides with the point at which the boundary layer depth meets the free surface (Henderson 1966, Keller & Rastogi 1975, Cain & Wood 1981). The entrainment mechanism appears to be associated with the emergence of streamwise vortices at the free surface. Such vortices would originate in the spillway crest region. A rough estimate of the concentration of air may be made using the following equations:

$$C = 0.743 \log(S_0/q^{1/5}) + 0.876 \tag{13.25}$$

where C = (volume of entrained air)/(volume of air and water) and S_0 is the bed slope, or

$$C = (x_I/y_I)^{2/3}/74 \qquad (13.26)$$

where x_I and y_I are measured from the point at which entrainment commences. C may vary considerably over the width of the spillway.

(iv) Cavitation arises when the local pressure in a liquid approaches the ambient vapour pressure. Such conditions may arise on spillways for a variety of reasons, especially where the velocity of flow is high (say $V > 30$ m/s). For example, if $H > 3H_d$, separated flow will probably arise at the crest.

Irregularities in the surface finish of the spillway may result in the generation of local regions of low pressure.

Cavitation has been observed in severely sheared flows (Kenn 1971, Kenn and Garrod 1981).

At the toe of the spillway, the flow will be highly supercritical. The flow must be deflected through a path curved in the vertical plane before entering the stilling zone or apron. This can give rise to very high thrust forces on the base and side walls of the spillway. A rough estimate of these forces may be made on the basis of the momentum equation.

The high velocity and energy of flow at the foot of the spillway must be dissipated, otherwise severe scouring will occur and the foundations at the toe will be undermined.

Siphon spillways

A siphon spillway is a short enclosed duct whose longitudinal section is curved as shown in Figure 13.12a. When flowing full, the highest point in the spillway lies above the liquid level in the upstream reservoir, and the pressure at that point must therefore be sub-atmospheric, this is the essential characteristic of a siphon. A siphon spillway must be self-priming.

The way in which this type of spillway functions is best understood by considering what happens as the reservoir level gradually rises. When the water level just exceeds the crest level, the water commences to spill and flows over the downstream slope in much the same way as a simple Ogee spillway. As the water level rises further, the entrance is sealed off from the atmosphere. Air is initially trapped within the spillway, but the velocity of flow of the water tends to entrain the air (giving rise to aeration of the water) and draw it out through the exit. When all air has been expelled, the siphon is primed (i.e. running full) and is therefore acting as a simple pipe.

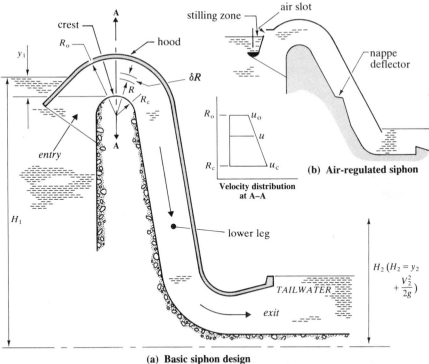

(a) **Basic siphon design**

Figure 13.12 Siphon spillway.

There are thus three possible operating conditions depending on upstream depth:

(a) gravity spillway flow;
(b) aerated flow;
(c) pipe ('blackwater') flow.

Operational problems with siphon spillways. The aerated condition is unstable and is maintained only for a short time while the siphon begins to prime, since (theoretically, at least) air cannot enter once the entry is covered. Therefore, in a simple siphon a small change in H produces a sharp increase or decrease in the discharge through the spillway. This can lead to problems if the discharge entering the reservoir is greater than the spillway flow but less than the blackwater flow, since the following cycle of events is set in train:

(a) If the spillway is initially operating with gravity flow, then the upstream (reservoir) level must rise;
(b) when the upstream level has risen sufficiently, the siphon primes and the spillway discharge increases substantially;
(c) the upstream level falls until the siphon de-primes and its discharge drops.

The cycle (a) to (c) is then repeated.

Obviously, this can give rise to radical surges and stoppages in the downstream flow. This problem can be overcome by better design.

Other potential problems encountered with siphon spillways are:

(a) blockage of spillway by debris (fallen trees, ice, etc.);
(b) freezing of the water across the lower leg or the entry before the reservoir level rises to crest level;
(c) the substantial foundations required to resist vibration;
(d) waves arising in the reservoir during storms may alternately cover and uncover the entry, thus interrupting smooth siphon action.

Problem (a) may be ameliorated by installing a trash-intercepting grid in front of the intake.

Discharge through a siphon spillway. Some of the principal discharge conditions may be estimated by combining some familiar concepts, i.e. weir flow ($Q = \text{constant} \times H^{3/2}$), pipe flow ($Q = \text{constant} \times H^{1/2}$) and free vortex flow.

Thus, if H_d coincides with the onset of blackwater flow, then the hydraulic losses may be equated to the head difference as for normal pipe flow:

$$H_1 - H_2 = \left(k_1 + k_2 + k_3 + k_4 + \frac{\lambda L}{d}\right)\frac{V^2}{2g} \qquad (13.27)$$

where k_1–k_4 are the loss coefficients for entry, first and second bends, and exit, respectively, and λ is the friction factor.

If the heads and coefficients are known, the corresponding velocity (or discharge) may be found. It must be emphasised that, due to the fact that the spillway is such a short pipe, the fully developed pipe flow condition is not attained and the estimates based on this equation must be regarded as rough approximations only.

For satisfactory siphon operation the pressure must nowhere approach the vapour pressure. It is therefore necessary to estimate the lowest pressure in the sipon. This will almost certainly occur in the flow around the first bend. Since the cross section of a siphon is large compared with the length and the head, the velocity and pressure distribution will be decidedly non-uniform. As a first approximation, the flow around the bend may be assumed to resemble a free vortex. The characteristic velocity distribution is then given by the equation

$$uR = \text{constant}$$

For this flow pattern, the lowest pressures coincide with the smallest radius.

Therefore, applying the Bernoulli equation,

$$H = \frac{p_c}{\rho g} + \frac{u_c^2}{2g}$$

(the subscript c refers to the spillway crest). Therefore

$$H - \frac{p_c}{\rho g} = h_c = \frac{u_c^2}{2g}$$

hence

$$u_c = \sqrt{2gh_c}$$

For the free vortex, $uR = \text{constant} = u_c R_c = \sqrt{2gh_c}R_c$, hence

$$u = \sqrt{2gh_c}R_c/R$$

Therefore the discharge through an elemental strip at radius R is

$$\delta Q = ub\,\delta R = \sqrt{2gh_c}(R_c/R)b\,\delta R$$

Thus the total discharge is

$$Q = \int_{R_c}^{R_o} \sqrt{2gh_c}(R_c/R)b\,\mathrm{d}R = \sqrt{2gh_c}R_cb\,\ln(R_o/R_c) \qquad (13.28)$$

ln denoting the natural (base e) logarithm.

If the required discharge and the bend radii are known, then the corresponding value of h_c may be found. As a guide, h_c should not be more than 7 m of water below atmospheric pressure at sea level to avoid formation of vapour bubbles.

The free vortex flow pattern will be modified in practice, due to the developing boundary layer in the spillway. The existence of the shear layer may tend to increase the likelihood of cavitation, as may any irregularities in the spillway surfaces, so great care is needed both in design and construction. Because of the form of the spillway rating curves, the maximum discharge is usually assumed to coincide with the onset of blackwater flow. At flows higher than this, H increases more rapidly for a given increase in Q, with the consequent danger of overtopping the dam. Therefore, for reasons of safety, siphons are often used in conjunction with an auxiliary emergency spillway.

Design improvements to siphon spillways. It has been pointed out that uncontrolled surging can be a problem. Two methods of improving the operation of a siphon installation are (a) the use of multiple siphons with differential crest heights and (b) air regulation.

For method (a), a series of siphons are installed. One siphon has the lowest crest height, a second has a slightly higher crest and so on. The siphons thus prime sequentially as the reservoir level rises and de-prime sequentially as the level falls. The system gives a much smoother $H–Q$ characteristic.

Method (b) involves a modification to the intake (Fig. 13.12b). The cowl is constructed with a carefully designed slot which admits controlled quantities of air when the head lies between the gravity flow and blackwater values. The operation of such a design is therefore as follows for a rising reservoir level (Fig. 13.13):

(a) Gravity flow for low heads.

(b) As the reservoir level rises further, air trapped inside the duct tends to be entrained and the pressure falls below atmospheric pressure. However, some air is drawn in through the air-regulating slot.

(c) A slight further increase in head leads to the entrainment of air trapped under the crown. Air drawn in at the air slot now mixes with the water to form a fairly homogeneous two-phase mixture, which is foamy in appearance.

(d) Blackwater flow sets in when the reservoir level seals off the air slot.

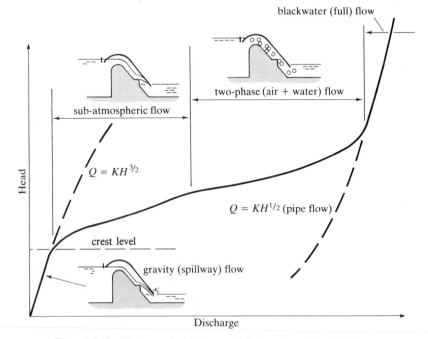

Figure 13.13 Discharge–head characteristic for air-regulated siphon.

If the air slot is well designed, a stable discharge of water can be maintained for conditions (b) and (c). Air regulation is thus a simple means of automatically matching the siphon discharge to the incoming discharge. Accurate prediction of the rate of air entrainment (and hence of the siphon discharge characteristic) is difficult. Model testing is a necessary prelude to design, though even tests do not give completely reliable results. Research is currently proceeding to clarify the reasons for the discrepancies. For an initial (and rough) estimate of entrainment, the equation of Renner may be used (see Proceedings of the Symposium on the Design and Operation of Siphons and Siphon Spillways 1975)

$$Q_{air}/Q = K\mathrm{Fr}^2 \qquad\qquad (13.29)$$

The value of K appears to depend on the angle between the deflected nappe and the hood, but an average value is 0.002. The Froude Number refers to the flow at the toe of the deflector.

A large amount of information regarding design, modelling and operation of siphon spillways has been published. The reader may refer to the Proceedings of the Symposium on the Design and Operation of Siphons and Siphon Spillways (1975) and also to Head (1975), Ervine (1976), Ali and Pateman (1980) and Ervine and Oliver (1980).

Shaft (morning glory) spillways

This type of structure consists of four parts (Fig. 13.14a):

a circular weir at the entry;
a flared transition which conforms to the shape of a lower nappe for a sharp-crested weir;
the vertical drop shaft;
the horizontal (or gently sloping) outlet shaft.

The discharge control may be at one of three points depending on the head:

(a) When the head is low the discharge will be governed at the crest. This is analogous to weir flow and the discharge may therefore be expressed as

$$Q = CLH^{3/2}$$

where L is usually referred to the arc length at the crest. C is a function of H and the crest diameter, it is therefore not a constant. The magnitude of C is usually in the range 0.6–2.2, in SI units (1.1–4.0 in foot-sec units).

Below the crest, the flow will tend to cling to the wall of the transition and of the drop shaft. The outlet shaft will flow only partially full and is therefore, in effect, an open channel.

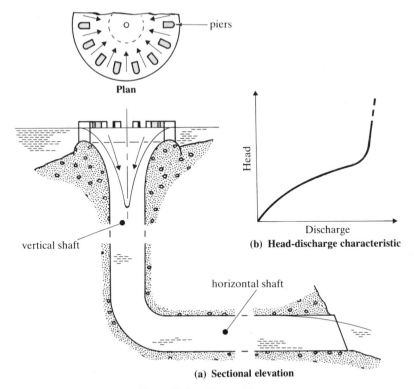

Plan

(b) **Head-discharge characteristic**

(a) **Sectional elevation**

Figure 13.14   Shaft spillway.

(b) As the head increases, so the annular nappe must increase in thickness. Eventually, the nappe expands to fill the section at the entry to the drop shaft. The discharge is now being controlled from this section, and this is often referred to as 'orifice control'. The outlet tunnel is not designed to run full at this discharge.

(c) Further increase in head will induce blackwater flow throughout the drop and outlet shafts. The $Q–H$ relationship must now conform to that for full pipe flow, and the weir will in effect be 'submerged'. The head over the weir rises rapidly for a given increase in discharge, with a consequent danger of overtopping the dam.

The complete discharge–head characteristic is therefore as shown in Figure 13.14b. The design head is usually less than the head required for blackwater flow. This is done to leave a margin of safety for exceptional floods. Even so, the discharge which can be passed by a shaft or siphon spillway is limited, so great care must be exercised at the design stage. An auxiliary emergency spillway may be necessary. It is also worth noting that as the flow enters the transition it tends to form a spiral vortex. The vortex pattern must be minimised in order to maintain a smoothly converging flow, so anti-vortex baffles or piers are often positioned around the crest.

It may be undesirable for sub-atmospheric pressure to occur anywhere in the system, since this can lead to cavitation problems. To avoid such problems the system may (a) incorporate vents, (b) be designed with an outlet shaft which is large enough to ensure that the outlet end never flows full or (c) have an outlet shaft with a slight negative slope, sufficient to ensure that the outlet does not flow full (and can therefore admit air).

Some recent shaft spillways incorporate a bank of siphon spillways around the crest to reduce the rise in reservoir level for a given discharge. Trash grids must be installed around the entrance to a shaft spillway to exclude large items of debris.

13.6 Energy dissipators

The flow discharged from the spillway outlet is usually highly supercritical. If this flow were left uncontrolled, severe erosion at the toe of the dam would occur, especially where the stream bed is of silt or clay. Therefore, it is necessary to dissipate much of the energy, and to return the water to the normal (subcritical) depth appropriate to the stream below the dam. This is achieved by a dissipating or 'stilling' device. Typical devices of this nature are:

(a) stilling basin;
(b) submerged bucket; and
(c) ski jump/deflector bucket.

These are now described (see Fig. 13.15).

The stilling basin

A stilling basin consists of a short, level apron at the foot of the spillway (Fig. 13.15a). It must be constructed of concrete to resist scour. It incorporates an integrally cast row of chute blocks at the inlet and an integral sill at the outlet. Some designs also utilise a row of baffle blocks part way along the apron.

The function of the basin is to decelerate the flow sufficiently to ensure the formation of an hydraulic jump within the basin. The jump dissipates much of the energy, and returns the flow to the subcritical state. The chute blocks break the incoming flow into a series of jets, alternate jets being lifted from the floor as they pass over the tops of the blocks. The sill (or baffle blocks and sill) provide the resistance required to reduce energy and control the location of the jump. Baffle blocks are not usually used where the velocity of the incoming flow exceeds 20 m/s, due to the likelihood of cavitation damage.

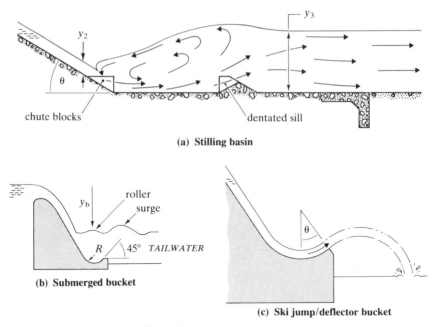

(a) Stilling basin

(b) Submerged bucket

(c) Ski jump/deflector bucket

Figure 13.15 Energy dissipators.

Provided that the Froude Number of the incoming flow (Fr_2) exceeds 4.5, a stable jump can be formed. However, if $2.5 < Fr_2 < 4.5$, then the jump conditions can be less well defined and disturbances may be propagated downstream into the tailwater. Based on a combination of operational and empirical data, a number of standard stilling basin designs have been proposed. A typical set of designs were published by the US Bureau of Reclamation.

US Bureau of Reclamation stilling basins

(a) Type II stilling basin (for $Fr_2 > 4.5$, $V_2 > 20$ m/s), which has a length of $4.3y_3$;
(b) Type III stilling basin (for $Fr_2 > 4.5$, $V_2 < 20$ m/s), which has a length of $2.7y_3$;
(c) Type IV stilling basin (for $2.5 < Fr_2 < 4.5$), which has a length of $6.1y_3$.

It is usually assumed that air entrainment makes little difference to the formation of the hydraulic jump in a stilling basin.

The stage–discharge characteristic of the tailwater is a function of the channel downstream of the dam. The stilling basin should therefore give an hydraulic jump with a downstream depth–discharge characteristic which matches that of the tailwater.

The submerged bucket

A submerged bucket is appropriate when the tailwater depth is too great for the formation of an hydraulic jump. The bucket is produced by continuing the radial arc at the foot of the spillway to provide a concave longitudinal section as shown in Figure 13.15b. The incoming high velocity from the spillway is thus deflected upwards. The shear force generated between this flow and the tailwater leads to the formation of the 'roller' motions. The reverse roller may initially slightly scour the river bed downstream of the dam. However, the material is returned towards the toe of the dam, so the bed rapidly stabilises.

The relationship between the depths for the incoming and tailwater flows cannot readily be derived from the momentum equation. Empirical relationships of an approximate nature may be used for initial estimates, e.g.

$$y_b \simeq 0.0664(h + E_{S1})\left(\frac{q}{g^{1/2}(h + E_{S1})^{3/2}}\right)^{0.455}$$

(where h is the dam height (Fig. 13.9)
and

$$\text{tailwater depth} \simeq 1.25 y_b$$

Some submerged buckets are 'slotted'. This improves energy dissipation, but may bring about excessive scour at high tailwater levels.

Detailed design data may be obtained from the appropriate publications of the US Bureau of Reclamation (1964, 1977).

The ski jump/deflector bucket

This type of dissipator has a longitudinal profile which resembles the submerged bucket (Fig. 13.15c). However the deflector is elevated above the tailwater level, so a jet of water is thrown clear of the dam and falls into the stream well clear of the toe of the dam. Spillways may be arranged in pairs, and it is then usual for the designer to angle the jets inwards so that they converge and collide in mid-air. This breaks up the jets, and is a very effective means of energy dissipation.

Example 13.5   Spillway and stilling basin design

Design a gravity spillway and stilling basin to pass a design flood of 2100 m^3/s. Dam base elevation = 268 m, dam crest elevation = 300 m, reservoir design water level = 306 m and $C = 2.22$ $m^{1/2}/s$.

Solution

$$2100 = 2.22bH_d^{3/2}$$

For a first trial, assume $H_d = 306 - 300 = 6$ m

$$2100 = 2.22b6^{3/2}$$

Therefore $b = 64.36$ – say 65 m. Applying the specific energy equation between the spillway crest (Station 1) and the foot (or toe) of the spillway (Station 2):

$$306 - 268 = 38 = \frac{2100^2}{2g \times 65^2 \times y_2^2} + y_2$$

so $y_2 = 1.2$ m. Therefore

$$V_2 = \frac{2100}{65 \times 1.2} = 26.9 \text{ m/s}$$

and $Fr_2 = 7.84$. A US Bureau of Reclamation Type II stilling basin is appropriate. If y_3 is the depth downstream of the hydraulic jump, then

$$y_3/y_2 = \tfrac{1}{2}(\sqrt{1 + (8 \times 7.84^2)} - 1) = 10.6$$

therefore $y_3 = 12.72$ m. The length of the basin is $4.3 \times 12.72 = 54.7$ m – say 55 m. The depth y_3 should be compared with the tailwater rating for this flow. It may be necessary to adjust the elevation of the spillway apron to meet tailwater requirements.

13.7 Control gates

Whenever a discharge has to be regulated, some form of variable aperture or valve is installed. These are available in a variety of forms. When gates are installed for channel or spillway regulation, they may be designed for 'underflow' or 'overflow' operation. The overflow gate is appropriate where logs or other debris must be able to pass through the control section. Some typical gate sections are shown in Figure 13.16.

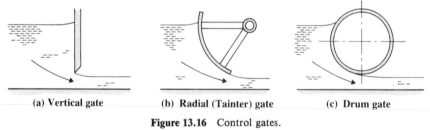

(a) Vertical gate   (b) Radial (Tainter) gate   (c) Drum gate

Figure 13.16   Control gates.

The choice of gate depends on the nature of the application. For example, the vertical gate (Fig. 13.16a) has to be supported by a pair of vertical guides. The gate often incorporates roller wheels on each vertical side, so that the gate moves as smoothly as possible in the guides. Even so, once a hydrostatic load is applied, a considerable force is needed to raise or lower the gate. Furthermore, in severe climates, icing may cause jamming of the rollers. The radial (Tainter) gate (Fig. 13.16b) consists of an arc-shaped face plate supported on braced radial arms. The whole structure rotates about the centre of arc on a horizontal shaft which transmits the hydrostatic load to the supporting structure. Since the vector of the resultant hydrostatic load passes through the shaft axis, no moment is applied. The hoist mechanism has therefore only to lift the mass of the gate. Tainter gates are economical to install, and are widely used in overflow and underflow formats. Other gates (such as the drum or the roller gate) are expensive, and have tended to fall out of use.

The hydraulic characteristics of gates are related to the energy equation, so for free discharge (Fig. 13.17)

$$y_1 + \frac{V_1^2}{2g} = y_2 + \frac{V_2^2}{2g}$$

which can be rearranged in terms of Q:

$$Q = by_1 y_2 \sqrt{\frac{2g}{y_1 + y_2}}$$

Now $y_2 = C_c y_G$. Furthermore, it is usual to express the velocity in terms of $\sqrt{2gy_1}$:

$$Q = bC_c y_G \sqrt{2gy_1 \frac{y_1}{y_1 + y_2}} \tag{13.30}$$

or

$$Q = bC_d y_G \sqrt{2gy_1} \tag{13.31}$$

where

$$C_d = \frac{C_c}{\sqrt{1 + (C_c y_G / y_1)}} \tag{13.32}$$

The quantity $\sqrt{2gy_1}$ should be regarded purely as a 'reference velocity', and not as an actual velocity at any point in the system.

(a) Sectional elevation through vertical sluice gate

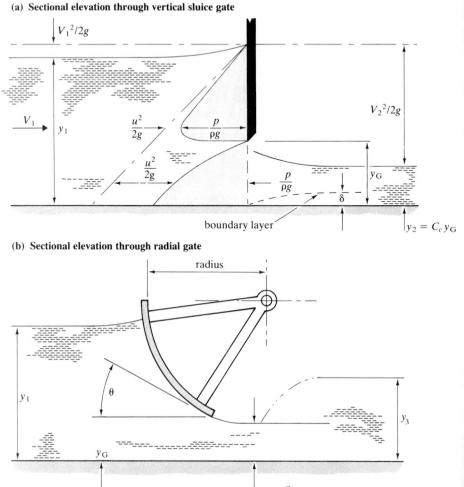

(b) Sectional elevation through radial gate

Figure 13.17 Flow past control gates.

Thus, the magnitude of C_d depends on the gate opening (y_G) and on the contraction of the jet, which in turn is a function of the gate geometry. For a vertical sluice gate under free discharge conditions, it has been found that $C_c = 0.61$ for $0 < (y_G/E_{S1}) < 0.5$. However, when a radial gate is raised or lowered, the lip angle is altered, and the value of C_c must alter correspondingly. An empirical formula for C_c has been evolved for free discharge through underflow radial gates, based on the work of a number of investigators:

$$C_c = 1 - 0.75\left(\frac{\theta}{90}\right) + 0.36\left(\frac{\theta}{90}\right)^2 \qquad (13.33)$$

where θ is measured in degrees.

Clearly, under certain conditions, the outflow from an underflow gate may be submerged (Fig. 13.18), i.e. when the downstream depth exceeds the conjugate depth to y_2. An approximate analysis may then be made as follows, assuming that all losses occur in the expanding flow downstream of the gate between Stations 2 and 3. From the energy equation,

$$y_1 + \frac{Q^2}{2gb^2y_1^2} = y + \frac{Q^2}{2gb^2y_2^2} \tag{13.34}$$

Note that the hydrostatic term on the right-hand side is represented by the downstream depth y, not by y_2. Between Station 2 and Station 3, the momentum equation is applicable:

$$\frac{y^2}{2} + \frac{Q^2}{gb^2y_2} = \frac{y_3^2}{2} + \frac{Q^2}{gb^2y_3} \tag{13.35}$$

In most practical situations y_1 and y_3 will be known, whereas y_2 will have to be estimated from known values of y_G and C_c.

Control gates are used for a range of applications, of which typical examples are regulation of irrigation flows and spillway flows.

Regulation of irrigation flows. It is worth pointing out that for free discharge, supercritical conditions may occur under and downstream of the gate, so a protective apron with a stilling arrangement may be required to protect the stream bed.

Spillway flows. While it is cheapest and simplest to have an unregulated spillway, it is not always the best system. This is the case where a reservoir is to be used for flood attenuation as well as storage. Consider a reservoir sited above low-lying farmland. During heavy rainstorms, the overflow down an unregulated spillway might add to flooding problems downstream. However, a system of control gates may be installed along the spillway crest. During the storm, the gates would be closed, which would increase the reservoir storage capacity and shut off the spillways. This additional

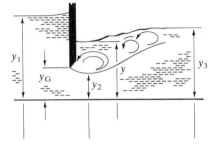

Figure 13.18 Submerged flow through an underflow sluice gate.

storage would be available to accommodate all or part of the incoming flood flow. The excess capacity may then be released in a controlled fashion after the effects of the storm have subsided.

Example 13.6 Discharge control by sluice gate

An irrigation scheme is fed from a river by a diversion channel. The discharge into the system is controlled by an underflow vertical sluice gate. The irrigation channel is 4 m wide, and is roughly faced with cemented rubble giving an estimated value of 0.028 for Manning's n. The bed slope is 0.002. Estimate the required gate opening (y_G) and the flow condition downstream of the gate if (a) the depth of flow in the river is 2 m and the irrigation demand is 11 m³/s and (b) the depth of flow in the river is 3 m and irrigation demand has fallen to 5 m³/s.

Solution

(a) $Q = 11$ m³/s and y_1 (assume same as river depth) = 2 m. Using Manning's equation, estimate $y_n = 1.8$ m = y_3.

1st trial. Assume gate opening $y_G = 1.4$ m, then $y_2 = C_c \times y_G = 0.61 \times 1.4 = 0.854$ m. Hence

$$V_2 = 3.22 \text{ m/s and } Fr_2 = \frac{3.22}{\sqrt{9.81 \times 0.854}} = 1.11$$

Check conjugate depth relationship:

$$\frac{y_3}{y_2} = \frac{y_3}{0.854} = \tfrac{1}{2}(\sqrt{1 + (8 \times 1.11^2)} - 1) = 1.147$$

Conjugate value for $y_3 = 0.98$ m, so submerged conditions exist downstream of gate.

From the specific energy equation (13.34) (between Stations 1 and 2)

$$2 + \frac{11^2}{(4 \times 2)^2 \times 2g} = y + \frac{11^2}{(4 \times 0.854)^2 \times 2g}$$

Therefore $y = 1.568$ m.

From the momentum equation (13.35) (between Stations 2 and 3)

$$\frac{1.568^2}{2} + \frac{11^2}{g \times 4^2 \times 0.854} = \frac{1.8^2}{2} + \frac{11^2}{g \times 4^2 \times 1.8}$$

$$2.132 = 2.048$$

2nd trial. Assume $y_G = 1.25$ m. Then

$$y_2 = C_c \times y_G = 0.61 \times 1.25 = 0.7625 \text{ m}$$

From the specific energy equation,

$$2 + \frac{11^2}{(4 \times 2)^2 \times 2g} = y + \frac{11^2}{(4 \times 0.7625)^2 \times 2g}$$

Therefore $y = 1.433$ m.
    From the momentum equation,

$$\frac{1.433^2}{2} + \frac{11^2}{g \times 4^2 \times 0.7625} = \frac{1.8^2}{2} + \frac{11^2}{g \times 4^2 \times 1.8}$$

$$2.038 = 2.048$$

which is sufficiently accurate.
    Therefore $y_G = 1.25$ m, and outflow is submerged.
    (b) $Q = 5$ m^3/s and $y_1 = 3$ m. $Q = C_d b y_G \sqrt{2gy_1}$. Assume $C_d = 0.6$ for first approximation:

$$5 = 0.6 \times 4 \times y_G \sqrt{2g \times 3}$$

Therefore $y_G \approx 0.27$ m and

$$C_d = \frac{0.61}{\sqrt{1 + [(0.61 \times 0.27)/3]}} = 0.594$$

Hence,

$$Q = 0.594 \times 4 \times 0.27 \sqrt{2g \times 3} = 4.92 \text{ m}^3/\text{s}.$$

From Manning's formula, $y_n = 1.05$ m $= y_3$.
    Check whether discharge is free, using the conjugate depth relationship:

$$y_2 = C_c \times y_G = 0.61 \times 0.27 = 0.165 \text{ m}$$

$$\text{Fr}_2 = V_2/\sqrt{gy_2} = 5.86$$

Therefore

$$y_3/y_2 = \tfrac{1}{2}(\sqrt{1 + (8 \times 5.86^2)} - 1) = 7.8$$

so

$$y_3 = 7.8 \times 0.165 = 1.287 \text{ m}$$

which is greater than y_n, so free discharge must occur.
    The hydraulic jump will be a short distance downstream of the gate. An adequate length of apron would be required to prevent scour, or a stilling arrangement could be used to control jump location.

13.8 Lateral discharge structures

Equations for lateral flow

There are a few circumstances when it is necessary to discharge water into
or out of a stream (Fig. 13.19), typical instances being side weirs in sewers
and side spillways. However, this presents the problem of a varying
discharge in the stream. In turn, this implies that none of the equations
encountered up to now is appropriate. The equations for such problems are
usually based on the momentum principle, and the approach is now
outlined.

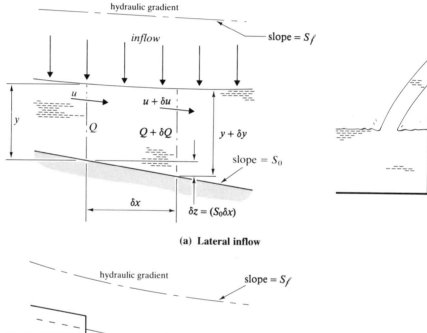

(a) Lateral inflow

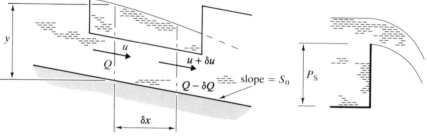

(b) Lateral outflow

Figure 13.19   Lateral discharge.

The first case considered will be that of a lateral inflow of fluid (Fig. 13.19a). The incoming fluid is assumed to enter the channel in a direction perpendicular to the direction of flow of the stream. A force must therefore be applied to the incoming fluid to accelerate it. The total momentum in the stream must therefore be changed. The forces applied to the stream in the direction of flow are:

(a) hydrostatic pressure force $= -\rho g A\, \delta y$ (where A is the cross-sectional area of the stream);
(b) gravity force component $= \rho g A S_0\, \delta x$;
(c) friction force $= \rho g A S_f\, \delta x$.

The discharge, Q, in the stream will increase by δQ as it passes through a longitudinal element of length δx. Therefore, the rate of change of momentum will be

$$\frac{\delta M}{\delta t} = \rho (Q + \delta Q)(u + \delta u) - \rho Q u$$

$$= \rho Q\, \delta u + \rho u\, \delta Q$$

(ignoring products of small quantities). Now $u = Q/A$ and $(u + \delta u) = (Q + \delta Q)/(A + \delta A)$, so

$$\delta u = \frac{A\, \delta Q - Q\, \delta A}{A^2}$$

and

$$\frac{\delta M}{\delta t} = \rho Q \left(\frac{2\, \delta Q}{A} - \frac{Q\, \delta A}{A^2} \right)$$

Force equals the rate of change of momentum, so

$$-\rho g A\, \delta y + \rho g A S_0\, \delta x - \rho g A S_f\, \delta x = \rho Q \left(\frac{2\, \delta Q}{A} - \frac{Q\, \delta A}{A^2} \right)$$

Now, as $A = by$, $\delta A/A \simeq \delta y/y$. Substituting for δA in the momentum equation and dividing throughout by gA, taking the limit as $\delta x \to 0$, and rearranging,

$$\frac{dy}{dx} = \frac{S_0 - S_f - (2Q/gA^2)\, dQ/dx}{1 - (Q^2/gA^2 y)} \tag{13.36}$$

The equation for the case of lateral outflow (Fig. 13.19b) may be developed in a similar manner. There is no loss of momentum in the fluid remaining in the stream, so the equation is

$$\frac{dy}{dx} = \frac{S_0 - S_f - (Q/gA^2)\, dQ/dx}{1 - (Q^2/gA^2 y)} \tag{13.37}$$

The only difference between (13.36) and (13.37) is the coefficient of the third term of the numerator.

The reader will have noticed that the above equations bear some resemblance to the gradually varied flow equation (5.31).

If the value of the coefficient (β) departs significantly from unity, then the equation must be modified accordingly:

$$\frac{dy}{dx} = \frac{S_0 - S_f - (2Q/gA^2)\, dQ/dx(2\beta - 1)}{1 - (\beta Q^2/gA^2 y)} \tag{13.38}$$

There is clearly a difficulty in producing a solution to this equation, since Q is a variable as well as y. A number of methods have been proposed. Two solutions for flow over a side weir are now briefly outlined. Solutions for other gradually varying discharge problems are analogous (see Chow 1959).

Ven te Chow's method. This method is in many ways the simplest to understand. It is suitable for computation on a calculator, and may be laid out in tabular form. The solution used by Chow is based on a variant of (13.36) expressed in finite difference form:

$$\delta(y + z) = \alpha \frac{Q_1(u_1 + u_2)}{g(Q_1 + Q_2)} \delta u \left(1 - \frac{\delta Q}{2Q_1}\right) - S_f\, \delta x \tag{13.39}$$

where u_1 and Q_1 are, respectively, the velocity and discharge at the upstream end of the finite (x) element, u_2 and Q_2 are corresponding values for the downstream end of the element, $\delta u = (u_1 - u_2)$, $\delta Q = (Q_1 - Q_2)$ and S_f may be estimated from the Manning or Colebrook–White equations. The steps in solving the equation are as follows:

(1) divide the longitudinal section at the weir into a convenient number of x-increments;
(2) for the first increment, guess a value for δy;
(3) using this value, compute y_2, A_2, $y_{mean}(=(y_1 + y_2)/2)$;
(4) using y_{mean}, compute the flow (δQ) over the weir for the first increment, using an appropriate weir discharge equation;
(5) compute Q_2 $(=Q_1 \pm \delta Q)$ and u_2 $(=Q_2/A_2)$;
(6) substitute for Q_1, u_1, Q_2, u_2, etc., in (13.39), and compute δy – if

this value differs excessively from the initial guess, then the computation should be repeated until the two values are considered to have converged sufficiently;

(7) the next increment is then subjected to the same computation.

Balmforth and Sarginson's (B&S) method. This method was outlined in Balmforth and Sarginson (1978). The method relies on a simultaneous solution of two equations (the gradually varying discharge equation and the weir flow equation) by a Runge–Kutta fourth-order procedure. It is therefore suitable for computer application.

The steps in the solution are:

(1) use a suitable equation to obtain a value of $\delta Q/\delta x$. B&S proposed:

$$\frac{\delta Q}{\delta x} = \frac{2}{3} C_d \sqrt{2g} (y - P_S)^{3/2}$$

(the value of C_d may be estimated with good accuracy from any of the standard equations);

(2) substitute for $\delta Q/\delta x$ in the varying discharge equation

$$\delta y = \left(\frac{S_0 - S_f + \dfrac{u}{gA} \dfrac{\delta Q}{\delta x} (2\beta - 1)}{1 - \dfrac{\beta Q^2 b}{gA^3}} \right) \delta x \qquad (13.40)$$

(3) compute the Runge–Kutta coefficients by solving the equation using y_1, u_1, Q_1, etc.;

(4) compute δy;

(5) proceed to the next increment.

No explicit mention has been made in this section of the nature of the water surface profile in the vicinity of the lateral inflow or outflow structure. Frequently, the flow will be gradually varying, but this is not invariably the case. It is necessary to check the effect of the slope and of any other channel controls on the surface profile. Potentially, the same variety of profiles can exist as in any other channel flow problem.

Example 13.7   Side weir

A rectangular channel of width 2.5 m incorporating a side weir is to carry water. The channel has a slope of 0.002 and a Manning's n value of 0.015. The sill height is 0.9 m above the bed at its downstream end, and the weir discharge coefficient is 0.7. If the maximum depth of flow at the downstream end of the weir is 1.2 m, estimate the length of the weir if it is to draw off 1.5 m^3/s from the channel.

Table 13.1 Solution to Example 13.7.

x	δx	z	$(y+z)_n$	Estim'd $\delta(y+z)$	$(y+z)_{n+1}$	y_{mean}	R_{mean}	A_{mean}	Q_n	u_n	δQ	δu	$u_n + u_{n+1}$	$Q_n + Q_{n+1}$	$S_f\,\delta x$	Calc'd $\delta(y+z)$
0	0.5	0	1.2	0.02	1.22	1.2095	0.62	3.025	6.45	2.15	0.178	0.058	4.358	13.078	negl.	0.013
	0.5	0.001	1.213	0.013	1.213	1.213	0.615	3.03	6.45	2.127	0.18	0.059	4.315	13.08	negl.	0.013
0.5	0.5	0.002	1.226	0.013	1.226	1.218	0.62	3.05	6.627	2.173	0.187	0.061	4.432	13.44	negl.	0.013
1.0	0.5	0.003	1.240	0.014	1.240	1.23	0.62	3.08	6.814	2.212	0.197	0.064	4.51	13.826	negl.	0.014
1.5	0.5	0.004	1.254	0.014	1.254	1.244	0.623	3.11	7.011	2.24	0.208	0.067	4.58	14.23	0.001	0.014
2.0	0.5	0.005	1.270	0.015	1.270	1.257	0.632	3.173	7.219	2.275	0.22	0.069	4.57	14.507	0.001	0.015
2.5	0.5	0.006	1.286	0.016	1.286	1.273	0.632	3.183	7.439	2.337	0.227	0.071	4.746	15.105	0.001	0.016
3.0	0.5	0.007		0.017	1.303	1.288	0.634	3.22	7.666	2.381	0.25	0.075	4.839	15.582	0.001	0.017
3.5									7.916							

Total weir discharge = $7.916 - 6.45 = 1.466\ \mathrm{m^3/s}$. Therefore weir length required = 3.5 m.

Solution

The solution is presented in tabular form in Table 13.1.

13.9   Outlet structures

Outlet structures are designed to release controlled volumes of water from a reservoir into the water supply, irrigation or other system. Such structures may be conveniently divided into groups according to either the form of the structure or the function of the system. For example, a given installation could be described equally well as a 'gate controlled conduit' or as a 'river outlet', and so on. Such structures may also be used to lower the reservoir level to permit dam maintenance, etc. Some typical cases are as follows:

(a) For low heads, it is possible (and economical) to use an open channel with a control gate. The upstream end usually incorporates fish/trash grids (this is true for virtually all outlet works).

(b) For low to medium heads, a simple conduit may be used with a gate valve to control discharge. Velocities at the outlet end may be great enough to warrant a stilling system.

(c) For higher pressures, a more robust conduit is required (often steel lined). The conduit may consist of a horizontal shaft, or may incorporate a drop shaft entry. If a drop shaft is used, it is economical to arrange two intakes (for the spillway and outlet systems) concentrically on one drop shaft. The spillway tunnel and outlet conduit are then run separately.

Where the outlet system feeds a hydroelectric or pumping scheme, it is imperative that air entrainment is avoided. Large air bubbles in the flow may cause severe damage to hydraulic machines.

13.10   Concluding remarks

A range of hydraulic structures has been considered. The range is by no means exhaustive. However, the various principles which have been applied will usually be capable of adaptation to other appropriate problems. It cannot be too strongly emphasised that our knowledge is still limited in many areas. Most of the solutions given above are approximate. In many instances, it will be important to commission model tests to confirm (or otherwise!) initial estimates.

References and further reading

Ackers, P., W. R. White, J. A. Perkins and A. J. M. Harrison 1978. *Weirs and flumes for flow measurement*. Colchester: Wiley.

Ali, K. H. M and D. Pateman 1980. Theoretical and experimental investigation of air-regulated siphons. *Proc. Instn Civ. Engrs, Part 2* **69**, 111–38.

Balmforth, D. J. and E. J. Sarginson 1978. A comparison of methods of analysis of side weir flow. *Chartered Municipal Engr* **105**, 273–9.

Bos, M. G. (ed.) 1978. *Discharge measuring structures*, 2nd edn. Wageningen: Int. Instn for Land Reclamation and Improvement.

Bradley, J. N. and A. J. Peterka 1957. The hydraulic design of stilling basins. *Am. Soc. Civ. Engrs, J. Hydraulics Divn* **83**(HY5), 1401–6.

Cain, P. and I. R. Wood 1981. Measurements of self-aerated flow on a spillway. *Am. Soc. Civ. Engrs, J. Hydraulics Divn* **107**(HY11), 1425–44.

Chow, V. T. 1959. *Open channel hydraulics*. New York: McGraw-Hill.

Ervine, D. A. 1976. The design and modelling of air-regulated siphon spillways. *Proc. Instn Civ. Engrs, Part 2* **61**, 383–400.

Ervine, D. A. and G. S. C. Oliver 1980. The full scale behaviour of air regulated siphon spillways. *Proc. Instn Civ. Engrs, Part 2* **69**, 687–706.

Head, C. R. 1975. Low-head air-regulated siphons. *Am. Soc. Civ. Engrs, J. Hydraulics Divn* **101**(HY3), 329–45.

Henderson, F. M. 1966. *Open channel flow*. New York: Macmillan.

Keller, R. J. and A. K. Rastogi 1975. Prediction of flow development on spillways. *Am. Soc. Civ. Engrs, J. Hydraulics Divn* **101**(HY9), 1171–84.

Kenn, M. J. 1971. Protection of concrete from cavitation damage. *Proc. Instn Civ. Engrs*. Tech. Note TN48 (May).

Kenn, M. J. and A. D. Garrod 1981. Cavitation damage and the Tarbela tunnel collapse of 1974. *Proc. Instn Civ. Engrs, Part 1* **70**, 65–89.

Proceedings of the Symposium on the Design and Operation of Siphons and Siphon Spillways 1975. London: British Hydromechanics Research Assoc.

Rao, N. S. L. 1971. Self aerated flow characteristics in wall region. *Am. Soc. Civ. Engrs, J. Hydraulics Divn* **97**(HY9), 1285–303.

US Army Waterways Experimental Station 1959. *Hydraulic design criteria*. Washington, DC: US Dept of the Army.

US Bureau of Reclamation 1964. *Hydraulic design of spillways and energy dissipators*. Washington, DC: US Dept of the Interior.

US Bureau of Reclamation 1967. *Water measurement manual*, 2nd edn. Washington, DC: US Dept of the Interior.

US Bureau of Reclamation 1977. *Design of small dams*. Washington, DC: US Dept of the Interior.

14

Canal and river engineering

14.1 Introduction

This chapter draws together various strands of knowledge which have been developed separately in the text. Open channel flows of the steady, varying and unsteady types, hydrology and sediment transport all play a part in river and canal engineering. The vast body of literature which has been produced over the last 100 years bears witness to the fascination, complexity and controversial nature of the subject. There is no complete theoretical approach to many of the problems, though various numerical models can help the engineer to make rough predictions. Above all, the engineer concerned with river hydraulics must be a careful observer who uses his knowledge to co-operate with natural laws as far as is possible. Before any major modifications are made to the course of a river, it is usual to institute physical model studies as a check on the viability of the proposals.

Channels may be lined or unlined, artificial or natural (or may consist of various combinations of these).

14.2 Optimisation of a channel cross section

In designing an artificial channel, cost is a prime consideration, so the engineer must use a channel whose geometry minimises the cost of excavation and lining. The uniform flow equation can be used to give some indication of the optimum section shape from an hydraulic viewpoint. Take the case of a channel of prismatic section (Figure 14.1). For uniform flow,

$$Q = \frac{1}{n} \frac{A^{5/3}}{P^{2/3}} S_0^{1/2}$$

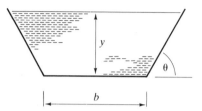

Figure 14.1   Trapezoidal channel.

(equation 5.9). If n and S_0 can be assumed to be constant, then for an economic section the value of $A^{5/3}/P^{2/3}$ must be a maximum for a given value of A. Put another way, the requirement is for P to be the minimum value for a given value of A. From the geometry of the section,

$$A = by + y^2/\tan\theta \qquad (14.1)$$

$$P = b + 2y/\sin\theta \qquad (14.2)$$

From (14.1)

$$b = \frac{A}{y} - \frac{y}{\tan\theta}$$

therefore

$$P = \left(\frac{A}{y} - \frac{y}{\tan\theta}\right) + \frac{2y}{\sin\theta}$$

For the minimum P,

$$\frac{dP}{dy} = -\frac{A}{y^2} - \frac{1}{\tan\theta} + \frac{2}{\sin\theta} = 0$$

hence

$$A = -\frac{y^2}{\tan\theta} + \frac{2y^2}{\tan\theta}$$

and so

$$b = \frac{A}{y} - \frac{y}{\tan\theta} = -\frac{2y}{\tan\theta} + \frac{2y}{\sin\theta}$$

Therefore, for the optimum section,

$$R = \frac{A}{P} = \frac{-y^2/\tan\theta + 2y^2/\sin\theta}{[-(2y/\tan\theta) + (2y/\sin\theta)] + 2y/\sin\theta} = \frac{y}{2} \qquad (14.3)$$

Several considerations arise regarding the above:

(a) It is readily shown that the optimum trapezoidal section may be defined as that shape which approaches most nearly to an enclosed semi-circle whose centre of radius lies at the free surface.

(b) The above development took into account only hydraulic consider-ations. Thus the area A is the area of flow, not the cross-sectional area of the excavation (the latter will naturally be greater than the former).

(c) If the channel is unlined and runs through erodible material, con-siderations of bank stability (Section 14.3) must also enter into the calcula-tions. Furthermore, the bed slope S_0 may not coincide with the natural stable slope in that material for the given flow.

(d) Where the channel is lined, lining methods and costs also play a deter-mining role in deciding section shape.

(e) The slope of the local topography cannot be ignored, and this may be neither constant nor identical with the desired value for S_0.

The points above serve to emphasise the fact that estimates based on (14.3) represent an optimum section shape only from a very restricted viewpoint.

14.3   Unlined channels

Unlined channels may be natural or man-made, and may pass through rocky and/or particulate geological formations. Given a fast enough current, any material will erode (as witness the action of fast-flowing streams in gorges). Where the stream boundaries are both particulate and non-cohesive (e.g. a river traversing an alluvial plain) erosion and deposi-tion occur. The fundamental ideas relating to sediment transport have been developed in Chapter 9. These ideas find an application here. In under-standing the processes which govern channel formation, the starting point must be the current patterns which occur in channels.

Current patterns in channels

Some mention has already been made of secondary flows (Section 5.4), but in view of their importance in determining the paths of natural watercourses, a further look at certain aspects is justified. In a straight watercourse, the pattern of secondary currents is usually assumed to be symmetrical. Consequently, the distribution of shear stress around the boundary should also be symmetrical and this, in turn, should theoretically give a symmetrical pattern of erosion (or deposition) and lead to a symmetrical boundary shape in an erodible bed material. However, when the flow traverses a bend, the secondary current patterns are changed. The velocity of an element of water near the channel centre at the surface will be higher than that of a second element near the channel bed (Fig. 14.2). The centrifugal force on the first element is higher than that on the second, which implies that the force pattern is not in equilibrium. The first element therefore tends to migrate outwards towards the bank, displacing other elements as it does so. This mechanism sets up a spiral vortex (Fig. 14.3) around the outside of the bend. The vortex causes an asymmetrical flow pattern which tends to erode material from the region near the outside of the bend and deposit it near the inside. This implies that the bend will tend to migrate outwards over a period of time. From observations, it would seem that secondary currents are seldom symmetrical, even in a straight channel. Furthermore, ground conditions are rarely homogeneous. Hence the problem of predicting, let alone controlling, events. Even in an apparently straight channel, the line of maximum depth may not be straight. The actual current patterns are more complicated than this outline suggests, nevertheless it gives an indication of the type of mechanism by which the alignment of an unlined watercourse changes with time. The centrifugal effects at the bend also cause a differential elevation of the water surface (superelevation) with higher levels at the outside of the bend than at the inside.

Stable channels and the 'regime' concept

An unlined channel exhibits multiple degrees of freedom, whereas the lined channel has only one (the depth). The unlined channel is subject to erosion and deposition, which may change the bed slope and the channel cross section and alignment. It is interesting that an artificially straight channel in an erodible bed is seldom stable. The flow will usually compel the course to 'meander' (see Section 14.4). Therefore the degrees of freedom of an erodible channel are depth, breadth, slope and alignment. A system with one degree of freedom will rapidly stabilise in a steady state (normal depth) for a given discharge. Stabilisation of a channel with multiple degrees of freedom may take a long time and, indeed, may in some cases never be quite attained. In other cases, 'armouring' (Section 9.4: 'Concluding notes') will assist in the long term stabilisation of the channel.

Where channels run through regions of fine-grained soils, in which irrigation may be vital, it is also vital to avoid excessive erosion of the precious topsoil. Such conditions are widespread throughout the Middle East and the

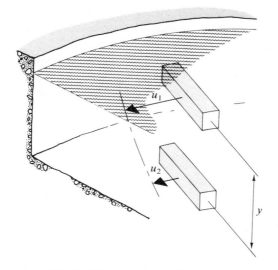

Figure 14.2 Variation of velocity with depth, in channel.

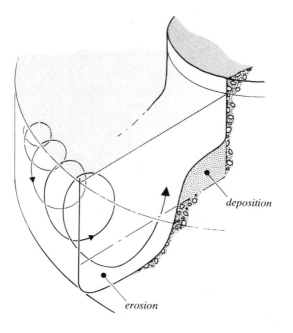

Figure 14.3 Secondary current pattern.

Indian subcontinent, and, indeed, it was in India that some of the early work on the design of stable channels was initiated. A channel which does not exhibit long term changes in geometry and alignment, and in which scour and deposition are in equilibrium, is said to be stable (or 'in regime'). Two approaches have been developed to the design of stable channels with particulate boundaries: the 'regime' approach and the 'tractive force' approach.

The 'regime' approach

This was probably originated by R. G. Kennedy towards the end of the 19th century. Kennedy studied a number of irrigation channels in the Punjab, and developed a formula for the mean velocity V_{CR} required for a stable channel:

$$V_{CR} = K y^n \quad \text{(or } K R^n \text{)}$$

Unfortunately, n is not a constant, but varies roughly between the limits $0.5 < n < 0.73$. In common with many other 'regime' formulae, the numerical coefficient K is not dimensionless, but is purely empirical. Some engineers have criticised the regime approach on precisely the grounds that it lacks any coherent theoretical framework.

Further work has been done since Kennedy's formulae were published. Major contributors include Lacey (in a series of papers published between 1929 and 1958), Blench (in the 1950s and 1960s) and Simons and Albertson.

Blench suggested that the regime approach was primarily applicable to channels having the following characteristics:

(a) steady discharge and with Froude Number less than unity;
(b) steady sediment load (the sediment is therefore small, i.e. $D_{50} \ll y$);
(c) straight alignment;
(d) cross section in which $B > 3y$ and the bank angle approximates to the 'natural' angle;
(e) bed and banks are hydraulically smooth;
(f) sufficiently long established to ensure that equilibrium has been attained and that the channel is stable.

In a sense, (f) begs many questions and, indeed, it is questionable whether it is possible to predict the timescale for its fulfilment when designing an artificial channel.

Simons and Albertson extended the data base for regime work using the information from India, adding data gathered from a number of North American rivers. They concluded that five channel types could be distinguished:

Type 1 – channels with sandy boundaries
Type 2 – channels with sandy bed and cohesive banks

Type 3 – channels with cohesive boundaries
Type 4 – channels with coarse non-cohesive boundary material
Type 5 – Type 2 channel with heavy load of fine silty sediment
A selection of their equations (in metric units) is given below, (a complete list may be found in Henderson (1966)):

$$B = 0.9\,P \qquad P = K_1 Q^{1/2}$$

$$R = K_2 Q^{0.36} \qquad S_0 = \frac{1}{R^2}\left(\frac{V}{K_3}\right)^{1/n}$$

$$y_n = \begin{cases} 1.21R, & R < 2.1\text{ m} \\ 0.61 + 0.93\,R, & R > 2.1\text{ m} \end{cases}$$

The values for the various constants depend on the channel type. For example, a Type 2 channel would use

$$K_1 = 4.71 \qquad K_2 = 0.484 \qquad K_3 = 10.81 \qquad n = 0.33$$

In spite of the efforts of its protagonists, other researchers remain doubtful about the general applicability of regime methods. The approach is not universally accepted.

The 'tractive force' approach

This makes use of the tractive force equations for sediment transport (see Section 9.4). Most of the early work was based on the Du Boys equation; subsequently, the Shields equations were adopted. If the Du Boys equation is used, with the shear stress evaluated by the Chézy Formula ($\tau = \rho g R S_0$), then

$$q_S = K_3 \tau_0 (\tau_0 - \tau_{CR})$$

$$= K_3\,\rho g R S_0 (\rho g R S_0 - \rho g R S_{CR}) \qquad (14.4)$$

Note that S_{CR} is the slope estimated to give the critical shear stress, it has no necessary connection with the slope S_c which would give critical flow (i.e. a uniform flow with Fr = 1).

Equation (9.1b) may be used in conjunction with (14.4) to provide a simple tractive force design criterion.

Shearing action affects the banks as well as the bed. However, in Chapter 9, no explicit account was given of the effects of shear stress on the stability of channel banks. Clearly, this is an important aspect of channels with particulate boundaries, so a simple numerical model is now developed. In

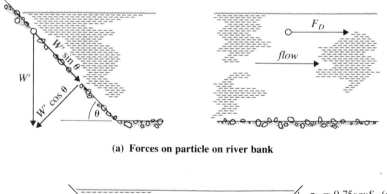

(a) Forces on particle on river bank

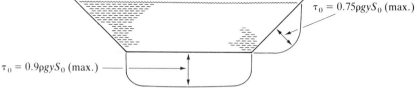

(b) Shear stress distribution

Figure 14.4 Force and shear stress in a channel.

estimating the shear stress at the bank, account should be taken (if possible) of the secondary currents in the channel, especially in the vicinity of bends. Consideration of the forces acting on a particle resting on a bank (Fig. 14.4a) leads to the following:

$$\text{force perpendicular to bank} = W' \cos \theta$$

therefore

$$\text{resistance to motion} = W' \cos \theta \tan \phi$$

Forces parallel to bank are (1) due to the immersed weight of the particle ($= W' \sin \theta$) and (2) due to the drag force of the fluid on the particle ($= F_D$). The angle ϕ is the angle of repose.

All of the above forces can be expressed in terms of the fluid shear force. At the threshold of movement,

$$\tau_{CR} D^2 / A_p = W' \tan \phi$$

therefore

$$W' \cos \theta \tan \phi = (\tau_{CR} D^2 / A_p) \cos \theta$$

$$W' \sin \theta = \tau_{CR} \frac{D^2}{A_p} \frac{\sin \theta}{\tan \phi}$$

$$F_D = \tau_0 D^2 / A_p$$

Therefore, at the threshold of movement (which constitutes the limit of bank stability) $\tau_0 \to \tau_{bc}$ and

$$\tau_{CR}^2 \cos^2 \theta = \tau_{bc}^2 + \left(\tau_{CR} \frac{\sin \theta}{\tan \phi}\right)^2$$

therefore

$$\left(\frac{\tau_{bc}}{\tau_{CR}}\right)^2 = \cos^2\theta - \frac{\sin \theta}{\tan \phi}$$

and so

$$\frac{\tau_{bc}}{\tau_{CR}} = \cos \theta \left[1 - \left(\frac{\tan \theta}{\tan \phi}\right)^2\right]^{\frac{1}{2}}$$

Since both drag (or shear) and self-weight ($W' \sin \theta$) components are combining to dislodge a grain, the shear (τ_{bc}) required to move the grain is less than τ_{CR}. In order to use the equation, it is necessary to estimate the shear stress at the bank. Figure 14.4b shows the shear distribution typical of a trapezoidal channel. The maximum bank shear stress is shown as $0.75 \rho g y S_0$. The coefficient 0.75 is not a constant, but is a function of θ. The magnitude 0.75 departs significantly from 0.75 only when $B/y < 2$ and $45° < \theta < 90°$.

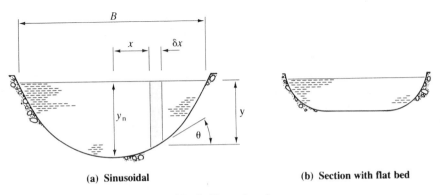

(a) Sinusoidal   (b) Section with flat bed

Figure 14.5 Stable section shape.

The principles outlined above have been refined and extended; for example, by Lane and the US Bureau of Reclamation (Lane 1953) in America. Using a section through a canal (Fig. 14.5a), the following result was derived:

$$\frac{y}{y_n} = \cos\left(\frac{x\tan\phi}{y_n}\right)$$

Thus, the stable bank profile has a sinusoidal form.

The banks may be separated by a flat base (Fig. 14.5b), but the shape of the bank is practically unaffected. This treatment may be further developed to yield estimates of the principal geometrical characteristics for a channel design. For example, with no flat base,

$$A = 2y_n^2/\tan\phi$$

$$P = \frac{2y_n}{\sin\phi}\int_0^{\pi/2}(1-\sin^2\phi\sin^2\theta)^{1/2}\,d\theta$$

For channels of low to moderate width, (9.1b) can then be used to determine the bed slope appropriate to a given sediment size.

The tractive force equations tend to be used for channels through coarse non-cohesive sediments for which the threshold criterion is significant (i.e. there may be little or no transport at low flows).

14.4 Morphology of natural channels

This section covers the study of the processes which are involved in the formation of natural channels. A large number of variables may be involved, not all of which may be known. The first, and most obvious, point is that the discharge in a river may be highly variable with time, being a function of the climate, geology and topography of the area. This, in turn, implies a wide variability in the nature of the sediment transport processes. Floods may denude topsoil in some areas, deposit it elsewhere, and so on. The river engineer is usually concerned with planning for the medium term, and the quality of the information available to assist in planning has improved in recent years.

The discharge

As there is no unique value of Q for a river, the question arises as to what value of Q is appropriate for making engineering estimates. A number of approaches are possible, for example:

(a) the use of a numerical model for predicting the discharge hydrograph over a long period. Such an approach requires detailed information, and access to a computer to perform the analysis.

(b) the use of a single value of Q, known as the 'dominant discharge'. This is defined in a number of ways by different researchers. For example, Ackers and Charlton define it as the steady discharge which would cause the same meander (bend) effect as occurs in the actual river. Alternatively, Henderson (following Leopold & Wolman) proposes that it is the discharge which would maintain a channel at its present cross section, and which is not exceeded sufficiently often for berm build-up to occur on the banks.

(c) the use of the discharge which occurs when the channel runs full – the 'bank-full' discharge.

'Braiding' and 'meandering'

The combined effect of currents and sediment transport is to modify the watercourse. Leopold and Wolman (1957) proposed that channels should be classed as straight, meandering (Fig. 14.6a) or braided (Fig. 14.6b).

For the straight and meandering cases, a single watercourse is present, whereas the braided stream is subdivided by spits. Initially straight channels with particulate boundaries will meander only if the slope and discharge exceed certain critical values, and will braid only beyond yet higher critical values. Note that the values of S_0 and Q are not independent variables here. Broadly speaking, meandering is characteristic of lowland rivers with small slopes, and braiding is characteristic of the steeper upland reaches. Braiding appears to be a natural mechanism for dissipating energy, and is associated with a degree of armouring.

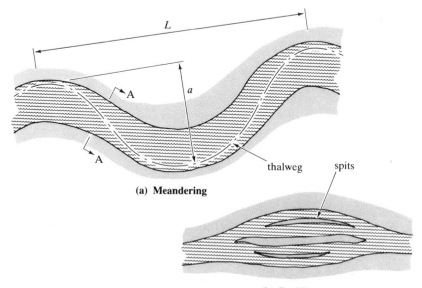

(a) Meandering

(b) Braiding

Figure 14.6 Meandering and braiding in natural channels.

It should not be assumed that braiding and meandering are always separate occurrences. A few rivers both braid and meander at certain points.

Meanders may be characterised by their length, L, and amplitude, a (or radius r). They are probably brought into being (or at least strongly influenced) by the action of the asymmetrical spiral secondary current (see Section 14.3) which scours the outer bank and deposits at the inner bank at a bend. The river section is therefore asymmetrical at the bend (Fig. 14.7a) and symmetrical at the intersection between one bend and the next (Section A–A in Fig. 14.6a). This means that the point of greatest depth does not lie at the centreline of the channel, but swings towards the outside of the bends. This line is known as the 'thalweg'.

This simple explanation of meanders does not command universal acceptance. Some engineers have postulated that it is the influence of small surge waves superimposed on the main flow which initiates the formation of sediment bars at certain points. The deflection of the flow by the bar then leads to meander formation.

Meander patterns are often present even in apparently straight channels, since the thalweg is often found to swing from one side of the channel to the other, thus forming a series of 'hidden bends'. Furthermore, the depth is not necessarily uniform. Measurements of depth often reveal alternating deeper and shallower zones ('pools' and 'riffles'), which exhibit much the same frequency characteristics as meanders.

Several researchers have developed relationships for the geometrical characteristics of meanders. Some examples are:

Leopold and Wolman (1960) $7 < L/B < 11,   2 < r/B < 3$;
Charlton and Benson (1966) $L \propto Q^{0.515} D^{-0.285}$;
Anderson (1967) $L/A^{\frac{1}{2}} = 72\mathrm{Fr}^{\frac{1}{2}},   L = 46Q^{0.39}$.

These equations apparently indicate a series of stable relationships (between L and Q, for example), but this disguises the complexities found in nature. Meander patterns may vary throughout the year at any one point, or they may vary from point to point due either to changes in the geological formation of the bed or to the entry of a tributary. Recent work has tended to emphasise the stochastic nature of the phenomenon (see, for example, Einstein 1971). Other researchers have also emphasised the influence of stream power (or energy) on channel formation (for example, Chang 1984

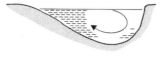

(a) Section at bend   (b) Section at intersection between bends

Figure 14.7   River sections.

and Lewin 1981, pp. 106–21). Henderson (1966) proposed a criterion for the slope of a single watercourse ('straight' or meandering) in a coarse alluvial bed. In its metric form, the equation is

$$S_0 = 0.517 \, D_{50}^{1.14} \, Q^{-0.44}$$

where D_{50} is in metres and Q is in m^3/s. Braided channels have bed slopes greatly in excess of the values derived from this equation.

Braiding is a feature of channels with steeper slopes, where flows have high energy. Therefore, the particulate banks are vulnerable to attack. It seems likely that the surplus energy is dissipated by erosion of the banks, which leads to a wider, shallower watercourse with sediment spits. Braiding may also occur where there is a heavy sediment load, or where the watercourse has (over a long timespan) varied seasonally. The latter type is observable on shingle deltas at the foot of a mountain range, where high discharges (snow melt) take a shorter route than low (summer) discharges.

14.5 River engineering

Many rivers are used by mankind for one or more purposes: water supply, irrigation, drainage and navigation, to name but four. Left to themselves, rivers will usually exhibit features which are inconvenient to man. A river with a strongly meandering course will present a high resistance to a flood wave. The flood may therefore take a 'short cut' across the bends, which will be unfortunate for towns built along the banks. Some aspects of flooding and flood waves have already been alluded to in Sections 5.11 and 10.7.

River engineering, therefore, consists of a range of constructional techniques, training, installation of levees, construction of docks, locks, dams, reservoirs, etc. Some of these techniques will now be reviewed.

Training

Training is the technique of confining or realigning a river to a straight and more regular course than that which occurs in nature (Fig. 14.8). In order to ensure long term stability, the banks usually have to be protected or artificially armoured. A realignment usually presents a problem regarding the bed slope, since the requisite change in level now has to take place over a shorter river length. The river will to some extent readjust itself in response to the training scheme. By protecting the banks, the number of degrees of freedom has been reduced, so the adjustment process will mainly affect the slope and the depth. The shortening of the channel path often results in erosion upstream and deposition downstream, as the channel

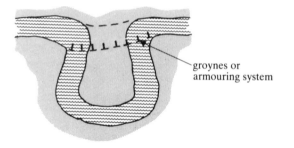

Figure 14.8　River training.

seeks to restore its natural slope. If the engineer wishes to maintain the steeper slope, it may be necessary to armour the bed artificially. Armouring often consists of the dumping of stones, though on the outer banks short groynes, plastic sheeting or sheet piling may be installed. The engineer may use regime or tractive force equations (depending on the nature of the river-bed sediments) to estimate the required geometrical characteristics for a given bank-full discharge. A number of examples of river training schemes are illustrated in Jansen *et al.* (1979).

Flood protection

Flood problems are usually most acutely felt along the lower reaches of rivers. In nature, these reaches often traverse vegetated floodplains. The floodplain provides a natural temporary storage reservoir. However, once the plain becomes populated, steps must be taken to control the extent of flooding. This may be achieved by various methods, such as:

(a) the construction of levees (Fig. 14.9);
(b) the construction of embankment walls (Fig. 14.9);
(c) the construction of flood storage reservoirs (on or off the river line).

The first and second of these methods increase river storage capacity, but they also raise backwater levels during the passage of a flood wave. The engineer must therefore beware of solving a problem in one place only to create another elsewhere. The use of reservoirs for routing purposes has

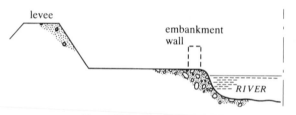

Figure 14.9　Flood protection.

been considered in Section 10.7. However, great care must be taken if a dam is to be constructed across the path of the watercourse. For example, if the incoming flow is carrying a heavy sediment load, much of the sediment will settle in the reservoir (cf. Example 9.2), progressively reducing its effective capacity. This may also imply erosion of the river bed further downstream. The dam will also have considerable implications for the backwater levels. Reservoir systems often combine a water storage or hydroelectric scheme with the routing function to maximise economic viability. Design calculations for any of these (or other) systems start from hydrological data and information regarding flood waves in the existing channel. Incidentally, river training schemes or the introduction of a designed reduction in channel roughness may help to reduce flooding problems.

14.6 Concluding note

The design of canal and river works is still something of an art. However, the array of analytical tools is being continually researched and improved. Within the confines of this chapter, omissions are inevitable – for example, the design of desilting structures has not been considered. Readers who wish to study this area more deeply should consult the references. As a useful starting point, Novak (1983) contains a helpful introduction to sediment transport problems in irrigation canals, while Jansen *et al.* (1979) gives a more detailed account of river engineering techniques.

References and further reading

Anderson, A. G. 1967. *On the development of stream meanders.* Proc. 12th Congr. IAHR 1, Fort Collins. Paper A46, 370–8.

Blench, T. 1957. *Regime behaviour of canals and rivers.* London: Butterworths.

Blench, T. 1966. *Mobile bed fluviology.* Alberta: T. Blench.

Chang, H. H. 1984. Analysis of river meanders. *Am. Soc. Civ. Engrs, J. Hydraulic Engng* 110(1), 37–50.

Charlton, F. G. and R. W. Benson 1966. Effect of discharge and sediment charge on meandering of small streams in alluvium. Symp. CWPRS II, Poona, 285–90.

Einstein, H. A. 1971. *Probability; statistical and stochastic solutions.* Proc. First Symp. Stochastical Hydraulics, University of Pittsburgh.

Henderson, F. M. 1966. *Open channel flow.* New York: Macmillan.

Jansen, P., L. v. Bendegom, Jvd Berg, M. de Vries and A. Zanen 1979. *Principles of river engineering.* London: Pitman.

Lacey, G. 1953. Uniform flow in alluvial rivers and canals. *Proc. Instn Civ. Engrs* 237, 421.

Lane, E. W. 1953. *Progress report on studies on the design of stable channels by the Bureau of Reclamation.* Proc. Am. Soc. Civ. Engrs. No. 280 (September).

Leopold, L. B. and M. G. Wolman 1957. River channel patterns: braided, meandering and straight. *US Geol Survey, Prof. Paper* 282-B.

Leopold, L. B. and M. G. Wolman 1960. River meanders. *Bull. Geol Soc. Am.* **71**, 769–94.

Lewin, J. (ed.) 1981. *British rivers.* London: George Allen & Unwin.

Novak, P. (ed.) 1983. *Developments in hydraulic engineering—1.* London: Applied Science Publishers.

Simons, D. B. and M. L. Albertson 1963. Uniform water conveyance in alluvial material. *Trans. Am. Soc. Civ. Engrs* **128**(1), 65–167.

15

Coastal engineering

15.1 The action of waves on beaches

This may seem a strange point from which to start discussing coastal engineering works. However, it is appropriate because an understanding of natural coastal defence mechanisms gives insights into suitable forms of engineering structures for coastal protection and sea defence.

Beaches form a natural coastal protection system. It is only when these are inadequate that further measures should be contemplated. Examples of such inadequacies include erosion of the coastline (which will result in encroachment on existing buildings and roads, etc.) and insufficient beach height (resulting in flooding by overtopping).

The action of waves on beaches depends on the type of wave and the beach material. For simplicity, wave types are generally categorised as storm waves or swell waves (refer to Ch. 8), and beach materials as sand or shingle.

In general terms, swell waves cause an onshore movement of beach material, resulting in accretion of the beach and a steep beach profile. Conversely, storm waves cause an offshore movement of beach material, resulting in depletion and much flatter beach slopes. Sand beaches respond more drastically (and quickly) to storm and swell waves than do shingle beaches. Typical beach slopes resulting from storm waves are $1:25$ to $1:50$ for sand and $1:7$ to $1:10$ for shingle.

The mechanisms of on- and offshore movement may be explained as follows. Under swell conditions, the wave heights are small and their period long. When the waves break, material is thrown into suspension and carried up the beach (as bed and suspended load) in the direction of movement of the broken wave (the uprush). The uprush water percolates into the beach, so the volume and velocity of backwash water is reduced. Sediment is deposited by the backwash when the gravity forces predominate. The net result is an accumulation of material on the beach. In addition, the beach material is naturally sorted, with the largest particles being left highest on the beach and a gradation of smaller particles seaward. Under storm conditions, the waves are high and steep-fronted, and have shorter periods. Consequently, the volumes of uprush are much larger, and the beach is quickly saturated. Under these conditions the backwash is much more severe, causing rapid removal of beach material. Also, an hydraulic jump often forms when the backwash meets the next incoming wave. This puts more material into suspension, which is then dropped seaward of the jump. The net result is depletion of the beach.

On sand beaches, the material moved offshore is often deposited seaward of the breaker line as a sand bar. During storm conditions, the formation of such a bar has the effect of causing waves to break at a greater distance from the beach, thus protecting the beach head from further attack. The subsequent swell waves then progressively transport the bar material back on to the beach in readiness for the next storm attack.

From the foregoing, it can be appreciated that beaches are an excellent means of coastal protection. Providing that sufficient beach material is available, and that the building line is kept behind the upper limit of beach movement, no further defence is necessary. However, these remarks are only applicable to stable or accreting beaches. In locations where beaches are depleting, or where no natural beach exists, other measures are necessary.

15.2 Littoral drift

Whether beaches are dynamically stable or unstable depends not only on the relative volumes of on- and offshore movements, but also on any movement along the shore (this type of movement is generally referred to as littoral drift).

Waves usually approach a shoreline obliquely, and their direction varies throughout the year. However, there is often a predominant swell direction, generated from remote ocean regions, whereas local storm waves come from a variety of directions. In the transition zone, waves have an effect on the sea bed and may move sediment along the sea bed or bring it into suspension. In the case of swell waves, there is a net movement at the sea bed in the direction of the wave fronts. Conversely, storm waves cause a net movement at the sea bed of opposite direction to the wave fronts. Hence, swell waves 'sweep along' material on the sea bed, bringing it to the shore at an oblique angle. The component of this sweeping action along the shore constitutes littoral drift. Storm waves have a similar effect, but with reversed direction. For coastlines subject to a predominant swell direction, there will be a net movement of material along the coastline from 'upcoast' to 'downcoast' (as storm waves generally occur for only a small part of the year). In the case of shingle beaches, shingle movement normally only takes place on the beach in the surf zone, and littoral drift is largely restricted to this zone. As the uprush is in the direction of the wave fronts and the backwash is normal to the beach slope, littoral drift takes place by a zig-zag movement along the beach.

Littoral drift, in itself, does not cause beach accretion or depletion. These only occur when the supply of material upcoast is greater or less than that downcoast. For example, the construction of a large breakwater (to protect an harbour) or installation of a new groyne system may reduce the upcoast littoral drift, resulting in downcoast depletion and the loss of beaches. Littoral drift in relation to shoreline protection is discussed by Bijker and Van de Graaf in *Shoreline protection* (ICE 1982).

Estimating littoral drift

An appreciation of the importance of littoral drift and beach processes is fundamental to good engineering design of coastal protection projects. Unfortunately, quantitative estimation is extremely difficult. Changes in beach volumes may be calculated from ground or aerial surveys. If surveys are carried out over several years, a trend of accretion or depletion may be ascertained. This is not a measure of littoral drift, but a measure of the imbalance of upcoast and downcoast movements. However, where marine structures are constructed which cut off the upcoast supply, comparisons of beach volumes before and after construction can give a good indication of littoral drift.

Direct measurement of littoral drift has been attempted using tracer methods (e.g. injection of radioactive material or dyes) and using artificial sediments. None of these methods is entirely satisfactory or in general use.

Prediction of littoral drift using various sediment transport theories has been attempted, but our understanding of the transport mechanisms is far from complete. One equation is given here (due to Scripps & Komar) which relates the annual volume of littoral drift to wave power and angle of attack:

$$Q = \frac{KP \sin 2\alpha}{(\rho s - \rho)g} \tag{15.1}$$

where Q is the volumetric transport (m^3/s), P is the wave power of the breaking wave (cf. Equations (8.7)–(8.9)), α is the angle between the breaking wave front and the beach, ρs is the sediment density (kg/m^3) and K is a non-dimensional constant.

The value of K is different for sand and shingle beaches. For shingle beaches, the Hydraulics Research Station (1973) used K = 0.035. This was found to give good agreement with studies of a beach at Bournemouth (UK). However, for some other beaches studied, the transport rate was an order of magnitude less. For sand beaches a value of K = 0.385 was derived by Komar using H_{rms} for wave height.

To determine the annual transport rate using (15.1) requires an estimation of the relevant wave heights and directions throughout the year at the breaker line. One method of dong this (in the absence of direct measurement) is to start from a frequency record of wind speed and direction, convert to deep water wave heights and directions and determine the inshore regime by refraction and shoaling analysis.

For more details of methods for quantifying littoral drift, see Komar (1976), and Muir Wood and Fleming (1981).

15.3   Natural bays

Where an erodible coastline exists between relatively stable headlands, a bay will form. The shape of such bays is determined by the wave climate and the amount of upcoast littoral drift. Silvester (1974) investigated the shape of naturally stable bays by a series of wave tank model experiments. His results indicated that, in the absence of upcoast littoral drift, a stable bay will form in the shape of a half heart. These are called crenulated bays. The reason why crenulated bays are stable is that the breaker line is parallel to the shore along the whole bay. Littoral drift is therefore zero. Silvester also gives a method of predicting the stable bay shape and its orientation to the predominant swell direction (see Section 15.8).

These results have several significant implications. For example, the ultimate stable shape of the foreshore, for any natural bay, may be determined by drawing the appropriate crenulated bay shape on a plan of the natural bay. If the two coincide, then the bay is stable and will not recede further. If the existing bay lies seaward of the stable bay line, then either upcoast littoral drift is maintaining the bay, or the bay is receding. Also, naturally stable bays act as 'beacons' of the direction of littoral drift. Silvester has scrutinised the coastlines of the world to identify crenulated bays, and hence has determined the directions of predominant swell and littoral movements. Where such movements converge at a point, large depositions and accumulation of coastal sediments may be expected. Finally, the existence of crenulated bays suggests a method of coastal protection in sympathy with the natural processes, by the use of artificial headlands. This is discussed further in Section 15.8.

15.4 Sea defence or coast protection?

The term 'sea defence' is normally used to describe schemes which are designed to prevent flooding of coastal regions under extremes of wind and tide. By contrast, the term 'coast protection' is normally reserved to describe schemes designed to protect an existing coastline from further erosion. The distinction between the two may appear to be artificial, but it is nevertheless an important one, as they are often the responsibility of different authorities. For example, in England, the Regional Water Authorities are generally responsible for sea defence schemes, and the local authorities for coast protection schemes. In addition, it is often the case that different sources of (government) finance are available for the two types of scheme. From the point of view of the engineer planning protective measures for a particular coastline, this state of affairs is unsatisfactory, since several different authorities with conflicting requirements may be involved.

Before any new scheme can proceed, it is often necessary to carry out a cost–benefit study. The purpose of such a study is to ascertain whether the cost of the proposed scheme can be justified in terms of the benefits it will provide over the life of the scheme. For a coastal protection scheme, such benefits might include the replacement costs of roads and buildings, etc. For a sea defence scheme, flood damage, loss of income and loss of life would be the prime considerations. A discussion of the economics of shoreline protection is given by Parker and Penning-Rowsell in *Shoreline protection* (ICE 1982).

15.5 Sea walls

At first sight, sea walls appear to offer a sure means of defence and protection. They are very common. However, with a little forethought, it should be evident that they are far from satisfactory, and should only be used when all other measures have been considered and discounted.

Vertical sea walls will reflect virtually all of the incident wave energy (refer to Ch. 8). This sets up a short-crested wave system adjacent to the wall, doubling the wave heights and causing severe erosive action at the sea bed. Consequently, the wall foundations will be quickly undermined unless very substantial toe protection is provided. Immediately downcoast of the sea wall, the short-crested wave system will cause further erosion, and this leads to a temptation to extend the sea wall. Any beach material in front of the sea wall will be rapidly removed downcoast under storm attack, removing any natural defence mechanism and allowing larger waves to attack the wall.

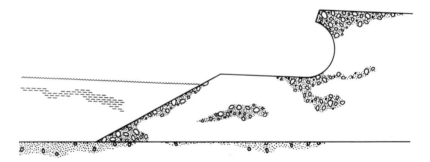

Figure 15.1 Typical form of a sea wall.

The forces exerted on a sea wall by wave action can be considered to be composed of three parts: the static pressure forces, the dynamic pressure forces and the shock forces. The shock forces arise due to breaking waves trapping pockets of air, which are rapidly compressed. As a result, very high localised forces will exist. Thus, sea walls must have a high structural strength, and their construction materials must be able to withstand the shock forces.

Modern designs of sea walls have tended to alleviate some of the problems associated with vertical sea walls (refer to Fig. 15.1). First, the wall is given a sloping face to reduce reflection. Secondly, a curved wave wall is placed at the top of the sea wall to deflect waves downward, and thus dissipate reflected wave energy by turbulence. Finally, one of the various forms of rip-rap protection to the toe of the wall are provided. Such sea walls are expensive to construct, and the efficacy of the initial design should be tested by physical modelling before the design is finalised. A review of the hydraulic design of sea walls is given by Owen in *Shoreline protection* (ICE 1982).

15.6 Groynes and beach replenishment

Groynes are often used in an attempt to stabilise an eroding beach. They consist of wall-like structures built at right angles to the beach line, and their function is to accumulate beach material by reducing the littoral drift along the beach. These structures are commonly constructed of wooden piles driven into the beach, with wooden planks attached between the piles (see Fig. 15.2a). Such groynes are termed 'permeable', as they allow water to flow through them. Impermeable groynes, constructed in concrete or stone, are also used.

The layout of a groyne system is governed by their length, spacing, height and orientation to the beach line (see Fig. 15.2b). Currently, the interrelation of these parameters is largely based on empirical rules or observation.

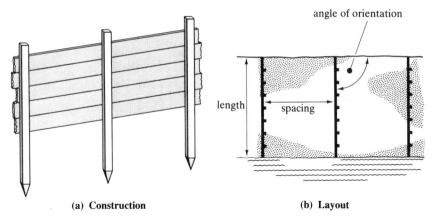

Figure 15.2   Typical groyne details.

As a general guide, the following empirical rules may be used:

(a) the length is normally taken as the distance between beach head and the low water line (for ease of construction and low cost);
(b) the height is between 0.5 and 1.0 m, or sufficient to accommodate the difference in height between storm and swell profiles;
(c) the spacing is between one and three times the length;
(d) the orientation is at right angles to the beach head.

More detailed guidance is given by Berkeley Thorn and Roberts (1981) and Muir Wood and Fleming (1981).

The effectiveness of groynes in reducing littoral drift is a subject of widespread debate. Some authors (for example, Silvester) suggest that they may actively promote littoral drift by inducing rip currents. The question is a difficult one to answer. For example, merely measuring beach accretion or depletion on an existing groyned beach does not necessarily give a good indication of the effectiveness of the groyne system. These measurements need to be compared with conditions before construction of the groynes. In addition, to compare the relative merits of different layouts, such measurements need to be taken for many beaches, in conjunction with estimates of the prevailing wave climates and littoral drift and a knowledge of the beach materials. At the time of writing, such research is being undertaken in the UK sponsored by the Construction Industry Research and Information Association (CIRIA). The interested reader may refer to Summers and Fleming (1983).

Beach replenishment may be used as an alternative to, or in conjunction with, the installation of a groyne system. Its purpose is to renourish the beach which forms the natural defence and protection. It is not usually necessary to sort the material, or to place it to a particular gradient, since wave action will sort and distribute it along the beach. The economics of replenishment schemes will depend on the rate of beach depletion (as distinct from the rate of littoral drift) and the source (and hence cost) of the supply material. In the cost–benefit study, due account must be taken of the recurrent annual costs in maintaining the beach material. Beach monitoring and shingle recharge is described by Foxley and Shave in *Shoreline protection* (ICE 1982).

15.7 Permeable breakwaters and rip-rap

Breakwaters are often used in harbour works to form the primary means of protection from storm attack. They are also used offshore to protect a coastline by absorbing the wave energy before it reaches the beach. Their primary mode of action is to absorb wave energy through friction and turbulence. Hence, they do not suffer the major faults inherent in sea walls. Also, they are a relatively cheap form of construction, since the breakwater may be formed from the land by progressive dumping of material to sea. Many breakwaters are constructed using large blocks of rock (the 'armour units') placed randomly over suitable filter layers. More recently, rock has been replaced by numerous shapes of massive concrete blocks (for example, dolos, tetrapod and cob). Many of these shapes have been patented. A typical breakwater is shown in Figure 15.3a, and concrete armour units in Figure 15.3b and Photograph 8.

The preliminary design of a breakwater may be based on the Hudson formula. This formula was originally derived for rock, but it was later extended to concrete armour units. The formula is:

$$W = \frac{\omega_r H^3}{K_D(S_r - 1)\cot \theta}$$

where W is the required weight of the armour unit, ω_r is the unit weight of armour unit, H is the design wave height, S_r is the specific gravity of the armour unit, θ is the slope angle of the breakwater and K_D is a non-dimensional constant.

The value of K_D is determined by the type of armour unit. For rock, its value is approximately 1–3 whereas for the 'best' concrete armour units, it may be in excess of 10 (consequently reducing the required weight). However, the validity of the Hudson formula has been questioned in recent years, particularly when used for concrete armour units, as there have been some instances of major failures. Designs should not be finalised until extensive physical model studies have been undertaken. For more detailed guidance, the reader should refer to *Breakwaters – design and construction* (ICE 1983).

Photograph 8 Breakwater with 'SHED' armour units, St Helier, Jersey.

A rip-rap revetment consists of rock or stone placed on filter layers of finer material and, as such, is similar to a breakwater. It is normally used to stabilise a shoreline where a beach does not exist. Also rip-rap is often used as toe protection to sea walls. Details of their design (e.g. slopes, stone sizing, filter layers, wave run-up) are given by Thompson and Shuttler (1976).

15.8 Artificial headlands

Silvester's experimental and field work on the stability of natural bays has already been discussed (see Section 15.3). Figure 15.4 summarises his results in graphical form. For any given angle of wave approach (β), there is a fixed ratio of bay indentation (a) to bay length (b), which is known as the indentation ratio (a/b).

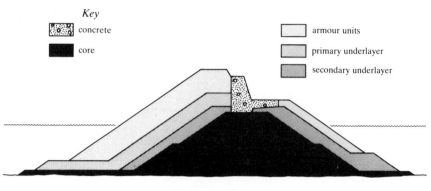

Key

▨ concrete

■ core

▢ armour units

▨ primary underlayer

▨ secondary underlayer

(a) Cross section

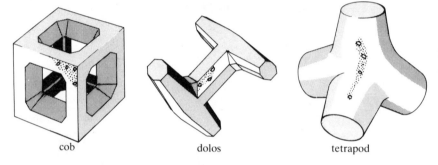

cob   dolos   tetrapod

(b) Armour blocks

Figure 15.3   A typical breakwater.

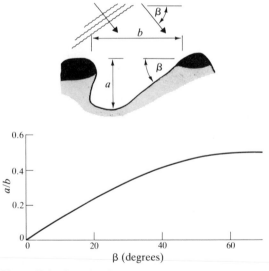

Figure 15.4   Crenulated bays and the indentation ratio.

These results suggest a method of coastal protection on a grand scale. If two artificial headlands are formed on an eroding coastline, then a new stable bay should form between them. Thus, by this simple expedient, whole sections of coastline may be stabilised in one scheme, without recourse to sea walls, groynes or beach replenishment.

Such schemes have been constructed – for example, in Singapore (see Silvester and Ho 1972). In the UK, Muir Wood and Fleming (1981) describe another scheme at Barton-on-Sea, where artificial headlands have been used.

However, for a variety of reasons this method of coastal protection has not gained great popularity. Perhaps the most significant of these reasons is the uncertainty of the bay stability in conditions where there is not a very predominant swell wave direction, and where storm attack comes from a variety of directions.

References and further reading

Berkeley Thorn, R. and A. G. Roberts 1981. *Sea defence and coast protection works*. London: Thomas Telford.

Hydraulics Research Station 1973. *Review of shingle sea defences*. Wallingford: HRS.

Institution of Civil Engineers 1982. *Hydraulic modelling in maritime engineering*. London: Thomas Telford.

Institution of Civil Engineers 1982. *Shoreline protection*. London: Thomas Telford.

Institution of Civil Engineers 1983. *Breakwaters – design and construction*. London: Thomas Telford.

Komar, P. D. 1976. *Beach processes and sedimentation*. New Jersey: Prentice–Hall.

Muir Wood, A. M. and C. A. Fleming 1981. *Coastal hydraulics*, 2nd edn. London: Macmillan.

Silvester, R. 1974. *Coastal Engineering, 1 & 2*. Oxford: Elsevier.

Silvester, R. and S. K. Ho 1972. *Use of crenulate shaped bays to stabilise coasts*. Proc. 13th Conf. Coastal Eng **2**

Summers, L. and C. A. Fleming 1983. *Groynes in coastal engineering: a review*. CIRIA Tech. Note 111. London: CIRIA.

Thompson, D. M. and R. M. Shuttler 1976. *Design of riprap slope protection against wind waves*. CIRIA Rep. 61. London: CIRIA.

Postscript

In this book we have tried to juxtapose theoretical concepts and empirical results in a manner which is useful to civil engineers. We would strongly emphasise here that there are other aspects which must be considered if a knowledge of hydraulics is to be applied to the solution of engineering problems.

Engineering is essentially concerned with the provision of technological means for satisfying the physical requirements of mankind. Hence, engineers are not principally concerned with ascertaining scientific truths, *per se*, but in applying them to the benefit of mankind. To achieve this aim, the successful engineer will possess a thorough theoretical grounding, coupled with the ability to translate this into schemes which are practical and economical.

A good civil engineer must observe and understand those natural processes which may be relevant to a given scheme (there is no excuse for shoddy analysis here), must appreciate the limitations of the theoretical concepts which are to be applied and must have the breadth of vision to distinguish between the general (i.e. universal principles) and the particular (that which is valid only for one case) in any scheme.

These qualities are not acquired merely by reading textbooks, or even by obtaining an engineering degree. Experience is also essential, though useful experience is not to be measured in years, but in the growth of one's ability to draw out general principles from each project, and to apply these appropriately in the future. Good engineers thus develop a sense of intuition and a judgemental capacity which enables them to weigh alternatives and reject unsuitable proposals before embarking on a detailed analysis. Such qualities are priceless when one has to break new ground (in research, design or construction), or when investigating the occasional major failure where application of current knowledge has proved to be insufficient.

It is our hope that student readers will be inspired to become good engineers, and that this book will have helped them on their way.

Problems for solution

CHAPTER 1

1. A pipe contains oil (density = 850 kg/m^3) at a gauge pressure of 200 kN/m^2. Calculate the piezometric pressure head (a) in terms of the oil and (b) in terms of water.

$$[24 \text{ m}, \ 20.4 \text{ m}]$$

2. A sloping tube manometer has the capacity for a maximum scale reading $R_p = 150$ mm. If it is to measure a maximum pressure of 400 N/m^2, what must be the angle θ at which the tube is set? Assume a fixed zero position. The gauge fluid has a density of 1800 kg/m^3.

$$[8.69°]$$

3. For the manometer in Question 2, the horizontal cross section of the tank is 40 times the tube cross section. What will be the percentage error in the indicated pressure reading due to the fall in the level in the tank?

$$[2.43\%]$$

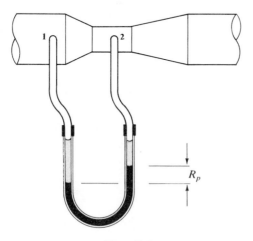

Figure P.1

4. A mercury manometer is connected to a flow meter in a pipeline (Fig. P.1). The gauge pressure at Point 1 is 38 kN/m^2 and at Point 2 the vacuum pressure is − 50 kN/m^2. The fluid in the pipeline is water. Calculate the manometer reading R_p.

[712 mm]

5. An inverted tube has its upper end sealed and the air has been evacuated to give a vacuum (Fig. P.2). The lower end is open and stands in a bath of mercury. If the air pressure is 101.5 kN/m^2, what will be the height of the mercury in the tube?

[760 mm]

6. A gas holder is sited at the foot of a hill (Fig. P.3) and contains gas at an absolute pressure of 103 000 N/m^2. If the atmospheric pressure is 101 500 N/m^2, calculate the gauge pressure in head of water. The gas holder supplies gas through a main pipeline whose highest point is 150 m above the gas holder. What is the gauge pressure at this point? Take density of air as 1.21 kg/m^3 and density of gas as 0.56 kg/m^3.

[150 mm water, 250 mm water]

7. A large drain is 1 m square in cross section. It discharges into a sump whose wall is angled at 45° (Fig. P.4). At the outlet end of the drain there is a steel flap valve hinged along its upper edge. The mass of the flap is 100 kg, and the centre of gravity lies at its geometrical centre. The flap gate is also held shut by a weight of 250 kg on a 500 mm cantilever arm. To what level will the water rise in the drain before the valve lifts?

[Roof of drain]

8. A mass concrete dam has the section shown in Figure P.5 and spans a channel 200 m wide. Estimate the magnitude of the resultant force, its angle to the horizontal, and the point at which its line of action passes through the base line.

[630.5 MN, 13.5°, 36.6 m from O]

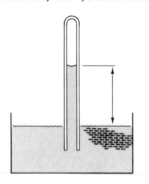

Figure P.2

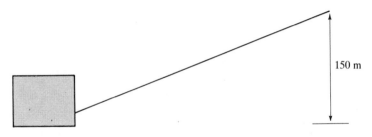

Figure P.3

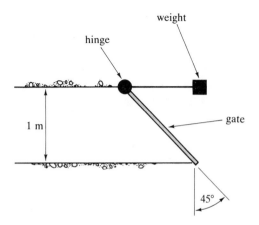

Figure P.4

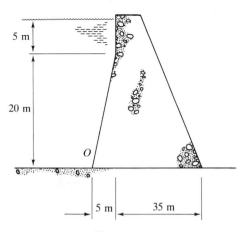

Figure P.5

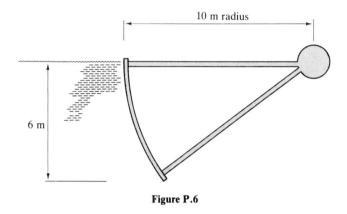

Figure P.6

9. A radial gate is to be used to control the flow down a spillway (Fig. P.6). The gate is 10 m in radius and 12 m wide, and is supported on two shaft bearings. Calculate the load on each bearing. Prove that the resultant hydrostatic force passes through the axis of the bearings.

[1.163 MN]

10. A pontoon is to be used as a working platform for diving activities associated with a dockyard scheme. The pontoon is to be rectangular in both plan and elevation, and is to have the following specification:

width = 6 m;
mass = 300 000 kg;
metacentric height $\geqslant 1.5$ m;
centre of gravity = 0.3 m above geometrical centre;
freeboard (height from water level to deck) $\geqslant 750$ mm.

Estimate the overall length, L, and overall height, h, of the pontoon if it is floating in fresh water (density = 1000 kg/m^3)

[$L = 36.25$ m, $h = 1.44$ m]

CHAPTER 2

1. For the pipeline shown in Fig. P.7, estimate the velocity of flow $V (= Q/A)$ at Section 1, and the pressure p_1. The fluid is water, and the pressure head at entry is 2 m of water. Assume that there are no losses in the pipe itself. The only loss of energy is due to the dissipation of the kinetic energy at the exit (Section 2). Explain why this solution is physically impossible

[10.85 m/s, -239 kN/m^2]

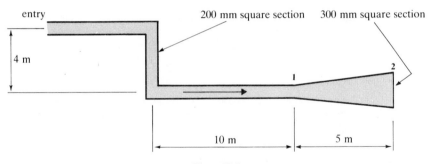

200 mm square section   300 mm square section

4 m

10 m   5 m

Figure P.7

2. A jet of water 50 mm in diameter is directed vertically upwards. The initial (datum) velocity of the jet is 14 m/s. Determine the jet velocity and diameter 2.5 m above datum. At what height above datum would the jet come to rest?

[12.1 m/s, 53.8 mm, 10 m]

3. Calculate the velocity (α) and momentum (β) coefficients for the case of laminar pipe flow if the velocity distribution is given by

$$u_r = K(R^2 - r^2)$$

where u_r is the velocity at radius r, R is the pipe radius and K is a constant. (*Hint*. Fig. 4.1 may be of assistance.)

[$\alpha = 2$, $\beta = 4/3$]

4. A reducing pipe bend is shown in Fig. P.8. If the discharge is 500 l/s and the upstream pressure is 1000 kN/m^2, find the magnitude and direction of the force on the bend.

[174.4 kN, 8.26°]

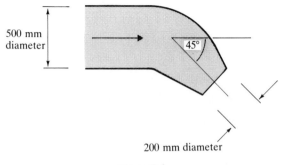

500 mm
diameter

45°

200 mm diameter

Figure P.8

5. For the pipe junction shown in Fig. P.9, estimate the magnitude and direction of the required resistance force to prevent movement of the junction for the following conditions: $p_1 = 69$ kN/m^2, $Q_1 = 570$ l/s and $Q_3 = 340$ l/s

$$[F_R = 6.93 \text{ kN}, \ \theta = 25.1°]$$

6. A duct of diameter 0.8 m carries gas ($\rho = 1.3$ kg/m^3). At one point, the duct diameter reduces to 0.74 m. Starting from Bernoulli's equation, estimate the velocity of flow in the reduced section, and the mass flow rate ($= \rho Q$) through the duct if the pressure difference between a point upstream of the reduction and another point at the reduction is equivalent to 30 mm of water.

$$[41.1 \text{ m/s}, 23.0 \text{ kg/s}]$$

7. A venturi meter is installed in a 300 mm diameter water main to measure the discharge as shown in Fig. 2.9a. If the throat diameter is 200 mm, the manometer reading is 15 mm (using mercury as a gauge fluid (density 13.6×10^3 kg/m^3)) and $C_d = 0.98$, what is the discharge?

$$[66.2 \text{ l/s}]$$

8. Starting from first principles, derive the discharge equation for a 90° V-notch weir. (*Hint.* Fig. 2.11 is a good starting point.)

$$[Q_{\text{ideal}} = \tfrac{8}{15} \tan 45 \sqrt{2g} h^{5/2}]$$

9. A source, having a discharge of 4 m^3/m s occurs at the origin $(0, 0)$ of Cartesian co-ordinates. This is superposed on a uniform rectilinear flow of 5 m/s parallel to the x-axis. Plot the flow net. Determine the co-ordinates of the stagnation point.

$$[127 \text{ mm}, 0]$$

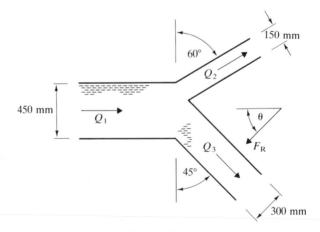

Figure P.9

10. An ideal (potential) flow pattern is made up from a combination of:

a source with a discharge of 3.8 m^3/s;
a forced vortex of 300 mm diameter rotating at 100 rad/s;
a free vortex surrounding the forced vortex.

Sketch the stream function diagram which represents the above pattern. Estimate the pressure and velocity at a radius of 250 mm.

[40.66 N/m^2, 9.32 m/s]

CHAPTER 3

1. Determine whether the following flows are laminar or turbulent:
    (a) A flow of water ($\rho = 1000$ kg/m^3, $\mu = 1.2 \times 10^{-3}$ kg/m s) through a pipe of square cross section. The section is 500 × 500 mm and the mean velocity of flow is 3 m/s.
    (b) A flow of air ($\rho = 1.24$ kg/m^3, $\mu = 1.7 \times 10^{-5}$ kg/m s) through a pipe of diameter 25 mm. The mean velocity is 0.1 m/s.

2. A laminar boundary layer forms over a plate which moves at 3 m/s through water. Estimate the boundary layer thickness 5 m downstream of the leading edge. Assume $\rho = 1000$ kg/m^3 and $\mu = 1 \times 10^{-3}$ kg/m s. The velocity distribution is

$$\frac{u}{U_\infty} = \left(\frac{2y}{\delta} - \left(\frac{y}{\delta}\right)^2\right)$$

[7.1 mm]

3. Define the terms boundary layer, displacement thickness and momentum thickness. A rectangular surface 500 mm wide and 1 m long is immersed in water. The water is flowing at a free stream velocity of 20 m/s parallel to the surface. Evaluate the shear force of the fluid on the surface, assuming that the flow is turbulent throughout the boundary layer. For water, $\rho = 1000$ kg/m^3 and $\mu = 1.2 \times 10^{-3}$ kg/m s. The velocity distribution in the boundary layer $(u/U_\infty) = (y/\delta)^{1/7}$, $\tau_0 = 0.0225\rho U_\infty^2 (\rho U_\infty \delta/\mu)^{-1/4}$.

[119 N]

4. A chimney for a new chemical plant is 30 m tall and 1 m in diameter, and is to be fabricated from stainless steel. It is to be designed for a maximum wind speed of 80 km/h. Calculate the total wind loading on the structure. If the chimney has a natural frequency of 0.5 Hz, determine whether a problem could arise due to self-induced oscillation and, if so, at what wind speed this would occur. Assume $C_D = 1.0$. The density and viscosity of air are 1.24 kg/m^3 and 1.7×10^{-5} kg/m s, respectively.

[9183 N, yes, at 9 km/h]

CHAPTER 4

1. Find the maximum discharge in a 12 mm diameter domestic plumbing system for which the flow is laminar. For this discharge find the head loss per metre run, the maximum velocity and the friction factor. Take $\mu = 1.14 \times 10^{-3}$ kg/m s.

$$[Q = 0.0215 \text{ l/s}, \ h_f = 4.9 \text{ mm/m}, \ V_{max} = 0.38 \text{ m/s}, \ \lambda = 0.032]$$

2. Write an essay describing the historical development of turbulent pipe flow theory. The essay should include the contributions of all those named in Table 4.1, and should both describe their work and explain its relevance to current practice.

3. Write a computer program which will calculate the necessary data to reproduce the Moody diagram. The program calculations should be based on the Colebrook–White transition equation.

4. Write a computer program to calculate the discharge in a simple pipe system including local losses. Hence, or otherwise, find the discharge for the following two cases:

   (a) Pipe length = 10 000 m;
       pipe diameter = 500 mm;
       pipe roughness $(k_S) = 0.03$ mm;
       static head = 150 m;
       local loss coefficient $(k_L) = 0$.
   (b) Pipe length = 50 m;
       pipe diameter = 300 mm;
       pipe roughness $(k_S) = 0.6$ mm;
       static head = 10 m;
       local loss coefficient $(k_L) = 10$.

   Take $\mu = 1.14 \times 10^{-3}$ kg/m s.

$$[\text{(a) } Q = 0.68 \text{ m}^3/\text{s}, \text{ (b) } Q = 0.265 \text{ m}^3/\text{s}]$$

5. Water flows vertically down a pipe of diameter 150 mm at 2.4 m/s. The pipe suddenly enlarges to 300 mm diameter. Find the local head loss. If the flow is reversed and the coefficient of contraction is 0.62, find the new local head loss.

$$[h_L = 0.165 \text{ m}, \ h_L = 0.11 \text{ m}]$$

6. A sewer of diameter 1200 mm is laid to a gradient of 1 m in 100 m. Using the HRS tables or charts with $k_S = 0.6$ mm find:

   (a) the full bore discharge and velocity;
   (b) the flow depth and velocity for $Q = 0.5$ m^3/s.

   Why might this pipe size be considered unsuitable?

$$[\text{(a) } Q = 4.5 \text{ m}^3/\text{s}, \ V = 3.75 \text{ m/s}; \text{ (b) } d = 0.28 \text{ m}, \ V = 2.5 \text{ m/s}]$$

7. A concrete pipe 750 mm in diameter is laid to a gradient of 1 in 200. The estimated value of Manning's n is 0.012, and the pipe-full discharge is estimated to be 0.85 m^3/s. (a) Calculate the discharge for a proportional depth of 0.9 using Manning's equation. (b) Explain why the discharge in (a) is greater than the pipe-full discharge.

$$[Q = 0.91 \text{ m}^3/\text{s}]$$

CHAPTER 5

1. To solve Manning's equation for depth of flow given the discharge (Q) requires an iterative procedure. Prove that for a wide rectangular channel a good estimate of the depth (y) is given by

$$y = y_1(Q/Q_1)^{0.6}$$

where y_1 and Q_1 are initial estimates of y and Q. Hence, write a computer program to solve Manning's equation for depth given discharge for any trapezoidal channel, such that the solution for y gives a discharge to within 1% of Q. Use this program to find the depth of flow for the following conditions:

(a) $b = 2$ m, side slope $1:1$, $n = 0.015$, $S_0 = 2$ m/km, $Q = 21$ m^3/s;
(b) $b = 10$ m, side slope $1:2$, $n = 0.04$, $S_0 = 1$ m/km, $Q = 95$ m^3/s

$$[(a) \ y = 1.846 \text{ m}, \ (b) \ y = 3.776 \text{ m}]$$

2. Produce a graphical solution to Example 5.4, drawn accurately to scale, as shown in Figure 5.10.

3. The normal depth of flow in a rectangular channel (2 m deep and 5 m wide) is 1 m. It is laid to a slope of 1 m/km with a Manning's $n = 0.02$. Some distance downstream there is a hump of height 0.5 m on the stream bed. Determine the depth of flow (y_1) immediately upstream of the hump.

$$[y_1 = 1.27 \text{ m}]$$

4. Starting from first principles, show that the following equation holds true for a hydraulic jump in a trapezoidal channel:

$$\rho g \left(\frac{by^2}{2} + \frac{xy^3}{3} \right) + \frac{\rho Q^2}{(b + xy)y} = \text{constant}$$

where y is the depth of flow, b is the bottom width, Q is the discharge and x is the side slope (1 vertical to x horizontal). Hence draw the force–momentum

diagram for the following conditions and determine the initial depth if the
sequent depth is 0.2 m: $Q = 50$ l/s, $b = 0.46$ m, $x = 1$.

[Initial depth $= 0.038$ m]

5. Suppose that a stable hydraulic jump forms between the two sluice gates shown
in Figure 5.15. If the flow depth downstream of a sluice gate is 61% of the gate
opening (y_g), determine the downstream gate opening (y_{g2}) for the following
conditions: $y_1 = 0.61$ m, $Q = 15$ m³/s, $b = 5$ m. (*Hint.* Assume no loss of energy
through a sluice gate)

[$y_{g2} = 1.115$ m]

6. A rectangular concrete channel has a broad-crested weir at its downstream end
as shown in Figure 5.16. The channel is 10 m wide, has a bed slope of 1 m/km
and Manning's n is estimated to be 0.012. If the discharge is 150 m³/s,

(a) calculate the minimum height of the weir (Δz) to produce critical flow;
(b) if $\Delta z = 0.5$ m, calculate the upstream head (H_1) if $C_d = 0.88$.

[(a) $\Delta z = 0.179$ m, (b) $H_1 = 5.14$ m]

7. A trapezoidal channel of bed width 5 m and side slopes 1:2 has a flow of
15 m³/s. At a certain point in the channel the bed slope changes from 10 m/km
(upstream) to 50 m/km (downstream). Taking Manning's n to be 0.035, determine the following:

(a) the normal depth of flow upstream and downstream
(b) the critical depth of flow upstream and downstream
(c) the Froude Number upstream and downstream
(d) the depth at the intersection of the two slopes.

Sketch the flow profile.

[(a) $y_n = 0.953$ and 0.61 (m), in (b) $y_c = 0.86$ (m), in (c) Fr $= 0.84$ and 1.77,
(d) $y = y_c$, sketch: see Figure 5.18a]

8. A lake discharges directly into a rectangular concrete channel. If the head of
water in the lake above the channel bed is 3 m and the channel is 6 m wide with
Manning's $n = 0.015$, find:

(a) the discharge for a channel bed slope of 100 m/km
(b) the discharge for a channel bed slope of 1 m/km.

[(a) $Q = 53.15$ m³/s, (b) $Q = 42$ m³/s]

9. Using one of the computer programs given in Appendix B, or otherwise, verify
that the three mild slope profiles given in Figure 5.20 are correct.

10. For the situation described in Problem 8(a), find the distance over which the
depth reduces from critical depth to $y = 0.8$ m and the normal depth. Why is
the solution unlikely to be very accurate?

[For $Q = 53.15$ and $y_c = 2.0$ m, $y = 0.8$ m after 57.5 m, $y_n = 0.642$ m; at $y = y_c$
flow is rapidly varied and for steep slopes refer to Section 13.5]

11. Starting from the references at the end of Chapter 5, write a computer program
to solve the gradually varied unsteady flow equations using an implicit finite
difference scheme. Hence re-solve Example 5.10 and compare the results.

CHAPTER 6

1. A steel pipeline is 2.5 km long and 500 mm in diameter, and carries water. A control valve is sited at the downstream end. Calculate:

   (a) the celerity of the shock wave in the fluid;
   (b) the maximum discharge if the surge pressure following instantaneous shut down is limited to 2900 kN/m^2.

   Take $K = 2.11 \times 10^9$ N/m^2.

   [1450 m/s, 390 l/s]

2. The penstock to a turbine is 1 km long and 150 mm in diameter. The pipe is designed for a maximum pressure of 5600 kN/m^2. The operating discharge is 44 l/s, and the corresponding pressure at the governor valve is 2450 kN/m^2. What is the minimum time for complete shut down? Take $c = 1400$ m/s.

   [1.7 s]

3. A water supply main is 3 km long and carries water. The velocity of the water is 2.5 m/s. Calculate the surge pressure for instantanteous closure if $K = 2.13 \times 10^9$ N/m^2.

   [372 m head]

4. A steel pipeline is 750 m long and 100 mm in diameter, and has a wall thickness of 7.5 mm. It carries oil at a velocity of 1 m/s and has a control valve at the downstream end. The valve shuts in 1 s. Determine (a) the speed of sound in the medium, (b) whether the 1 s closure is 'rapid' or 'instantaneous' and (c) the surge pressure. Take $E = 205 \times 10^9$ N/m^2 for steel. For the oil, $\rho = 950$ kg/m^3 and $K = 0.67 \times 10^9$ N/m^2.

   [(a) 801 m/s, (c) 761 kN/m^2]

5. A large hydroelectric scheme is fed through a lined tunnel. The tunnel is of 2.1 km overall length. For the first 1 km, the tunnel diameter is 4 m, and for the remaining length it is of diameter 3 m. The governor valve is at the downstream

end, and its closure characteristics are given below. If the maximum design discharge is 20.5 m³/s estimate the surge pressure at the governor, 3 s after the commencement of closure.

friction factor $\lambda = 0.012$;
celerity of sound in 4 m diameter tunnel = 1000 m/s;
celerity of sound in 3 m diameter tunnel = 1100 m/s;
head at governor = 157.7 m for the design discharge.

Closure characteristics

Time (s)	0	1	2	3	4	5	6
$A_R(=A/A_0)$	1.0	0.87	0.7	0.5	0.3	0.1	0

[277 m]

CHAPTER 7

1. A centrifugal pump is designed for a discharge of 1 m³/s of water. The water enters the pump casing axially and leaves the impeller with an absolute velocity of 11.5 m/s at an angle of 16.5° to the tangent at the impeller periphery. The impeller is 500 mm in diameter, and rotates at 710 rev/min. Estimate the exit vane angle and the hydraulic power delivered.

[25°, 102 kW]

2. A pump is to deliver 60 l/s through a 200 mm diameter pipeline. The pipe is 150 m long and rises 2 m. The friction factor $\lambda = 0.028$. Two pumps are being considered for the duty. Select the more suitable pump, and estimate the power required to drive the pump and the specific speed. Should the pump be of radial, axial or mixed flow type?

Pump No. 1 performance data

Head (m)	8.6	8.35	7.56	6.35	4.95	3.7	2.3
Discharge (l/s)	0	18	39	60	75	88	100
Efficiency (%)	0	52	72	79	75	63	48

Pump No. 2 performance data

Head (m)	9.0	8.8	8.1	7.0	6.0	4.5	3.3
Discharge (l/s)	0	18	39	60	75	88	100
Efficiency (%)	0	52	75	76	67	58	46

[4.8 kW, 89, mixed flow]

3. Cavitation problems have been encountered in a mixed flow pump. The pump is sited with its intake 3 m above the water level in the reservoir and delivers $0.05 \text{ m}^3/\text{s}$. The total head at outlet is 31.7 m, and at inlet is -7 m. Atmospheric pressure $= p_A = 101.4 \text{ kN/m}^2$ and vapour pressure of water $= p_{vap} = 1.82 \text{ kN/m}^2$. Determine the Thoma cavitation number. A similar pump is to operate at the same discharge and head, but with $p_A = 93.4 \text{ kN/m}^2$ and $p_{vap} = 1.2 \text{ kN/m}^2$. What is the maximum height of the pump intake above reservoir level if cavitation must be avoided?

[0.081, 2.25 m]

CHAPTER 8

1. A deep water wave has a period of 8.5 s. Calculate the wave celerity, group wave celerity and wavelength in a transitional water depth corresponding to $d/L_0 = 0.1$.

[$L_0 = 112.8$ m, $C_0 = 13.27$ m/s, $c = 9.42$ m/s, $C_G = 7.63$ m/s, $L = 80.1$ m]

2. Re-solve Example 8.1 if the deep water wave is travelling at $30°$ to the shoreline.

[$H_B = 4.75$, $c = 7.0$ m/s, $d_B = 6.1$, $\alpha_B = 15.5°$]

3. Using Figure 8.12, determine the significant wave heights and periods corresponding to a wind speed of 20 m/s, a fetch length of 200 km and for wind durations of 1, 6 and 24 h.

[$t = 1$ h, $H_S = 2.1$ m, $T_S = 4.2$ s (duration limited)
$t = 6$ h, $H_S = 6.0$ m, $T_S = 7.4$ s (duration limited)
$t = 24$ h, $H_S = 6.3$ m, $T_S = 7.6$ s (fetch limited)]

CHAPTER 9

1. A river is 100 m wide and 8 m deep. The bed slope is 1 m per 2 km. The median sediment size is 10 mm. Estimate (a) the critical shear stress, (b) the sediment bed load using the Du Boys equation and (c) the minimum stable sediment size.

[9.06 N/m^2, $3.95 \text{ m}^3/\text{s}$, 37.5 mm]

2. Further downstream, the river in Problem 1 traverses an alluvial plain for which $D_{50} = 0.5$ mm. Estimate the total sediment load. River dimensions are unchanged.

[$0.32 \text{ m}^3/\text{s}$ approx]

CHAPTER 10

1. 'Any quantitative assessment of flood runoff should start from an appreciation of the catchment characteristics affecting runoff.'

   (a) State whether you agree or disagree with this statement, giving your reasons, and (b) describe the effects on runoff of the catchment characteristics mentioned in Section 10.3.

2. The data given below are the annual maxima floods for the River Don. Starting with linear graph paper, plot these data in a similar manner to that shown in Figures 10.2a, b and d (i.e. starting with a histogram, converting to a pdf and finally an EV1 probability plot). Hence estimate the 50 year return period flood and comment on the reliability of this estimate.

Year	56	57	58	59	60	61	62
Flood (m^3/s)	38.2	58.9	120.6	57.6	159.9	40.4	66.2

Year	63	64	65	66	67	68	69
Flood (m^3/s)	73.6	66.5	206.8	84.8	97.1	142.1	82.1

$$[Q_{50} = 214 \text{ m}^3/\text{s}]$$

3. Calculate the peak runoff rate resulting from the probable maximum precipitation at Ardingly Reservoir from the following data:

Time (h)	0	1	2	3	4	5	6
Rainfall (mm)		2	5	8	11	14	31

Time (h)	6	7	8	9	10	11	12	13
Rainfall (mm)		93	31	14	11	8	5	2

Percentage runoff = 68% (from catchment characteristics)
Baseflow = 1 m^3/s (from catchment characteristics)
10 mm, 1 hour unit hydrograph time to peak = 4 h
Catchment area = 21.82 km^2
(*Hint*. Use a synthetic UH with $T_p = 4$ h)

$$[140 \text{ m}^3/\text{s}]$$

4. Describe, in detail, what methods you would use to estimate the 50 year return period flood (both peak discharge and flood hydrograph) for the following two cases:

   (a) an ungauged catchment;
   (b) a gauged catchment with stream flow records of 15 years.

Suggest ways in which your estimate could be improved (presuming that your improved estimates are required within 1 year!).

5. If the reservoir in Example 10.2 was not full preceding the occurrence of the PMF, but required an inflow volume of 7056×10^3 m^3 to bring it to spillway crest level, determine the new peak outflow.

$$[115 \text{ m}^3/\text{s}]$$

6. Determine the necessary pipe diameters for the first three pipes of a drainage network using the modified rational method from the data given below.

Pipe no.	Length (m)	Gradient	Area (m^2)	t_e (min)	k_S (mm)	C_V
1.0	70	0.0175	1415	4	0.6	0.9
1.1	75	0.017	3275	4	0.6	0.9
1.2	92	0.0085	4085	4	0.6	0.9

Rainfall (mm/h)	55.3	54.3	53.4	52.5	51.7	50.8	50.0	49.3	48.5	47.8	47.1
Duration (min)	4.8	5.0	5.2	5.4	5.6	5.8	6.0	6.2	6.4	6.6	6.8

$$[1.0, \ d = 175 \text{ mm}; \ 1.1, \ d = 225 \text{ mm}; \ 1.2, \ d = 250 \text{ mm}]$$

CHAPTER 11

1. Use dimensional analysis to establish the dimensionless Π groups which are relevant to flow over a weir. Which group or groups are the most important for model analysis? Tests have been performed on a 1 : 50 scale model of a weir. The model discharge is 3.5 l/s. What is the corresponding discharge for the full scale weir?

$$[62 \text{ m}^3/\text{s}]$$

2. Show that the resistance F_R to the descent of a spherical particle through a liquid may be expressed as

$$F_R = \rho u^2 r^2 f\left(\frac{\rho u r}{\mu}\right)$$

where r is the radius of the sphere, u is the velocity of descent, and the other terms have their usual meaning. Hence prove that if F_R is proportional to u, then

$$F_R = K\mu r u$$

If $K = 6\pi$, $r = 0.01$ mm, $\rho = 1000$ kg/m^3 and $\mu = 1.1 \times 10^{-3}$ kg/m s, estimate the time taken for a particle to fall through a distance of 3 m. The density of the particle is 2500 kg/m^3.

[1.1 h]

3. A radial flow pump has been tested and the following results obtained:

Speed = 2950 rev/min;
discharge = 50 l/s;
head = 75 m;
efficiency = 75%

The pump impeller is 350 mm in diameter. Determine the size and speed of a dynamically similar pump for the following duty: discharge = 450 l/s, head = 117 m.

[940 mm, 1375 rev/min]

4. A new dock is to be constructed in an estuary. The estuary is 1.5 km wide at the seaward end and 7 km in length. The depth of water at the dock site ranges from 40 m at low tide to 50 m at high tide. The laboratory area is 20 m wide and 60 m long. Give principal model scales and the maximum flow.

[For $\lambda_x = 1/250$, tidal period = 45.5 min, $Q'' = 4$ l/s]

5. A flood alleviation scheme is required for a stretch of river 5 km long. The river is 60 m wide, and the average depth is 7.5 m, for the passage of the probable maximum flood discharge of 1700 m^3/s. The laboratory for the previous question is to be used, but the model length must be accommodated across the 20 m width. The maximum laboratory discharge is 20 l/s. Estimate the model scales.

[$\lambda_x = 1/300$, $\lambda_y = 1/50$, $Q'' = 16$ l/s]

APPENDIX A

The matrix approach to dimensional analysis

The indicial method of dimensional analysis has been discussed in Section 11.4, and the procedure applied in Example 11.1. A set of homogeneous linear equations involving unknown powers or indices were produced. The equations may be expressed in the form of a matrix, and it will be shown below that the dimensionless groups may be obtained in a methodical way by means of a standard matrix technique. To focus ideas, the indicial equations for an arbitrary physical problem will be formed and then expressed as a matrix, following Langhaar (1980).

Suppose that an engineering problem is thought to depend on seven variables (P, Q, R, S, T, U, V, say). Let these be assigned dimensions as follows:

$$\left.\begin{array}{ll} P & \stackrel{\mathrm{D}}{=} \quad M^2 L^1 T^0 \\ Q & \stackrel{\mathrm{D}}{=} \quad M^{-1} L^0 T^1 \\ R & \stackrel{\mathrm{D}}{=} \quad M^3 L^{-1} T^0 \\ S & \stackrel{\mathrm{D}}{=} \quad M^0 L^0 T^3 \\ T & \stackrel{\mathrm{D}}{=} \quad M^0 L^2 T^1 \\ U & \stackrel{\mathrm{D}}{=} \quad M^{-2} L^1 T^{-1} \\ V & \stackrel{\mathrm{D}}{=} \quad M^1 L^2 T^2 \end{array}\right\} \qquad (A.1)$$

Note that $\stackrel{\mathrm{D}}{=}$ means 'has the dimensions'.

A dimensionless Π group incorporating these variables will be in the form

$$\Pi \stackrel{\mathrm{D}}{=} P^{k_1} Q^{k_2} R^{k_3} S^{k_4} T^{k_5} U^{k_6} V^{k_7} \qquad (A.2)$$

where $k_1, \ldots, k_7$ are the unknown powers. Substituting from (A.1) into (A.2)

$$\Pi \stackrel{\mathrm{D}}{=} [M^2 L^1 T^0]^{k_1} [M^{-1} L^0 T^1]^{k_2} [M^3 L^{-1} T^0]^{k_3} [M^0 L^0 T^3]^{k_4} [M^0 L^2 T^1]^{k_5}$$
$$[M^{-2} L^1 T^{-1}]^{k_6} [M^1 L^2 T^2]^{k_7} \qquad (A.3)$$

But Π is dimensionless, i.e.

$$\Pi \stackrel{\mathrm{D}}{=} M^0 L^0 T^0 \qquad (A.4)$$

By combining powers of M, L, T in (A. 3) and (A. 4), three equations may be obtained which express the indices or powers for each fundamental dimension:

$$\left.\begin{array}{ll}(M) & 2\,k_1 - 1\,k_2 + 3\,k_3 + 0\,k_4 + 0\,k_5 - 2\,k_6 + 1\,k_7 = 0 \\ (L) & 1\,k_1 + 0\,k_2 - 1\,k_3 + 0\,k_4 + 2\,k_5 + 1\,k_6 + 2\,k_7 = 0 \\ (T) & 0\,k_1 + 1\,k_2 + 0\,k_3 + 3\,k_4 + 1\,k_5 - 1\,k_6 + 2\,k_7 = 0\end{array}\right\} \quad \text{(A.5)}$$

Equations (A.5) are the indicial equations and may be written in matrix form:

$$\begin{array}{c} \\ M \\ L \\ T \end{array} \begin{array}{ccccccc} P & Q & R & S & T & U & V \\ \left[\begin{array}{ccc|cccc} 2 & -1 & 3 & 0 & 0 & -2 & 1 \\ 1 & 0 & -1 & 0 & 2 & 1 & 2 \\ 0 & 1 & 0 & 3 & 1 & -1 & 2 \end{array}\right] \\ \underbrace{}_{\mathbf{a}} \quad\quad \underbrace{}_{\mathbf{b}} \end{array} \begin{bmatrix} k_1 \\ k_2 \\ k_3 \\ k_4 \\ k_5 \\ k_6 \\ k_7 \end{bmatrix} = \begin{bmatrix} 0 \\ 0 \\ 0 \end{bmatrix} \quad \text{(A.6a)}$$

or

$$\mathbf{M_D k} = \mathbf{0} \quad \text{(A.6b)}$$

where $\mathbf{M_D}$ is the dimensional matrix and $\mathbf{k}$ is the vector of unknown powers.

Equation (A.6a) is a set of rectangular homogeneous linear equations and, as such, has no unique solution since the values of k are not independent. However, it is possible to express some of the powers in terms of the others. Equation (A.6b) is manipulated by an appropriate partitioning of the matrix. $\mathbf{M_D}$ is partitioned into two submatrices, $\mathbf{a}$ and $\mathbf{b}$. It is essential that $\mathbf{a}$ be square and non-singular; that is, it must be invertible. An objection may be raised to the effect that it is not known in advance whether $\mathbf{a}$ is non-singular. However, as the manipulation proceeds, this will become apparent and appropriate steps may then be taken.

The column vector $\mathbf{k}$ is partitioned into an upper portion $\mathbf{k_U}$ and a lower portion $\mathbf{k_L}$, with $\mathbf{k_U}$ conformable for multiplication with $\mathbf{a}$. Equation (A.6a) then becomes

$$\mathbf{a k_U} + \mathbf{b k_L} = \mathbf{0} \quad \text{(A.7)}$$

Therefore $\mathbf{k_U} = -\mathbf{a}^{-1}\mathbf{b k_L}$, since $\mathbf{a}$ is assumed non-singular.

Also, $\mathbf{k_L} = \mathbf{I_4 k_L}$. This is an identity in which $\mathbf{I_4}$ is a 4×4 unit matrix. In general, $\mathbf{I}$ would be an '$m \times m$' unit matrix with $m = n - r$, where r is the rank of the indicial equation (Equation (A.6b)).

Regrouping,

$$\left[\begin{array}{c} \mathbf{k_U} \\ \hline \mathbf{k_L} \end{array}\right] = \left[\begin{array}{c} -\mathbf{a}^{-1}\mathbf{b} \\ \hline \mathbf{I_4} \end{array}\right] \mathbf{k_L} \quad \text{(A.8a)}$$

or

$$\mathbf{k} = \mathbf{M_k k_L} \quad \text{(A.8b)}$$

Equations (A.8) express $\mathbf{k}_U$ in terms of $\mathbf{k}_L$ through the operation of the matrix $\mathbf{M}_k$ which is the indicial (or 'powers') matrix. A version of $\mathbf{M}_k$ ($\mathbf{M}_k^*$) yields the dimensionless Π groups. Providing that $\mathbf{a}$ is non-singular, it may be inverted by any of the standard techniques. However, the Gauss–Jordan (G–J) approach is particularly appropriate since it automatically determines whether $\mathbf{a}$ is non-singular or not, and also provides for the necessary adjustments to be made to $\mathbf{a}$ if it is not. The procedure is now outlined.

Starting with the dimensional matrix $\mathbf{M}_D = [\mathbf{a} \vdots \mathbf{b}]$, the G–J technique is applied. The technique consists of a set of row manipulations which transforms $\mathbf{a}$ to $\mathbf{I}_3$, which is a 3×3 unit matrix. Simultaneously $\mathbf{b}$ is transformed to $\mathbf{b}^*$. The transformed matrix may therefore be expressed as

$$\mathbf{M}_D^* = [\mathbf{I}_3 \vdots \mathbf{b}^*]$$

Therefore, from (A.8a), $\mathbf{k} = \mathbf{M}_k^* \mathbf{k}_L$, where

$$\mathbf{M}_k^* = \left[\begin{array}{c} -\mathbf{I}_3 \mathbf{b}^* \\ \hline \mathbf{I}_4 \end{array}\right] \tag{A.9}$$

If the procedure is to be undertaken manually, the simplest method is to use matrix manipulation using a series of transformation matrices $\mathbf{T}$ in a Falk's array as shown below.

The $\mathbf{T}$ matrices for the row manipulations for Equation (A.6) are as follows:

$$\begin{array}{ccccccc} P & Q & R & S & T & U & V \end{array}$$

$$\left[\begin{array}{ccc|cccc} ② & -1 & 3 & 0 & 0 & -2 & 1 \\ 1 & 0 & -1 & 0 & 2 & 1 & 2 \\ 0 & 1 & 0 & 3 & 1 & -1 & 2 \end{array}\right]$$

$$\underbrace{\hphantom{aaaa}}_{\mathbf{a}} \qquad \underbrace{\hphantom{aaaa}}_{\mathbf{b}}$$

$$\begin{bmatrix} ½ & 0 & 0 \\ -½ & 1 & 0 \\ 0 & 0 & 1 \end{bmatrix} \left[\begin{array}{ccc|cccc} 1 & -½ & \tfrac{3}{2} & 0 & 0 & -1 & ½ \\ 0 & ½ & -\tfrac{5}{2} & 0 & 2 & 2 & \tfrac{3}{2} \\ 0 & 1 & 0 & 3 & 1 & -1 & 2 \end{array}\right]$$

$$\mathbf{T}_1$$

$$\begin{bmatrix} 1 & 1 & 0 \\ 0 & ② & 0 \\ 0 & -2 & 1 \end{bmatrix} \left[\begin{array}{ccc|cccc} 1 & 0 & -1 & 0 & 2 & 1 & 2 \\ 0 & 1 & -5 & 0 & 4 & 4 & 3 \\ 0 & 0 & ⑤ & 3 & -3 & -5 & 1 \end{array}\right]$$

$$\mathbf{T}_2$$

$$\begin{bmatrix} 1 & 0 & \tfrac{1}{5} \\ 0 & 1 & 1 \\ 0 & 0 & ① \end{bmatrix} \left[\begin{array}{ccc|cccc} 1 & 0 & 0 & \tfrac{3}{5} & \tfrac{7}{5} & 0 & \tfrac{9}{5} \\ 0 & 1 & 0 & 3 & 1 & -1 & 2 \\ 0 & 0 & 1 & \tfrac{3}{5} & -\tfrac{3}{5} & -1 & -\tfrac{1}{5} \end{array}\right]$$

$$\mathbf{T}_3 \qquad\qquad \underbrace{\mathbf{I}_3 \qquad\qquad\qquad \mathbf{b}^*}$$

$$\underbrace{\hphantom{aaaaaaaaaaaaaaaaaaaaaaaa}}_{\mathbf{M}_D^*}$$

Therefore,

$$\mathbf{M_k^*} = \begin{bmatrix} -\mathbf{b^*} \\ \mathbf{I_4} \end{bmatrix} \quad \text{from (A.9)}$$

and

$$\mathbf{k} = \begin{bmatrix} -\mathbf{b^*} \\ \mathbf{I_4} \end{bmatrix} \mathbf{k_L} \quad \text{from (A.8).}$$

Thus

$$\underbrace{\begin{bmatrix} k_1 \\ k_2 \\ k_3 \\ k_4 \\ k_5 \\ k_6 \\ k_7 \end{bmatrix}}_{\mathbf{k}} = \underbrace{\begin{bmatrix} -\frac{3}{5} & -\frac{7}{5} & 0 & -\frac{9}{5} \\ -3 & -1 & 1 & -2 \\ -\frac{3}{5} & \frac{3}{5} & 1 & \frac{1}{5} \\ 1 & 0 & 0 & 0 \\ 0 & 1 & 0 & 0 \\ 0 & 0 & 1 & 0 \\ 0 & 0 & 0 & 1 \end{bmatrix}}_{\mathbf{M_k^*}} \underbrace{\begin{bmatrix} k_4 \\ k_5 \\ k_6 \\ k_7 \end{bmatrix}}_{\mathbf{k_L}} \qquad (A.10)$$

Arbitrarily set

$$\begin{bmatrix} k_4 = 1 \\ k_5 = 0 \\ k_6 = 0 \\ k_7 = 0 \end{bmatrix} \quad \text{giving} \quad \begin{bmatrix} k_1 = -\frac{3}{5} \\ k_2 = -3 \\ k_3 = -\frac{3}{5} \\ k_4 = 1 \\ k_5 = 0 \\ k_6 = 0 \\ k_7 = 0 \end{bmatrix}$$

Then arbitrarily set

$$\begin{bmatrix} k_4 = 0 \\ k_5 = 1 \\ k_6 = 0 \\ k_7 = 0 \end{bmatrix} \quad \text{giving} \quad \begin{bmatrix} k_1 = -\frac{7}{5} \\ k_2 = -1 \\ k_3 = \frac{3}{5} \\ k_4 = 0 \\ k_5 = 1 \\ k_6 = 0 \\ k_7 = 0 \end{bmatrix}$$

To complete the procedure for k_6 and k_7:

$$\begin{bmatrix} k_4 = 0 \\ k_5 = 0 \\ k_6 = 1 \\ k_7 = 0 \end{bmatrix} \quad \text{giving} \quad \begin{bmatrix} k_1 = 0 \\ k_2 = 1 \\ k_3 = 1 \\ k_4 = 0 \\ k_5 = 0 \\ k_6 = 1 \\ k_7 = 0 \end{bmatrix} \quad \text{and} \quad \begin{bmatrix} k_4 = 0 \\ k_5 = 0 \\ k_6 = 0 \\ k_7 = 1 \end{bmatrix} \quad \text{giving} \quad \begin{bmatrix} k_1 = -\frac{9}{5} \\ k_2 = -2 \\ k_3 = \frac{1}{5} \\ k_4 = 0 \\ k_5 = 0 \\ k_6 = 0 \\ k_7 = 1 \end{bmatrix}$$

Substitution into (A.2) yields

$$\Pi_1 = P^{-\frac{3}{5}}Q^{-3}R^{-\frac{3}{5}}S^1$$
$$\Pi_2 = P^{-\frac{1}{5}}Q^{-1}R^{\frac{3}{5}}T^1$$
$$\Pi_3 = Q^1R^1U^1$$
$$\Pi_4 = P^{-\frac{9}{5}}Q^{-2}R^{\frac{1}{5}}V^1$$

Thus, the indices of the Π groups are simply the columns of $\mathbf{M}_k^*$. The accuracy of the calculation may be checked by substitution into (A.6), i.e.

$$\mathbf{M}_D\mathbf{k} = \mathbf{0}$$

Therefore

$$\mathbf{M}_D\mathbf{M}_k\mathbf{k}_L = \mathbf{0}$$

or

$$\mathbf{M}_D\mathbf{M}_k^* = \mathbf{0}$$

The whole operation can be neatly set out in the form of a 'tableau', and Example 11.1 will now be treated in this way:

$$\begin{array}{c} \\ M \\ L \\ T \end{array} \begin{array}{cccccc} \rho & U & L & F & \mu \\ \begin{bmatrix} 1 & 0 & 0 & 1 & 1 \\ -3 & 1 & 1 & 1 & -1 \\ 0 & -1 & 0 & -2 & -1 \end{bmatrix} \end{array} = \mathbf{M}_D$$

$$\mathbf{T}_1 = \begin{bmatrix} 1 & 0 & 0 \\ 3 & 1 & 0 \\ 0 & 0 & 1 \end{bmatrix} \quad \begin{bmatrix} 1 & 0 & 0 & 1 & 1 \\ 0 & 1 & 1 & 4 & 2 \\ 0 & -1 & 0 & -2 & -1 \end{bmatrix}$$

$$\mathbf{T}_2 = \begin{bmatrix} 1 & 0 & 0 \\ 0 & 1 & 0 \\ 0 & 1 & 1 \end{bmatrix} \quad \begin{bmatrix} 1 & 0 & 0 & 1 & 1 \\ 0 & 1 & 1 & 4 & 2 \\ 0 & 0 & 1 & 2 & 1 \end{bmatrix}$$

$$\mathbf{T}_3 = \begin{bmatrix} 1 & 0 & 0 \\ 0 & 1 & -1 \\ 0 & 0 & 1 \end{bmatrix} \quad \begin{bmatrix} 1 & 0 & 0 & 1 & 1 \\ 0 & 1 & 0 & 2 & 1 \\ 0 & 0 & 1 & 2 & 1 \end{bmatrix} = \mathbf{M}_D^*$$

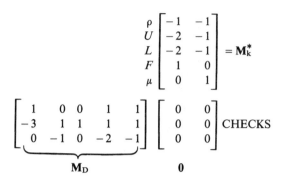

Thus, $\Pi_1 = F/\rho U^2 L^2$ and $\Pi_2 = \mu/\rho UL$, as before.

There is one question which remains, and which was alluded to at the outset: the non-singularity of **a**. It is not possible to know in advance of the calculation whether **a** will turn out to be non-singular. A small extension to the procedure will produce a non-singular **a** if it is attainable with the chosen variables. Returning to the arbitrary engineering problem which was solved above, it will be noticed that the variables P, Q and R occur in all of the Π groups (in Π_3 P is raised to the power of zero, and was therefore not actually written explicitly). Each Π group is completed by the presence of one of the other variables. These relationships are a direct result of the fact that P, Q and R formed the first three columns of $\mathbf{M_D}$. Thus, the variables which are placed in the first three columns automatically become the 'governing variables'. In real physical problems, the governing variables will sometimes be known in advance. Indeed, Barr (1982) has, in effect, pointed out that by careful selection of the governing variables $\mathbf{M_D}$ may be laid out in a form (row–echelon form) which minimises the amount of G–J manipulation.

Where the governing variables are not known, the columns of $\mathbf{M_D}$ could be set out in any order. Thus, for example, we could have written

$$
\begin{array}{c}
 \quad T \quad S \quad\; R \quad P \quad\;\; Q \quad\;\; U \quad V \\
\begin{array}{c} M \\ L \\ T \end{array}
\left[
\begin{array}{ccccccc}
0 & 0 & 3 & 2 & -1 & -2 & 1 \\
2 & 0 & -1 & 1 & 0 & 1 & 2 \\
1 & 3 & 0 & 0 & 1 & -1 & 2
\end{array}
\right]
\end{array}
$$

However, the row manipulation becomes impossible since the first column of $\mathbf{T_1}$ (the first transformation matrix) requires division by zero. To be able to continue with the calculation, columns T and R are interchanged. However, the following transformation matrix produces a further case of division by zero, so a further exchange of columns is required. The order of the governing variables is now R, T, S, and the resultant **a** is non-singular, as required. The advantage of the G–J procedure advanced in this Appendix should by now be self-evident. The column interchanges are now tabulated so that the reader can follow:

1st interchange

$$
\begin{array}{ccccccc}
R & S & T & P & Q & U & V
\end{array}
$$

$$
\begin{bmatrix}
3 & 0 & 0 & 2 & -1 & -2 & 1 \\
-1 & 0 & 2 & 1 & 0 & 1 & 2 \\
0 & 3 & 1 & 0 & 1 & -1 & 2
\end{bmatrix}
$$

$$
\begin{bmatrix}
\frac{1}{3} & 0 & 0 \\
\frac{1}{3} & 1 & 0 \\
0 & 0 & 1
\end{bmatrix}
\begin{bmatrix}
1 & 0 & 0 & \frac{2}{3} & -\frac{1}{3} & -\frac{2}{3} & \frac{1}{3} \\
0 & 0 & 2 & \frac{5}{3} & -\frac{1}{3} & \frac{1}{3} & \frac{7}{3} \\
0 & 3 & 1 & 0 & 1 & -1 & 2
\end{bmatrix}
$$

2nd interchange

$$
\begin{array}{ccccccc}
R & T & S & P & Q & U & V
\end{array}
$$

$$
\begin{bmatrix}
1 & 0 & 0 & \bigm| & \frac{2}{3} & -\frac{1}{3} & -\frac{2}{3} & \frac{1}{3} \\
0 & 2 & 0 & \bigm| & \frac{5}{3} & -\frac{1}{3} & \frac{1}{3} & \frac{7}{3} \\
0 & 1 & 3 & \bigm| & 0 & 1 & -1 & 2
\end{bmatrix}
$$

$$
\begin{bmatrix}
1 & 0 & 0 \\
0 & \frac{1}{2} & 0 \\
0 & -\frac{1}{2} & 1
\end{bmatrix}
\begin{bmatrix}
1 & 0 & 0 & \bigm| & \frac{2}{3} & -\frac{1}{3} & -\frac{2}{3} & \frac{1}{3} \\
0 & 1 & 0 & \bigm| & \frac{5}{6} & -\frac{1}{6} & \frac{1}{6} & \frac{7}{6} \\
0 & 0 & 3 & \bigm| & -\frac{5}{6} & \frac{7}{6} & -\frac{7}{6} & \frac{5}{6}
\end{bmatrix}
$$

$$
\begin{bmatrix}
1 & 0 & 0 \\
0 & 1 & 0 \\
0 & 0 & \frac{1}{3}
\end{bmatrix}
\begin{bmatrix}
1 & 0 & 0 & \bigm| & \frac{2}{3} & -\frac{1}{3} & -\frac{2}{3} & \frac{1}{3} \\
0 & 1 & 0 & \bigm| & \frac{5}{6} & -\frac{1}{6} & \frac{1}{6} & \frac{7}{6} \\
0 & 0 & 1 & \bigm| & -\frac{5}{18} & \frac{7}{18} & -\frac{7}{18} & \frac{5}{18}
\end{bmatrix}
$$

Some combinations of variables for M_D may conceivably produce a situation in which all possible interchanges had been tried, and no satisfactory solution achieved. In such a case, dependent indicial equations have been detected, i.e. the physical variables selected do not incorporate the correct 'amounts' of M, L and T to describe the problem. In terms of matrix methods, this is equivalent to saying that the column rank of M_D is less than the number of fundamental variables. The G–J procedure determines the row-rank (i.e. the number of equations for which the row transformations are possible) of M_D. Matrix theory states that row-rank equals column-rank. The procedure therefore determines automatically whether the selected physical variables are independent. It also inherently embraces Buckingham's proposals (Section 11.5):

(1) the statement that 'the rank of M_D must equal the number of fundamental variables' combines proposals (a) and (c);
(2) the form of M_k^* effectively embraces proposals (b) and (d);
(3) proposal (e) is necessarily embodied in the indicial formation of the original dimensional equation.

Computer application

In principle, the G–J manipulation is not too tedious to be carried out manually. However, there is clearly scope for arithmetical mistakes. Also, for some cases, it may be desirable to undertake a series of trials with different governing variables, which can become tedious. For these reasons, a computer program is a useful adjunct to model or similarity studies.

A program (in BASIC) to carry out the matrix manipulation is given in Appendix B. Data are entered into the DATA statements at the end of the program. The data comprise:

the size (number of rows × number of columns) of M_D;
the names (or symbolic letters) of the physical variables;
the numerical terms of M_D.

G–J elimination is then carried out. If zeros are encountered on the leading diagonal of **a**, then columns are interchanged. The output is then printed in the following order:

the matrix M_D^*, together with its rank;
the matrix M_k^*, (the appropriate variable names (or symbols) appear above their respective columns, the columns of M_k^* being the indices of the variables for each Π group).

The check for non-dimensionality is not carried out but can easily be incorporated into the program if the user requires it. As written, the program finally presents the user with the opportunity to change the order of the variables, so that various combinations of governing variables can be tried out.

References

For references, see Chapter 11.

APPENDIX B

Computer programs

All of the programs are in BASIC, numbers 1 to 3 and 7 were written for BBC BASIC, numbers 4 and 5 using Microsoft BASIC and number 6 using Microsoft BASIC with matrix algebra routines. The results of all of these programs are presented in tabular form, but in the authors' experience graphical presentation is often preferable. However, such graph plotting routines have not been included here, as the software tends to be machine-specific. More-sophisticated versions of these and other hydraulics programs for use in teaching and computer-aided learning are available from the authors on request.

Program no.	Brief details
1	Gradually varied flow using the direct step method. Refer to Example 5.7 for typical application
2	Gradually varied flow using the standard step method. Refer to Example 5.9 for typical application
3	Gradually varied unsteady flow using an explicit forward difference scheme. Refer to Example 5.11 for typical application
4	Extreme value analysis of annual maxima events. Refer to Example 10.1 for typical application
5	Convolution of net rainfall and unit hydrograph data using the matrix method. Refer to Example 10.1 for typical application
6	Dimensional analysis using matrix methods. Refer to Appendix A for full details
7	Sediment transport using the Ackers and White method. Refer to Example 9.2

Program 1

```
10 REM GVF USING DIRECT STEP METHOD
20 REM PRINT FORMAT 9 SPACES 4 D.P.
30 @%=&20409
40 DIM DL(100),Y(100)
50 INPUT"Discharge (cumecs)",Q
60 INPUT"Channel width (m)",B
70 INPUT"Manning's n",N
80 INPUT"Number of itervals(max 99)",M
90 INPUT"Slope (in decimals)",S
100 INPUT"SLOPE +OR-",SNG
110 S=S*SNG
120REM NORMAL DEPTH CALC
130 K=ABS(S)^.5/N
140 EYO=(Q/(B*K))^.6
150 YG=EYO:QG=Q
160 REPEAT
170 YG=YG*(Q/QG)^.6667
180 QG=K*(B*YG)^1.6667/(B+2*YG)^.6667
190 UNTIL ABS((QG-Q)/Q)<.001
200 YO=YG
210 REM CRIT DEPTH CALC
220 K2=Q^2/(B^2*9.81)
230 YC=K2^0.3333
240 PRINT"Normal Depth ";YO
250 PRINT   "Critical Depth ";YC
260 PRINT
270 INPUT"Initial depth",Y1
280 INPUT"Final depth",Y2
290 PRINT
300REM DIRECT STEP METHOD
310 PRINT:PRINT
320 DY=(Y2-Y1)/M
330J=1:Y(1)=Y1:X=0
340 PRINT"   y         A         P        Fr      (1-Fr^2)m   Sf      (So-Sf)m
X":PRINT
350REPEAT
360 Y(J+1)=Y(J)+DY
370 FR1=(Q/(B*Y(J)))/(9.81*Y(J))^.5
380 FR2=(Q/(B*Y(J+1)))/(9.81*Y(J+1))^.5
390 SF1=(N*Q/(B*Y(J)))^2*((B+2*Y(J))/(B*Y(J)))^1.333
400 SF2=(N*Q/(B*Y(J+1)))^2*((B+2*Y(J+1))/(B*Y(J+1)))^1.333
410 FRF=((1-FR1^2)+(1-FR2^2))/2
420 I=(SF1+SF2)/2
430DL(J)=DY*FRF/(S-I)
440 PRINT Y(J),B*Y(J),B+2*Y(J),FR1,"          ",SF1,"          ",X
450 PRINT"                    ",FRF,"          ",S-I
460 IF J=M THEN PRINT Y(J+1),B*Y(J+1),B+2*Y(J+1),FR2,"          ",SF2,"
",X+DL(J)
470 X=X+DL(J)
480J=J+1
490 UNTIL J=M+1
500 END
```

Program 2

```
10 REM GVF USING STANDARD STEP METHOD
20 REM PRINT FORMAT 9 SPACES 4 D.P.
30@%=&20409
40 INPUT"INPUT Q,N,SO,B,Y1,DX,L "Q,N,SO,B,Y1,DX,L
50K=(N*Q)^2
60
70PRINT"    Y          A          P        EG       SF      (SO-SF)M DE         EC"
80 FOR X=0 TO L STEP ABS(DX)
 90A1=B*Y1
100P1=B+2*Y1
110E1=Y1+(Q/A1)^2/19.62
120 SF1=K*P1^1.3333/A1^3.3333
130 IF X=0 THEN PRINT Y1,A1,P1,E1,SF1,"                    ",E1
140IF X=L THEN END
150 Y2=Y1+(SO-SF1)*DX
160REPEAT
170A2=B*Y2
180P2=B+2*Y2
190E2G=Y2+(Q/A2)^2/19.62
200SF2=K*P2^1.3333/A2^3.3333
210SFM=SO-(SF1+SF2)/2
220DE=DX*SFM
230E2=E1+DE
240IF (E2-E2G)>.001 THEN Y2=Y2+.001
250IF (E2G-E2)>.001 THEN Y2=Y2-.001
260UNTIL ABS(E2-E2G)<=.001
270PRINT"                                    ",SFM,DE
280PRINT Y2,A2,P2,E2G,SF2,"                    ",E2
290Y1=Y2
300NEXT X
310END
```

Program 3

Note. IMAX = no. of *x* grid points;
        DT   = time interval for computation (s);
        DX   = distance interval (m);
        CMAN = inverse of Manning's *n*;
        HO   = initial depth (m);
        SO   = bed slope (m/m);
        B   = channel width (m);
        DDT   = time interval for inflow hydrograph (s).

```
10REM EXPLICIT SOLUTION OF FLOOD PROPOGATION IN OPEN CHANNELS
20 REM PRINT FORMAT 8 SPACES 4D.P.
30 @%=&20408
40DIM Q2(100),Q1(100),H2(100),H1(100),QQ(100)
50INPUT"INPUT IMAX,DT,DX,CMAN,HO,SO,B,DDT",IMAX,DT,DX,CMAN,HO,SO,B,DDT
60IMAX1=IMAX-1
70 QO=CMAN*(B*HO)^1.667*SQR(SO)/(B+2*HO)^.667
80 PRINT "INITIAL DISCHARGE=";QO
90INPUT"INPUT NO. OF HYDROGRAPH VALUES",N
100PRINT"INPUT VALUES"
110FOR I=1 TO N
120INPUT QQ(I)
130 QQ(I)=QQ(I)+QO
140NEXT
150FOR I=1 TO IMAX
160H2(I)=HO
170H1(I)=HO
180Q2(I)=QO
190Q1(I)=QO
200 NEXT
210T=0:N1=0
220 REPEAT
230T=T+DT:N1=N1+1
240FOR I=1 TO IMAX1
250 H2(I)=H1(I)-(Q1(I+1)-Q1(I))/DX/B*DT
260NEXT
270L=INT(T/DDT+1)
280Q2(1)=QQ(L)+(QQ(L+1)-QQ(L))*(T-(L-1)*DDT)/DDT
290 IF L>=N THEN Q2(1)=QO
300FOR I=2 TO IMAX1
310HH=(H2(I)+H1(I)+H1(I-1)+H2(I-1))/4
320SF=Q1(I)^2/CMAN^2*(B+2*HH)^1.333/(B*HH)^3.333
330HH2=(H2(I)+H1(I))/2
340HH1=(H2(I-1)+H1(I-1))/2
350QQ2=(Q1(I+1)+Q1(I))/2
360QQ1=(Q1(I)+Q1(I-1))/2
370Q2(I)=(QQ2+QQ1)/2-DT*((QQ2^2/(HH2*B)-QQ1^2/(B*HH1))/DX-9.81*DT*HH*B*(HH2-HH
)/DX-9.81*HH*B*(SF-SO)*DT
380NEXT
390 Q2(IMAX)=B*(H2(IMAX1)+H1(IMAX1)-2*HO)/2*SQR(9.81*(H2(IMAX1)+H1(IMAX1))/2)
QO
400FOR I=1 TO IMAX
410Q1(I)=Q2(I)
420NEXT I
430FOR I=1 TO IMAX1
440H1(I)=H2(I)
450 NEXT
460 PRINT TAB(0,0)"T=",T
470 PRINT TAB(0,1)"      Q";"        H"
480FOR I=1 TO IMAX1
490 PRINT Q2(I),H2(I)
500 NEXT
510 PRINT Q2(IMAX)
520 UNTIL (T>=DDT*(N-1) AND Q2(IMAX)<=QO)
530END
```

Program 4

```
10 REM PROGRAM TO FIT AN EV1 TO AN ANNUAL MAXIMA SERIES
20 INPUT"INPUT GENERAL TITLE";T$
30 INPUT"INPUT DATA TITLE";Y$
40 INPUT"INPUT NO. OF YEARS OF DATA";N
50 DIM Y(N),X(N),YR(N)
60 PRINT"INPUT DATA AS : ANNUAL MAXIMA,YEAR"
70 FOR I=1 TO N
80 INPUT Y(I),YR(I)
90 NEXT I
100 REM CALCULATE MEAN,SD,CV
110 GOSUB 3000
120 REM SORT DATA
130 K=0
140 FOR I=1 TO N-1
150 IF(Y(I)>=Y(I+1)) THEN GOTO 230
160 TEMP=Y(I)
170 Y(I)=Y(I+1)
180 Y(I+1)=TEMP
190 TEMP=YR(I)
200 YR(I)=YR(I+1)
210 YR(I+1)=TEMP
220 K=K+1
230 NEXT I
240 IF(K>0) THEN GOTO 130
250 REM  ASSIGN EXCEEDENCE PROBABILITY TO SORTED DATA
260 FOR I=1 TO N
270 X(I)=(I-.44)/(N+.12)
280 NEXT I
290 REM CHECK FOR EQUAL VALUES
300 FOR I=2 TO N
310 IF(Y(I)=Y(I-1)) THEN X(I)=X(I-1)
320 NEXT I
330 REM PRINT SORTED DATA +STATS
340 LPRINT T$ :LPRINT :LPRINT
345 DIM  A$(16),B$(16),C$(16),Z$(16),V$(16),W$(16)
350 Z$="   ANNUAL MAXIMA" :V$="            YEAR" :W$="   RETURN PERIOD"
360 A$="\          \" : B$="£££££££££££££.£" : C$="£££££££££££££££££"
370 LPRINT USING A$;Z$,V$,W$
380 LPRINT USING A$;Y$
390 FOR I=1 TO N
400 LPRINT USING B$;Y(I),
410 LPRINT USING C$;YR(I),:LPRINT USING B$;1/X(I)
420 NEXT I
430 LPRINT :LPRINT
440 LPRINT "MEAN=";MEAN
450 LPRINT "  SD=";SD
460 LPRINT "  CV=";CV
470 REM FIT AN EV1
480 U=MEAN-.45*SD
490 B=.78*SD
495 REM CALCULATE XP FOR VARIOUS TR
500 DATA 100,50,25,10,5,2.33,1.01
510 DIM TR(7),XP(7)
520 FOR I=1 TO 7
530 READ TR(I)
540 XP(I)=U+B*(-LOG(-LOG(1-1/TR(I))))
550 NEXT I
560 REM PRINT PREDICTED VALUES
570 LPRINT :LPRINT
580 LPRINT" PREDICTED VALUES FOR AN EV1 "
585 LPRINT:LPRINT
590 LPRINT USING A$;W$;Y$
600 FOR I=1 TO 7
610 LPRINT USING B$;TR(I);XP(I)
630 NEXT I
640 END
3000 REM MEAN & STANDARD DEVIATION SUBROUTINE
3001 REM DATA IN ARRAY Y(N)
3002 REM CALCULATE MEAN
```

```
3005 S1=0  :S2=0
3006 FOR I=1 TO N
3007 S1=S1+Y(I) : S2=S2+Y(I)^2
3008 NEXT I
3009 MEAN=S1/N
3010 SD=SQR((S2-N*MEAN^2)/(N-1))
3011 CV=SD/MEAN
3012 RETURN
```

Program 5

```
10    REM *********************
20    REM CONVOLUTION BY MATRIX METHOD
30    REM *********************
40    JN=0
50  PRINT"   INPUT JOB TITLE"
60    INPUT JOB$
70    PRINT"INPUT TIME STEP (HRS), NO. OF NET RAINS, NO. OF  UH ORD'S"
80    INPUT DT,N,M
90    DIM P(N),UH(M),PMAT(M+N-1,M),Q(M+N-1)
100   PRINT"INPUT NET RAINS (MM)"
110   FOR I=1 TO N
120   INPUT P(I)
130   NEXT I
140   PRINT"INPUT UH ORDINATES (CU.M PER 10 MM. NET RAIN)"
150   FOR I=1 TO M
160   INPUT UH(I)
170   NEXT I
180   REM ******************
190   REM INFIL PMAT
200   REM ******************
210   FOR K=1 TO N
220   FOR J=1 TO M
230   D=J+K-1
240   PMAT(D,J)=P(K)
250   NEXT J
260   NEXT K
270   REM ******************
280   REM MATRIX MULTIPLICATION
290   REM ******************
300   FOR I=1 TO M+N-1
310   SUM=0
320   FOR J=1 TO M
330   SUM=SUM+PMAT(I,J)*UH(J)/10
340   NEXT J
350   Q(I)=SUM
360   NEXT I
370   REM ******************
380   REM PRINT OUT RESULTS
390   REM ******************
400   A$="££.£  "
410   B$="£££.£"
420   U$="££.££"
430   C$="\    \"
440   X$="     "
450   Z$="!  "
460   Y$="=  "
470   PRINT X$
480  LPRINT JOB$
490  LPRINT X$
500  LPRINT X$
510   LPRINT "   **** UNIT HYDROGRAPH ANALYSIS ****"
520   LPRINT X$
530   LPRINT "  MATRIX EQUATION FORM OF SOLUTION "
540   LPRINT X$
550   FOR I=1 TO M+N-1
560  LPRINT USING Z$;Z$,
570  FOR J=1 TO M
580  LPRINT USING A$;PMAT(I,J),
590  NEXT J
600  LPRINT USING Z$;Z$,Z$,
610  IF I>M THEN LPRINT USING C$;X$, ELSE LPRINT USING U$;UH(I)/10,
620  LPRINT USING Z$;Z$,
630  LPRINT USING Z$;Y$,
640  LPRINT USING Z$;Z$,
650  LPRINT USING B$;Q(I),
660  LPRINT USING Z$;Z$
670  NEXT I
680  LPRINT X$
690  LPRINT X$
```

```
700   LPRINT"   UNIT HYDROGRAPH ORDINATES (CUMECS PER 10 MM. NET RAIN)"
710   FOR I=1 TO M
720   LPRINT USING A$;UH(I),
730   NEXT I
740   LPRINT X$
750   LPRINT X$
760   LPRINT X$
770   LPRINT X$
780   IF UH(1)=0 THEN T=0 ELSE T=DT
790   LPRINT"  TIME   NET RAIN     RUNOFF"
800   LPRINT" (HRS)       (MM)   (CUMECS)"
810   FOR I=1 TO M+N-1
820   LPRINT USING D$;T,
830   IF I<=N THEN LPRINT USING E$;P(I), ELSE LPRINT USING G$;H$,
840   LPRINT USING F$; Q(I)
850   T=T+DT
860   NEXT I
870   END
```

Program 6

```
10      REM DIMENSIONAL ANALYSIS
20      DIM A(10,10),M(10,10),E(10)
30      DIM D(10,10),NORDER(10)
40      DIM VARBL$(10),FUNDL$(6)
50      DIM NVARD$(10),VARD$(10)
60      READ M,N
70      MAT READ FUNDL$(M),VARD$(N)
80      MAT READ D(M,N)
90      MAT A=D
100     MAT VARBL$=VARD$
110     PRINT "DIM. MAT."
120     PRINT
130     PRINT "                    ";
140     FOR I=1 TO N
150     PRINT USING "\      \";VARBL$(I);
160     NEXT I
170     PRINT
180     FOR I=1 TO M
190     PRINT FUNDL$(I),
200     FOR J=1 TO N
210     PRINT USING "££.££  ";A(I,J);
220     NEXT J
230     PRINT
240     PRINT
250     NEXT I
260     REM INITIALISE COLUMN ORDER RECORD
270     MAT E=ZER(N)
280     FOR I=1 TO N
290     LET E(I)=I
300     NEXT I
310     REM GAUSS JORDAN ELIMINATION
320     FOR K=1 TO M
330     LET S=A(K,K)
340     REM EXCH. COLS. IF NECESS.
350     IF S<>0   THEN 510
360     FOR Q=K+1 TO N
370     IF A(K,Q)<>0    THEN 400
380     NEXT Q
390     GOTO 610
400     FOR J=1 TO M
410     LET T=A(J,K)
420     LET A(J,K)=A(J,Q)
430     LET A(J,Q)=T
440     NEXT J
450     REM RECORD COLUMN EXCHANGES (IF ANY)
460     LET T1=E(K)
470     LET E(K)=E(Q)
480     LET E(Q)=T1
490     GOTO 330
500     REM CONTINUE ELIM.
510     FOR J=K TO N
520     LET A(K,J)=A(K,J)/S
530     NEXT J
540     FOR I=1 TO M
550     IF I=K THEN 600
560     LET S=A(I,K)
570     FOR J=K TO N
580     LET A(I,J)=A(I,J)-S*A(K,J)
590     NEXT J
600     NEXT I
610     NEXT K
620     PRINT
630     PRINT
640     PRINT
650     PRINT "DIM. MAT AFTER GAUSS JORDAN"
660     PRINT
670     FOR I=1 TO M
680     FOR J=1 TO N
690     PRINT USING "££.££  ";A(I,J);
```

```
700    NEXT J
710    PRINT
720    PRINT
730    NEXT I
740    REM DETERMINE RANK
750    LET R=0
760    FOR I=1 TO M
770    FOR J=1 TO N
780    IF A(I,J)<>0    THEN 810
790    NEXT J
800    GOTO 820
810    LET R=R+1
820    NEXT I
830    PRINT
840    PRINT "RANK=";R
850    REM INDEX MAT.
860    MAT M=ZER(N,N-R)
870    FOR I=1 TO R
880    FOR J=1 TO N-R
890    LET M(I,J)=-A(I,J+R)
900    NEXT J
910    NEXT I
920    FOR I=R+1 TO N
930    LET M(I,I-R)=1
940    NEXT I
950    PRINT
960    PRINT "INDEX MAT."
970    PRINT
980    PRINT "VARIABLE   ";
990    FOR I=1 TO N-R
1000   PRINT "      GP";I;
1010   NEXT I
1020   PRINT
1030   FOR I=1 TO N
1040   PRINT VARBL$(E(I)),
1050   FOR J=1 TO (N-R)
1060   PRINT USING "££.££      ";M(I,J);
1070   NEXT J
1080   PRINT
1090   PRINT
1100   NEXT I
1110   PRINT
1120   INPUT "DO YOU WISH TO CHANGE THE ORDER OF THE VARIABLES";ANS$
1130   A$=LEFT$(ANS$,1)
1140   IF A$="N" THEN 1390
1150   PRINT
1160   PRINT "ENTER THEM IN THEIR NEW ORDER"
1170   MAT INPUT NVARD$(N)
1180   FOR I=1 TO N
1190   FOR J=1 TO N
1200   IF NVARD$(J)<>VARD$(I) THEN 1230
1210   NORDER(J)=I
1220   GOTO 1240
1230   NEXT J
1240   NEXT I
1250   FOR I=1 TO M
1260   FOR J=1 TO N
1270   A(I,J)=D(I,NORDER(J))
1280   NEXT J
1290   NEXT I
1300   MAT VARBL$=NVARD$
1310   GOTO 110
1320   DATA 4,6
1330   DATA mu,L,T,M
1340   DATA mu,S,eps,e,D,T
1350   DATA 1,0,-1,-.5,0,0
1360   DATA 0,1,-2,.5,1,2
1370   DATA 0,-1,2,0,0,-2
1380   DATA 0,0,0,.5,0,1
1390   END
```

Program 7

```
10REM ACKERS AND WHITE SEDIMENT TRANSPORT METHOD
20REM
30PROCINPUT
40PROCGRAIN
50PROCPARAMETERS
60PROCMOBILITY
70PROCTRANSPORT
80PROCSEDIMENT
90END
100DEF PROCINPUT
110INPUT"surface width (m)",B
120 INPUT"area (sq.m)",A
130INPUT"perimeter (m)",P
140INPUT"bed slope (m/m)",SO
150INPUT"discharge (cumecs)",Q
160INPUT"grain size (D50) (mm)",D
170INPUT"sediment density (kg/m^3)",ROS
175INPUT"kinematic viscosity (m^2/s)",NU
180ENDPROC
190DEF PROCGRAIN
200 DGR=D/1000*((9.81*((ROS/1000-1))/(NU)^2))^(1/3)
210PRINT"Dgr=";DGR
220ENDPROC
230DEFPROCPARAMETERS
235 IF DGR<1 THEN PRINT"Dimensionless grain size too small":END
240IF DGR>60 THEN n=0
250IF DGR>60 THEN m=1.5
260IF DGR>60 THEN AGR=0.17
270IF DGR>60 THEN C=0.025
280IF DGR<=60 THEN n=1-0.56*LOG(DGR)
290IF DGR<=60 THEN m=1.34+9.66/DGR
300IF DGR<=60 THEN AGR=0.14+0.23/(DGR)^0.5
310IF DGR<=60 THEN C1=2.86*LOG(DGR)-(LOG(DGR))^2-3.53
315IF DGR<=60 THEN C=10^C1
320PRINT"n=";n
330PRINT"m=";m
340 PRINT"Agr=";AGR
350PRINT"C=";C
360ENDPROC
370DEF PROCMOBILITY
380VS=(9.81*(A/P)*SO)^.5
390V=Q/A
395 DM=A/B
400FGR=VS^n/((9.81*(D/1000)*(ROS/1000-1)))^0.5*(V/((32)^0.5*LOG(10*DM/(D/
1000))))^(1-n)
410PRINT"shear velocity =";VS
420PRINT"mean velocity =";V
430PRINT"hydraulic mean depth =";DM
440PRINT"mobility No. (Fgr) =";FGR
450ENDPROC
460DEFPROCTRANSPORT
465 IF FGR<AGR THEN PRINT"No sediment transport - threshold of motion not
exceeded":END
470GGR=C*(FGR/AGR-1)^m
480PRINT"transport parameter (Ggr) =";GGR
490ENDPROC
500DEFPROCSEDIMENT
510X=GGR*ROS/1000*D/1000/(DM*(VS/V)^n)
520QR=GGR/(DM/(D/1000)*(VS/V)^n)
530QS=QR*Q
540PRINT"dimensionless mass ratio (X) =";X
550PRINT"dimensionless volume ratio (QR) =";QR
560PRINT"sediment transport rate (cumecs) =";QS
570ENDPROC
```

APPENDIX C

Moments of area

One very common phenomenon in engineering is that of a pressure or stress which is distributed continuously, but not uniformly, over a surface. Typical examples are found in hydrostatics and in the bending of beams or columns. The intensity of pressure p (or stress σ) is often linearly distributed in one direction (Figs C.1a and b). That is to say, p (or σ) is proportional to distance (y, say) from a specified axis, so p (or σ) $= Ky$.

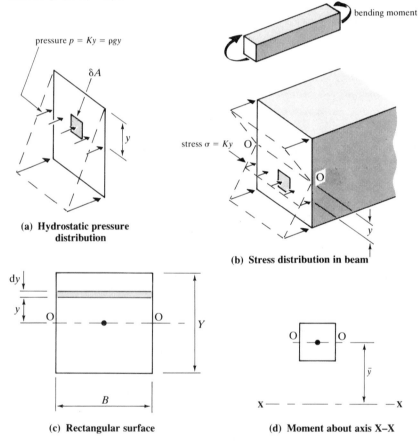

(a) Hydrostatic pressure distribution

(b) Stress distribution in beam

(c) Rectangular surface

(d) Moment about axis X–X

Figure C.1 Definition diagrams for moments of area.

The hydrostatic force, δF, acting on a small element of area is given by

$$\delta F = p\delta A = Ky\delta A$$

The force on the whole area is obtained by integration:

$$F = \int Ky\,\mathrm{d}A = K \times (\text{first moment of area}) = KA\bar{y}$$

where $\bar{y}$ is the distance from surface of liquid to the centroid.

The moment of the force about the axis O–O is obtained by taking the product of force and distance from O–O. For the element,

$$\delta M = \delta Fy = p\delta A\,y = Ky^2\delta A$$

Hence, the moment taken over the whole area is

$$M = \int Ky^2\,\mathrm{d}A = K \times (\text{second moment of area})$$

The second moment of area is denoted by I.

Both the first and second moments of area are simply functions of the geometry of the surface and position of the axis relative to the centroid. If the axis passes through the centroid, then the second moment has a particular value. I_0, for any given shape. Thus, for a rectangle (Fig. C.1c)

$$\text{area of element} = \delta A = B\,\delta y$$

$$\left.\begin{array}{l}\text{second moment of area of element} \\ \text{about axis through centroid}\end{array}\right\} = \delta A\,y^2 = B\,\delta yy^2$$

for whole area,

$$I_0 = \int By^2\,\mathrm{d}y = B\left[\frac{y^3}{3}\right]_{-Y/2}^{+Y/2}$$

$$= \frac{BY^3}{12} = BY\left(\frac{Y^2}{12}\right)$$

$Y^2/12$ may be regarded as the square of the length of a lever arm.

If the second moment of area about any other axis is required, then the above equations can be modified appropriately. For many common cases the arbitrary axis is parallel to O–O. Thus, for a rectangle (Fig. C.1d), the second moment of area of an element about axis X–X is given by

$$B\,\delta y(y+\bar{y})^2 = B\,\delta y\,y^2 + 2B\,\delta yy\bar{y} + B\,\delta y\,\bar{y}^2$$

Therefore

$$I_{X-X} = \int_{-Y/2}^{+Y/2} B y^2 \, dy + \int_{-Y/2}^{+Y/2} 2 B y \bar{y} \, dy + \int_{-Y/2}^{+Y/2} B \bar{y}^2 \, dy$$

The second term $\rightarrow 0$, so

$$I_{X-X} = I_0 + A \bar{y}^2$$

where $A = BY$.

This is a statement of the Parallel Axes Theorem.

HYDRAULICS IN CIVIL ENGINEERING

Computer Aided Learning Software

- Written by the authors specifically for the readers
- For use with a BBC B microcomputer (with disc drive)
- Self explanatory, interactive and user friendly
- Extensive use of colour graphics
- Pre-tested by undergraduate students
- Comprehensive error handling for trouble-free use
- Educational and fun to use

Two menu-driven packages available:

HYDRA   – Pipe flow
               – Channel flow
               – Rapidly varied flow
               – Gradually varied flow

HYDROL   – Flood frequency analysis
               – Unit hydrograph convolution
               – FSR design floods

Order Form:

Name ....

Address ....

        ....

        ....

        ....

Please send me:

...... copies of HYDRA at £10 (+ £1.50 p&p): cost ....

.... copies of HYDROL at £10 (+ £1.50 p&p): cost ....

                                     Total ....

Please enclose a cheque/PO made payable to A. J. Chadwick and send your order to:

A. J. Chadwick
Brighton Polytechnic
Department of Civil Engineering
Brighton BN2 4GJ
England

Index

Ackers, P. 262, 375, 377, 379, 381, 383, 388
Ackers & White formula 262–3, 265–6, 267
aeration 158, 394, 395
air chambers 367–8
air regulation 400
air valves 342
Anderson, A. G. 430
Archimedes 14
armouring 267, 422, 432
attenuation of flood wave 295

backwater profile 162, 165, 168
Bagnold, R. A. 260–2, 264
Bakhmeteff, B. A. 134, 161
Balmforth & Sarginson method 415
bankfull discharge 130, 429
Barr, D. I. H. 101, 107, 317, 348
baseflow 293
beaches 435–6
  replenishment of 441–2
Beaufort 229
bed load 247, 255, 259–61
  efficiency 260
  function 258
Bernoulli, D. xxi
  equation 27, 31, 34, 36, 39, 44, 54, 63, 66, 132, 340, 370, 395
Blasius, H. 98, 100
bores 170
boundary layer 74–83, 92, 386–8
  description of 74–5
  equations 76–81
Bourdon gauge 8
braiding 429–31
breaker line 233
breaking wave 233, 240
breakwater 442–3
bridge pier 158
broad crested weir *see* weirs, broad crested
Buckingham Pi Theory 312, 316–17
bulk modulus xxvi, 187

buoyancy 14
  centre of 14–16, 18, 19
  force 16
  force on quadrant gate 12
by-passing 366

Cain, P. 395
calibration of hydraulic model 329
catchment
  characteristics 271–3, 288, 294
  rural 273
  urban 303–6
cavitation 86, 186
  in hydraulic machines 223–4, 355, 366
  in sheared flows 393, 394
  in surge analysis 207
  on spillways 393–5, 402
Chang, H. H. 430
channel
  artificial 120, 419–21
  bank, stability of 425–8
  compound 130
  control points 158
  controls 143
  current patterns 421–2
  lined 421
  natural 120, 153
  section optimisation 419–21
  unlined 421, 422
channel flow, estimation of 128, 129, 130
channel routing 295, 299–301
characteristics, equations 198–204, 209
  for change in pipe geometry or material 202–4
  for dead end 205
  for reservoir 206
  for valve 205
characteristics, method of 193, 197–209
Charlton & Benson formula 430
Chézy formula 125, 251, 328, 425
choke 158
Chow, Ven te 161, 414

circulation 56
Clapotis Gauffré 238
closure of valve *see* valve closure
coast protection 439
Colebrook–White formula 88, 100–1, 104,
    107, 111, 112–13, 124, 348, 352
composite profile in channel flow 166
compressibility xxvi, 62, 183–6, 189
conservation of energy 23, 25
conservation of matter (mass) 23
conservation of momentum 23, 24, 29
continuity 24, 25, 32, 109
contracted weir 370, 374, 375
contraction
    coefficient of 44
    of nappe 373
    of spillway flow 392–4
control gates
    drum 407
    overflow 407
    radial 407
    roller 407
    sluice 410
    Tainter 407
    underflow 407
    vertical 407
control point 158, 166, 168
control volume 24
corresponding velocity 320
crenulated bays 438
critical bed slope 151–2, 153, 158
critical depth 139, 140, 151, 153–4, 157,
    166
    line 139, 154
    meter 148–51
critical flow in channels 137
critical flow, general equation of 137–9
critical velocity 139
Crump weir 381–2

Darbyshire & Draper 242
Darcy, H. xxii, 90, 97
Darcy–Weisbach equation 97, 107, 113,
    124, 197, 348, 349, 352
deflector bucket 405
design drought 269
design flood 269
diffraction of wave *see* wave diffraction
dimensional analysis 312
    Buckingham Pi Method 312, 316–17
    indicial method 312–16
    matrix method 316, 317, 318, 463–70
dimensional homogeneity xxviii, 311, 313
dimensionless groups 315, 318, 320
direct integration solution for gradually
    varied flow 161
direct step method 162

discharge
    coefficient 43
    measurement in pipes 41, 42
    measurement in channels 148–51, 379–91
displacement thickness 76
distortion in hydraulic models 326, 329,
    331
distribution pipe systems 346–53
dominant discharge 429
dominant fraction of sediment 267
drag 83–4, 252
    coefficient of 83
drawdown curve 165
drum gate *see* control gates
Du Boys, M. P. 252, 255, 425

effective head 375
effective width 375
Einstein, H. A. 256–9, 267, 430
energy
    coefficient of 30, 122
    dissipation in hydraulic jump 146–7
    dissipators 403–7
    line 340
    losses in pumps 215–16
    transfer in pumps 213–14
energy formulae for sediment transport
    260–6
entrainment function 249–50
estuaries, models of 332
Euler equation 29
extreme value distributions
    Type 1 277–8, 289, 294
    Type 2 278
    Type 3 278

Fleming, G. 271
flood
    design 285, 288, 302
    hydrograph 280
    prediction, rural catchment 270–1, 288
    prediction, urban catchment 303
    probable maximum 297, 302
    protection 432
    routing (*see also* channel routing and
        reservoir routing) 294
Flood Studies Report 270, 277, 278, 285,
    288, 293, 301, 302, 303
flotation, stability of 14
flow, classification in channels 156–8
flow net 22, 49
flow profiles in channels 156–61
flow separation 82
flow transition 136, 154–5
fluid, definition of xxv
flume 158, 384
    British Standard for 388
    design 386–91

Parshall 385
Venturi 131, 150–1, 384
Fox, J. A. 210
frequency analysis 273–80, 288
friction
  coefficient of 85
  factor 97
  velocity 73
Froude number 131, 141–3, 146, 155, 248, 321, 330, 332, 333, 401, 404, 424
  densimetric 332, 335
fully arisen sea 240, 243

Ganguillet 126
general extreme value distribution (*see also* extreme value distributions) 277, 278, 294
governing variables 313, 316, 318, 319, 468
gradually varied flow 151–69
  general equation of 155–6
  graphical solution for 160, 161
  numerical solution for 161, 162
  unsteady 173–8
Gringorten formula 276, 289
groundwater 280
group celerity of waves 229
groynes 440–1

Hagen, G. 90
Hagen–Poiseuille equation 95
Hamilton–Smith formula 375
Hardy–Cross method 348
Hazen–Williams formula 107
headlands, artificial 443–5
Henderson, F. M. 161, 327, 395, 425, 429, 431
hindcasting of wave climate 241–2
hydraulic analysis of pipe networks 347–53
hydraulic gradient 340–1, 390
hydraulic jump 131, 143, 159, 166, 403
  energy dissipation in 146
  equations 143–8, 166
  stability 147
  uses of 148
hydraulic machines 212–24, 322
hydraulic mean depth 121, 141
hydraulic models 326
hydraulic radius 112, 121, 123
Hydraulics Research Limited 115, 303
Hydraulics Research Station 101, 104, 105, 170, 303, 348, 350, 381, 438
hydraulic structures, models of 333
hydrograph
  convolution of 282, 284
  inflow 176, 177, 295, 296, 299, 301
  outflow 176, 295, 297, 299, 301
  synthetic unit 284, 288, 291, 294
  time base 284

time to peak 284
unit 280–5, 286, 294, 302

ideal flow *see* potential flow
ideal fluid 20, 46
infiltration 271
irrotational flow 49

JONSWAP wave spectrum 243

Kármán, Th. von xxiii, 99
Kindsvater–Carter formula 375
Kranenburg, C. 211
Kutter 126, 127

Lacey, G. 424
laminar flow (*see also* viscous flow) 66, 79, 90, 92–7, 98–9, 123, 322
laminar sublayer 80
Langhaar, H. L. 317, 463
Laplace equation 50
lateral discharge 412–17
  Balmforth and Sarginson solution 415
  Ven te Chow solution 414–15
Leopold and Wolman formula 430
levees 431, 432
Lewin, J. 431
lift 83, 256
littoral drift 228, 436–8
  estimation of 437–8
local head loss 108–11
long based weirs 379–84
loop method 348–51

Manning equation 126–30, 151, 154, 156, 161, 162, 172, 389, 390, 411, 414
Manning roughness coefficient 127, 128, 168, 170, 176, 410
manometer 6
meandering 429–31
metacentre 16, 19
mild slope 151, 153, 154, 157, 158
mobile bed models 326
modular flow 377
modular limit 377
moment of area 9, 17, 479–80
momentum
  coefficient of 30, 31, 122
  conservation of 23, 29
  equation 30, 35, 143, 195
  equation for pipe flow 90
  integral equation 77
  thickness 77–80
Moody, L. xxiii, 101
Moody diagram 101, 104
morphology of natural channels 428–31
Muir-Wood, A. M. and C. Fleming 228, 233, 238, 438, 445

Muskingum method 299, 300, 301
Muskingum–Cunge method 301, 306

nappe 371, 373
  contraction of 373, 374
natural bays 438
net positive suction head 224
Newton, I.
  2nd law of motion 24, 29
Newtonian fluid 65
Nikuradse, J. xxiii, 98, 99–100, 317
nodal method 351
non-uniform flow 21, 120
normal depth of flow 128, 151, 157, 166
numerical integration of gradually varied
    flow equation 161, 162

orifice, discharge through 44–6
outlet structures 417

Parmakian, J. 210
peak runoff 284
percolation 271
Pierson–Moskwitz wave spectrum 243, 245
piezometer 6
pipe
  commercial 100
  design of 339
  diameter, estimation of 106
  discharge, estimation of 111
  flow 88, 318
  flowing partially full 112
  forces on 35, 37
  jointing system for 341–2
  loop 348
  materials for 341
pipeline
  branched 344–6
  parallel 342, 344
  series 342, 343
  simple 342
pipe networks 353
pipe walls, straining of 191–2, 194
Pitot tube 41
Poiseuille 90
Poisson's ratio 192
pools 430
potential flow 46–60
  in curved path 53–4
  radial 53
  rectilinear 50–3
Prandtl, L. xxiii, 97, 100
  eddy model 72, 254, 255
pressure 3
  absolute 4
  gauge 4
  measurement 5
  on curved surface 11

on plane surface 9
on submerged bodies 14
transducer 8
proportional depth in pipes 114
pump
  axial 216–18, 220
  centrifugal 213, 214–17, 220
  failure 209, 358
  ideal head in 215
  impeller 214
  mixed flow 218, 220
  performance data 218–19
  radial flow 213, 214–16
  selection 219
  similarity 325
  specific speed 219–20
  type number 219
pumped storage 221
pumped mains 353–8
  economics 357
  hydraulic design 353–7
  matching of pump and pipe for 354
  with parallel pumps 355
  with series pumps 355

radial gate see control gates
rainfall
  effective 280
  net 280, 282, 284, 293, 304, 306
rapidly varied flow in channels 131
Rational Method 303
rectangular weirs 370–6
refraction 228, 231–3
  coefficient 234
reflected wave 236, 238
Regime formulae 424–5
region curves 278, 288, 294
Rehbock formula 374
relative roughness 98
reservoir routing 294–9
reservoirs
  safety of 302
return period 273, 274, 277, 278, 286
Reynolds, O. xxii, 67
  experiment 67, 90, 97
  number 68, 75, 84, 90, 98, 123, 128, 264,
    315, 319, 320, 321, 322, 323, 327, 373,
    377, 387
  stress 70–2
riffles 430
rip-rap 443
river models 326
rough turbulence 99, 114
roughness
  effective 100
  relative 98

Saint–Venant equations 176

sand bar formation 436
scale
  factor 310
  Schlichting, H. 84
Scripps and Komar 437
sea defence 439
sea wall 439–40
secondary current 122, 422
sediment
  concentration 254–5
  transport models of 332
separation of flow 82
sequent depth 143
sewer 112, 116
Shields, A. 250, 252, 256, 258, 259
shoaling 228, 230, 233–8
  coefficient 234
shock wave 184, 186, 193
significant wave height 241, 242
Silvester, R. 233, 238, 438, 445
similarity
  dynamic 311, 319
  geometric 310, 319
  kinematic 310
Simons and Albertson 424
sink 53
ski jump 403, 405
slope of channel
  critical 151–4, 158
  mild 151, 157, 158, 163
  steep 151, 158
sluice gate *see* control gates
SMB wave prediction method 242
source 53
specific energy 134–7, 148, 150, 151
  curve 136, 138, 151
specific speed 219–20, 223, 323–5
spillway 391–403, 409
  cavitation on 393, 395, 396, 403
  gravity (Ogee) 392–6
  model 334
  shaft (tunnel) 392, 401–3
  side channel 158, 392
  siphon 392, 396–401, 403
stage 121, 127, 296, 297
stagnation point 60
stagnation pressure 42
Standard step method
  for natural channels 162, 169
  for regular channels 162, 168
Stanton and Pannel 98
steady flow 21, 119–20
steep slope 151, 153, 154, 155, 158
stilling basin 158, 403, 404
streakline 22
stream function 47–9
streamline 21, 30, 32, 46–9, 59–61
stream power 260, 262, 264

stream tube 22, 24
Streeter, V. L. 210, 211
Strouhal number 84, 330
sub-critical flow 137, 140, 142, 143, 150, 154, 158, 166
submerged bucket 403
submergence 377
super-critical flow 137, 140, 142, 143, 147, 155, 158, 166, 171
surface roughness, effect on boundary layer 83
surface tension xxvi, 86, 374
surge in channels 170, 398
surge in pipelines 180–211
surge protection
  by-passing method 366
  for hydro-electric schemes 363, 365–6
  for pumped mains 366–8
  mechanical methods of 358, 366
surge tower 359–65
  equations of motion for 360–5
  numerical solutions to equations 362–5
superelevation 422
suspended load 247, 248, 255, 260–1
  efficiency 261

T-junction, forces on 39–41
terminal velocity 252
Thalweg 430
Thoma, D. 224
threshold of movement 247, 248, 250
thrust blocks 341
tidal phenomena, models of 330–3
time base 284
time of concentration 304
time to peak 284
total load 255, 260
  formulae 260–6
tractive force 425–8
training of rivers 431–2
transient pressure 181
transition from laminar to turbulent flow 68, 90, 98
transitional turbulence 99, 322
transitional water depth 228
translation of flood wave 295
turbines, hydraulic 212, 220–3
  axial 220
  Francis (radial) 220–2
  Kaplan 221
  Pelton 220, 223
  selection of 223
  specific speed 223
turbulent flow 68, 69–74, 80, 90, 122, 123

uncontracted weir 370, 374
undistorted models 326, 331
ungauged catchment 278

uniform flow 21, 120, 128
unit hydrograph *see* hydrograph, unit
United States Bureau of Reclamation 392, 404
unsteady compressible flow 183
unsteady flow 20, 120, 182
  in channels 170

valve closure
  instantaneous 189, 209
  rapid 183, 209
  slow 209
velocity
  coefficient 45, 131
  distribution 76, 80, 122
  measurement 41
  potential 48
  vectors in hydraulic machines 213–14
vena contracta 44
venturi flume *see* flume, venturi
venturi meter 42
vertical sluice gate *see* control gates
Villemonte's formula 377
viscosity xxvi, 20, 62–5
viscous flow (*see also* laminar flow) 65
volute 215
vortex 55
  forced 55
  free 56, 399
vorticity 56

wake 82
Wallingford hydrograph method 304, 306
washout 342
wave
  Airy 226–7
  celerity of 226, 228
  cnoidal 226
  deep water 226, 227
  diffraction 238–9, 331
  energy 229
  oscillatory 141, 170

periodic progressive 225, 226
power 229–30
shallow water 228
standing 236
Stokian 226
storm 240
surge 141, 170–3
swell 241
transitional 228
translatory 170
wave breaking 233
wave height
  maximum 244
  mean of highest 244, 245
  RMS 244
  significant 241, 244, 245
wave period
  of crests 244, 245
  of zero upward crossings 244, 245
wave phenomena, models of 330–3
wave prediction 239–43
wave records 243
wave spectrum 241, 243
Weber number 86, 377
Weibull formula 276
weirs
  broad crested 131, 149, 158, 379–81, 383
  compound 379
  Crump 382
  long based 379, 380–1
  rectangular 370, 371–6
  round nosed 381
  side 415
  Sutro 379
  thin plate 369–79
  trapezoidal 379
  vee 377
Weisbach, J. xxii, 90, 97, 371
wetted perimeter 121
White, W. R. 262, 268
White's formula 374
Wood, I. R. 395
Wylie, E. B. 210, 211